Primer on Flat Rolling

Primer on Flat Rolling

By

JOHN G. LENARD

ELSEVIER

Amsterdam • Boston • Heidelberg • London • New York • Oxford
Paris • San Diego • San Francisco • Singapore • Sydney • Tokyo

Elsevier
Linacre House, Jordan Hill, Oxford OX2 8DP, UK
84 Theobald's Road, London WC1X 8RR, UK

First edition 2007

British Library Cataloguing in Publication Data
Lenard, John G., 1937-
Primer on flat rolling
1. Rolling (Metal-work)
I. Title
671.3′2

Library of Congress Number: 2007927331

ISBN: 978-0-08-045319-4

Printed and bound in Great Britain
07 08 09 10 11 10 9 8 7 6 5 4 3 2 1

I dedicate this book to my wife, Harriet and my daughter, Patti.
Their active support, encouragement and love made
the writing possible and easy.

CONTENTS

PREFACE

I have been dealing with problems of the flat rolling process for the last 30 years. This included mathematical modelling, experimentation, consulting, publishing in technical journals, presenting my research at conferences and in industry, as well as lecturing on the topic at levels, appropriate for second and third year undergraduate students, graduate students and practicing engineers and technologists of aluminum and steel companies. The present book is a compilation of my experience, prepared for use by practitioners who work with metal rolling and who want to know about the "why"-s, the "what"-s and the interdependence of the material and process parameters of the rolling process. The book may also be useful for graduate students, researching flat rolling.

My interest in the process began while I spent a year at Stelco Research as an NSERC Senior Industrial Fellow, shortly after starting my academic career. I became aware of the tremendous complexity underlying the seemingly very simple process of metal rolling. I realized that while the process of flat rolling – that of two cylinders rotating in opposite directions and reducing the thickness of a strip as it passes between them – has not changed for centuries, its current sophistication places it at the top of the "high tech" activities. On return to academia, and as soon as research funds allowed, I designed and built a simple two-high experimental rolling mill and instrumented it to measure the important variables. The mill has been in use ever since to roll various metals – mostly aluminum and steel alloys – under a large variety of conditions. These conditions included dry and lubricated passes, use of neat oils and emulsions, high, low and intermediate temperatures, heated and non-heated rolls, speeds and reductions as high and low as the mill allowed. During these experiments, my students and I used smooth and rough roll surfaces, prepared by grinding or sand blasting.

In each of the tests, the roll separating forces, the roll torques, the entry and exit thickness, the rolling speed, the forward slip, the entry and exit temperatures of the strip, the roll's surface temperature, the amount of the lubricant, the flow rate and the temperature of the emulsion, the droplet size in the emulsion, the change of the width and the reduction of the strips were measured.

In addition to the experiments performed by myself, by academic visitors from China, Egypt, Germany, Hungary, India, Israel, Japan, Poland and South Korea, and by my graduate students, twice each year my undergraduate classes, typically 80–100 students strong, performed flat rolling tests, providing me with a very respectable collection of data.

Mathematical modelling of the process proceeded parallel to the experimental studies. The attention was on establishing the predictive abilities of the available models of the flat rolling process. The assumptions made in the derivation of the traditional 1D models were critically examined and were improved on by developing an advanced 1D model which makes use of as few arbitrary assumptions as possible. The use of finite-element models was also explored, in co-operation with Prof. Pietrzyk (University of Mining and Metallurgy, Krakow, Poland) and his colleagues and students.

During my academic career, I offered, once or twice a year, a specialist course on rolling, designed for technologists and engineers who work in the metal rolling industry. The educational level of the audience varied broadly, from those who completed high school to those with doctoral degrees. Each year I found two unchanging phenomena. The first was the shaky background my listeners possessed, essentially regardless of their education. When asked about the difference between engineering strains and true strains, about the difference between the plane-stress and the plane-strain conditions, the difference between static and dynamic recrystallization, and so on, the large majority of them betrayed serious ignorance. The second was the lack of a textbook that includes all I needed to develop the ideas in the course. The present book, resulting from the notes I used in these courses, attempts to compile, present and explain the disparate components, needed for a clear understanding of the topic.

The book contains 11 chapters. The first 10 of these deal with various aspects of the flat rolling process and the 11th presents a set of assignments and incomplete solutions, formulated to test the understanding of the reader of the material presented. Each chapter ends with a set of Conclusions.

The flat rolling process is defined in Chapter 1, the Introduction. The objectives are to give a very brief overview of the process. Details of the hot rolling process, using hot strip mills, are given. Continuous casting is described. The cold rolling process and cold mill configurations are presented next.

A general discussion of the rolling process is presented in Chapter 2. The components of a metal rolling system are defined. Reference is made to the rolling mill, designed by Leonardo da Vinci and the scale-model, built following his drawings. A description of the physical and the metallurgical events during the process is given, including the events as the strip to be rolled is ready to enter the roll gap, as it is partially reduced and as the process becomes one of steady-state. The independent variables of the system – the mill, the rolled metal and their interface – are listed. The minimum value of the coefficient of friction, necessary to commence the rolling process is given. Some of the simplifying assumptions that are usually made in mathematical models of the process of flat rolling are critically discussed: these include the idea of "plane-strain plastic flow" and "homogeneous compression of the strip". Microstructures of a fully recrystallized Nb steel, an AISI 1008 steel and a cold-rolled low carbon steel are presented.

Mathematical modelling of the rolling process is the topic of Chapter 3. Traditional and more advanced models are discussed in terms of their capabilities as far as their predictions are concerned. Models for both mechanical and metallurgical events are included. The chapter ends with the identification of three parameters, necessary for efficient, accurate and consistent modelling: the coefficients of heat transfer and friction and the resistance of the material to deformation.

Chapters 4 and 5 treat these in turn; material behaviour and tribology, respectively. In both, the emphasis is on how the concepts are to be used when combined with the models, presented in the previous chapter.

The objectives in preparing Chapter 6 are somewhat different. The chapter is entitled "Sensitivity studies" and in spite of some examination of the sensitivity of the predictions in previous chapters, some more calculations and applications are added.

Temper rolling is considered in Chapter 7. The differences between the usual flat rolling process and temper rolling are pointed out. Several mathematical models are given and the assumptions made in their development are discussed. The components that should make up a complete model of the process are listed.

The tenor of the book changes at that point. In each of Chapters 8, 9 and 10 – accumulative roll-bonding, cold-roll bonding and flexible rolling , respectively – a review of the literature is followed by the detailed descriptions of experimental work.

Chapter 11 contains two sections. In the first, problems are listed, for each of the chapters. Some of these require the direct application of the expressions and the formulas presented in the book. Some answers require Internet searches. Some require development of computer programs. Some are suggested topics for seminars or class discussions. In the second part, the solutions are given. Again, this is done in a variety of ways: in some cases detailed solutions are given while in some others, only the numerical answers are indicated. As well, in some instances, only a set of hints and recommended approaches are suggested.

I would like to acknowledge the contributions of my undergraduate and graduate students without whom my research would not have progressed. Also, I would like to thank the visiting scientists with whom co-operation was always most enjoyable.

LIST OF SYMBOLS

Ar_1, Ar_3	Austenite to ferrite transformation; stop and start temperatures, respectively
A, A_r	Apparent and true areas of contact, respectively
B	Material constant
C	Material constant, indicating strain-rate hardening; also half the distance between the roll centres in eq. 3.29
$[C], C_{eq}$	Carbon content, %; carbon equivalent
D	Diameter of the work roll
D_α, D_γ, D_r	Ferrite and austenite grain size; austenite grain size after recrystallization, respectively
D_{DRX}	Austenite grain size after dynamic recrystallization
D_{MD}	Austenite grain size after metadynamic recrystallization
E, E'	Elastic modulus; composite elastic modulus, respectively
F_I	Inertia force
H	Hardness
J^*	Externally supplied power
J_2	Second invariant of the stress deviator tensor
K	Material constant, strength coefficient
K_y	Grain boundary unlocking term
L	The contact length
M	Roll torque for both rolls per unit width
N	Roll rpm
P, P_{total}	Power required for plastic deformation and the total power needed to drive the rolling mill, respectively
P_n	Friction losses in four roll-neck bearings
P_r	The roll separating force per unit width
Q_p	The pressure intensification factor; a multiplier, accounting for the shape factor and the coefficient of friction
Q_{RX}	Activation energy for recrystallization
R', R	The radius of the flattened roll (by Hitchcock's equation) and the original radius of the roll, respectively
S, S_Γ, S_i	Surface; also mill stiffness in eq. 3.94; surfaces in the upper bound theorem, eq. 3.61, respectively
S_{ij}	Components of the stress deviator tensor
S_f, S_b	The forward and the backward slip
T, T_0	Temperature; roll temperature some distance below the surface

ΔT_{gain}, ΔT_{loss}	Temperature rise and loss of the strip in the pass, respectively
T_{strip}, T_{roll}	Temperature of the strip and the roll, respectively
T_{NRX}	Temperature above which recrystallization will occur
V	Volume
$\dot{W}$	The flattening rate
X, X_{DRX}	The recrystallized volume fraction; dynamically recrystallized volume fraction, respectively
X_{MD}	Metadynamically recrystallized volume fraction
Y	Material constant
Z	Zener-Hollomon parameter
a	Acceleration
a, b	Constants
c_p	Specific heat of the rolled strip
d	Diameter of the roll-neck bearing
f_t, f_n	Friction force, and the normal force, respectively
h	The thickness of the strip
h_{entry}	The thickness of the strip at the entry
h_{exit}	The thickness of the strip at the exit
h_{ave}	The average of the entry and the exit thickness
h_{min}	The minimum rollable thickness
h_{np}	The thickness of the strip at the neutral point
$h_{film,ave}$, h_s	Average oil film thickness; smooth oil film thickness
k	Yield strength of the material in pure shear
l_a	The lever arm
m, m_{ave}	Strain rate hardening exponent; also mass; average friction factor, respectively
n, n_1	Strain hardening exponent
p, p_{ave}	Roll pressure; average roll pressure
p, q, r, s	Material constants
p_s	Shakedown pressure
r	Reduction
t	time
v	Speed of the rolled strip; also Poisson's ratio in equation 3.5
v_r	Roll surface speed
w	Width of the strip
x	Distance along the direction of rolling, measured from the line connecting the roll centres
x_n	Location of the neutral point
α, $\overline{\alpha}$	Heat transfer coefficient; parameter in Hatta's equation
β^*	Correlation distance
γ	Austenite
δ_{ij}	Kronecker's delta
$\varepsilon_{0.5X}$	Strain for 50% recrystallization

$\varepsilon_c, \varepsilon_p$	The critical strain; strain at peak stress
$\varepsilon, \varepsilon_{max}$	Strain; strain the rolling pass, respectively
$\dot{\varepsilon}, \dot{\varepsilon}_{ave}$	Strain rate; average strain rate in the rolling pass
ε_r	Accumulated strain
ϕ	Angular variable around the roll, measured from the line connecting the roll centres
ϕ_1	The bite angle
ϕ_n	The neutral angle
φ	The Airy stress function
κ	Curvature of the asperity tip
Λ	Combined r.m.s roughness
ψ, ψ_s, ψ^*	Plasticity indices
η	Dynamic viscosity
η_m	Efficiency of the driving motor of the rolling mill
η_t	Efficiency of the transmission
μ, μ_c, μ_h	Coefficient of friction in the roll/strip contact; under steady state rolling; contribution of hydrodynamic action
μ_f, μ_b	Coefficient of friction in the forward and the backward slip regions;
μ_{min}	Minimum coefficient of friction required for successful entry
μ_n	Coefficient of friction in the roll-neck bearings
μ_{Hill}	Coefficient of friction by Hill
σ_c, σ	Dynamic, constrained yield strength of the rolled strip; yield strength of the softer metal;
σ_x	Stress in the direction of rolling
$\sigma_{entry}, \sigma_{exit}$	The external stresses at the entry and the exit, respectively
σ_{fm}	The average flow strength of the strip in the deformation zone
ρ	Density of the rolled strip
τ	The shear stress at the roll/strip contact
τ_0	Eyring shear stress

ADVICE FOR INSTRUCTORS

There are several topics mentioned in this book, the thorough understanding of which needs a broad and varied background. The instructor should be aware of the preparation of the audience and make sure that the following subjects are understood well before starting on the presentations of the book's contents. A brief quiz during the first lecture and the discussion of the results are often helpful in finding out what needs to be reviewed.

In the present writer's experience with rolling mill engineers, this background may have been there in the listeners' college or university days but not having used them daily for some considerable time, gaps are certain to exist. It is strongly recommended that at least the first six lectures be devoted to a review of the following.

The ideas involved with the strength of materials should be mastered first. These include the theory of elasticity and the analysis of stress and strain; the idea of equilibrium, static and dynamic; principal directions, principal stresses and strains need to be appreciated. Boundary conditions, surface and body forces should be clarified and it may be helpful to assign, and then discuss in class, some esoteric examples: such as the free-body diagram of a tooth while it is being extracted or the forces and torques acting on a rail-car wheel in motion. Identifying and sketching the loads on a bullet in flight would also pose a challenge. If these are well understood, their application during the course should become easy.

The difference between engineering and true stresses and strains should be made clear. Strain rates and the conditions under which they remain constant in a test, need to be mentioned.

The theory of plasticity is used throughout the book, without developing the basic ideas. Elastic–plastic boundaries, the yield and the flow criteria, the associated flow rules, the constancy of volume and the compatibility equations should be presented as part of the review. The stress and the strain tensors should be mentioned in addition to the tensor invariants.

Basic ideas from the field of metallurgy are needed. The grain structure of metals, the carbon equilibrium diagram, the hardening and the restoration mechanisms, the hot and cold response of metals to loading are all used in many of the developments in the course. It would be helpful for the students to have actually mounted, polished and etched a piece of metal for metallographical examination.

Some time should be devoted to a discussion of Tribology as well. Viscosity, Reynold's equation, lubricant and emulsion chemistry are all necessary here.

As a last comment to the instructors, nothing replaces the actual hands-on experimentation. Having a well-instrumented rolling mill and conducting some carefully designed experiments would lead to immeasurable benefits.

Some care needs to be exercised in assigning the problems from Chapter 11. Many of them are fairly straightforward and require the application of the ideas presented in the text. Many of them, however, require extensive reading and may well lead to some frustration. A discussion of the solution in class is often highly appreciated.

Seminars or class discussions are suggested when dealing with Chapters 7–10. These may require advance preparation so the discussions would not become professorial presentations. State-of-the-art reviews have been found helpful.

CHAPTER 1

Introduction

Abstract

The topic of the book that of the flat rolling of metals is introduced. The products of flat rolling, strips and thin plates are defined in terms of their geometry: the ratio of their thickness to width is much less than unity. Strips are substantially thinner than plates. Rolling of strips and plates is generally referred to as flat rolling. The objectives of the process – that of reducing the thickness of the work piece, increasing its length and thereby changing its mechanical and metallurgical attributes – are stated in this introductory chapter. The temperatures at which hot, warm and the cold rolling processes are performed are defined. Each of the processes is presented and discussed in general terms. The equipment: the hot strip mill, Steckel mills, mini-mills, Sendzimir mills, planetary mills and the cold rolling mill are shown and described, along with several mill configurations. The continuous casting process, as applied in the hot rolling industry, is also shown. New, recently developed equipment is described as well. A probably incomplete list of books and publications, dealing with the theory of plasticity, plastic forming of metals and specifically the flat rolling process, is followed by some general conclusions.

1.1 THE FLAT ROLLING PROCESS

The mechanical objective of the flat rolling process is simple. It is to reduce the thickness of the work piece, from the initial thickness to a pre-determined final thickness. This is accomplished on a rolling mill, in which two work rolls, rotating in opposite directions, draw the strip or plate to be rolled into the roll gap and force it through to the exit, causing the required reduction of the thickness. As these events progress, the material's mechanical attributes change. These in turn cause changes to the metallurgical attributes of the metal as well, which, arguably are of more importance as far as the product is concerned. A schematic, three-dimensional (3D) diagram of the back-up rolls and the work rolls is shown in Figure 1.1 where a single-stand, four-high mill is depicted; this may be a single stand roughing mill.

The figure shows the back-up rolls, the much smaller work rolls, the strip being rolled, the roll torques and the roll separating forces acting on the journals of the back-up roll bearings, keeping the centre-to-centre distance of

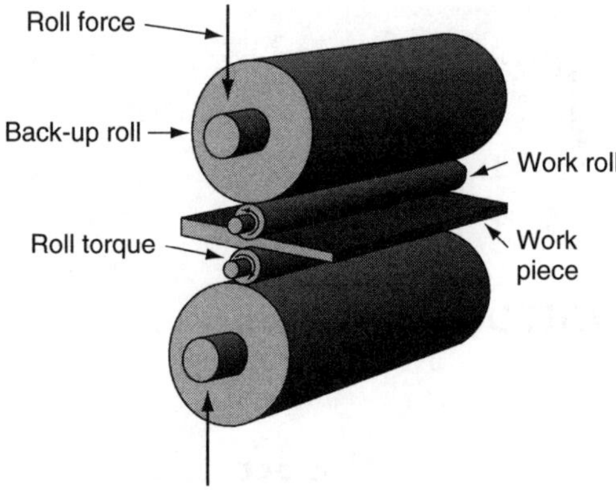

Figure 1.1 A schematic diagram of a single-stand, four-high set of rolls (See Plate 1).

the bearings as constant as possible[1]. As it will be demonstrated in Chapter 6, Sensitivity Studies, the energy requirements of the process may be decreased when small diameter work rolls are used. The drawback of that step is the reduced strength of the work roll, and this necessitates the use of the massive back-up rolls which minimize the deflections of the work roll.

1.1.1 Hot, cold and warm rolling

While the rolling process may be performed at temperatures above half of the melting point of the metal, termed hot rolling, or below that temperature in which case one deals with cold rolling, the division into these two categories should not be considered as cast-in-stone. There is a temperature range, beginning below and ending above the dividing line in between hot and cold rolling, within which the process is termed warm rolling and in some specific instances and for some materials this is the preferred process to follow, leading to mechanical and metallurgical changes of the attributes of the work piece, not possible to achieve in either the cold or the hot flat rolling regimes.

1.2 THE HOT ROLLING PROCESS

Hot rolling of metals is usually carried out in an integrated steel mill, on a "Hot Strip Mill", or since some changes were introduced in the last couple of decades, on mini-mills[2]. Both have advantages and disadvantages, of course,

---

[1] Mill stretch will be discussed in Chapter 3.
[2] See Section 1.4 in this Chapter for a brief discussion of mini-mills.

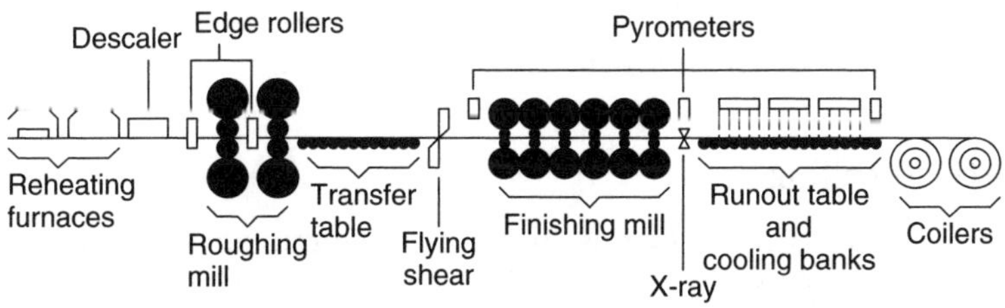

Figure 1.2 The schematic diagram of a traditional hot strip mill.

concerning capital costs, flexibility, quality of the product and danger to the environment.

A schematic diagram of a traditional hot strip mill (HSM) is depicted in Fig. 1.2, showing the major components.

There are several basic components in the hot strip rolling mill. In what follows, they are discussed briefly.

1.2.1 Reheating furnace

These constitute the first stop of the slab after its delivery from the slab yard. The slab is heated up to 1200–1250°C in the furnace to remove the cast dendrite structures and dissolve most of the alloying elements. The decisions to be made in running the reheat furnace in an optimal fashion concern the temperature and the environment within. If the temperature is high, more chemical components will enter into solid solution but the costs associated with the operation become very high and the thickness of the layer of the primary scale will grow. If the temperature is too low, not all alloying elements will enter into solid solution, affecting the metallurgical development of the product and the likelihood of hard precipitates remaining in the metal increases. As well, thinner layers of scale will form, a fairly significant advantage. A judicial compromise is necessary here and is usually based on financial considerations. The cost savings associated with a one-degree reduction of the temperature within the furnace can be calculated without too many difficulties and the changes to the formation of solid solutions may be estimated, however, the annual savings may well be significant.

Primary scales of several millimetre thickness form on the slab's surface in the reheat furnace. The thickness of the scale may be reduced by providing a protective environment within the furnace, albeit at some increased cost; an approach that is rarely followed. As the furnace doors open and the hot slab slides down on the skids to the conveyor table, the instant chilling caused by the water-cooled skids creates marks that are often noticeable on the finished product. As well, fast cooling of the surfaces and especially of the edges is also immediately noticeable, indicating a non-uniform distribution of the temperature within the slab and leading to possibly non-homogeneous dimensional, mechanical and metallurgical attributes.

1.2.2 Rough rolling

Before the rolling process begins, the scale is removed by a high-pressure water sprays and/or scale breakers. The slab is then rolled in the roughing stands in which the thickness is reduced from approximately 200–300 mm to about 50 mm in several passes, typically four or five. The speeds in the rougher vary from about 1 m/s to about 5 m/s. In the roughing process, the width increases in each pass and is controlled by vertical edge rollers. The vertical edgers compress and deform the slab somewhat, causing some thickening which is corrected in the subsequent passes. A large variety of roughing mill configurations is possible, from single-stand reversing mills to multi-stand, one-directional mills, referred to as roughing trains. These usually have a scale breaker as the first stand where the mill deforms the slab sufficiently just to loosen the scale, which is then removed by the high-pressure water jets. Roughing scale breakers are usually vertical edgers, capable of reducing the width of the slab by up to 5-10 cm and causing stresses at the steel surface-scale layer interface which then separate the scales. Roll diameters are near 1000 mm. The rolls are usually made of cast steel or tool steel. Roughing stands are either of the two-high or four-high configurations. At the end of the rough rolling process, the strip is sent to the finishing mill along the transfer table where it is referred to as the "transfer bar". The temperature of the slab in the roughing stands is high enough so the transfer bar is fully recrystallized, containing strain free, equiaxed grains. In general, though, the grain structure at the end of the rough rolling process seems to have little influence on the structure by the time the strip has passed through several stands of the finishing mill.

1.2.3 Coil box

Not shown in Figure 1.2 is a device – an invention by the Steel Company of Canada and first installed in the early 1970s in Stelco's Hilton Works – called the coil box, placed in between the roughing mill and the finishing train, in place of the transfer table. Since its introduction several integrated steel companies have installed the coil box in their hot strip mills. A photograph of the coil box is shown in Figure 1.3 below.

When the words "Coil Box" are entered into Google[3], a plethora of information is found, including the possibility of watching a video of the coil box in motion. A detailed description of the events when the steel arrives to the

[3] It is acknowledged that the contents of websites on the Internet are changed and updated regularly.

Figure 1.3 The coil box (courtesy The Steel Company of Canada) (See Plate 2).

coil box and when it is within the coil box are also given in Google, an edited version of which follows.

The transfer bars, exiting from the roughing stand, are formed into coils at the coil box which consists of two entry rolls, three bending rolls, a forming roll, two sets of cradle rolls, coil stabilizers, peeler, transfer arm and pinch rolls. The adoption of a coil box configuration has several advantages:

- it reduces the overall length of the mill line;
- it increases the productivity;
- it enlarges the strip width and the length to be rolled and
- it eliminates the thermal rundown along the strip length when compared to the conventional HSM.

Thus, uniform temperature and constant rolling speed conditions are maintained. On uncoiling from the coil box, the transfer bars are end-cut, processed through high-pressure descaling sprays, and then they are ready to enter the finishing stands. With the introduction of advanced high-strength steels such as Hot Roll Dual Phase steel[4], the benefits of the coil box are even more significant in providing uniform mechanical properties throughout the length of the coil[5].

---

[4] To be discussed in Chapter 4.

[5] The first coil box was installed in the Hilton Works shortly before the writer spent a year as a Senior Industrial Fellow at the Research Department of the Steel Company of Canada. At that time the information concerning the coil box was proprietary and so carefully guarded that no permission was obtained to read any of the reports written on the performance or the analysis of the equipment.

1.2.4 Finish rolling

When the transfer bar, now coiled up in the coil box, reaches the appropriate temperature, it is uncoiled and it is ready to enter the last several stands of the strip mill, the finishing train. The crop shear prepares the leading edge for entry, and the transfer bar enters the first stand, assisted by edge rollers. Its velocity is in the range of 2.5–5 m/s [6]. The finishing train in the strip mill is traditionally composed of five to seven tandem stands. The roll configuration is usually four-high, employing large diameter back-up rolls and smaller diameter work rolls. The entry of the strip into the first stand is carefully controlled and it is initiated only when the temperature is deemed appropriate, according to the draft schedule, which was prepared using sophisticated off-line mathematical models. These determine the reductions and the speeds at each mill stand as well as predicting the resulting mechanical and metallurgical attributes of the finished product. After entry into the first stand, the strip is continuously rolled in the finishing mill. At the entry to the finishing mill, the temperature of the strip is measured and at the exit, both temperature and thickness are measured; the thickness at the exit from each intermediate stand is estimated using mass conservation[7]. In some modern mills there are several optical pyrometers placed along the finishing train. The Automatic Gauge Control (AGC) system uses the feedback signals from several transducers to control the exit thickness of the strip. The finishing temperature may also be controlled by changing the rolling speed; however, only small variations of that are possible without causing tearing – if the speed of the subsequent mill stand is too high – or buckling, referred to as "cobble" of the strip, if the speed there is too low. On some newer and more modern strip mills inter-stand cooling and/or heating devices have been installed, which minimize the temperature variation across the rolled strip and thereby increase the homogeneity and the quality of the product. As the thickness is reduced the speed must increase, as demanded by mass conservation, and the speeds in the last stand may be as high as 10–20 m/s. The rolls of the finishing mill are cooled by water jets, strategically placed around the rolls. Without cooling, the surface temperature of the work rolls would rise to unacceptable levels. It has been estimated that when in contact with the hot strip, the roll surface temperatures may rise to as high as 500°C at a very fast rate. Of course, the roll surface would cool during its journey as it is turning around and is subjected to water cooling but the thermal fatigue it experiences accelerates roll wear and is, in fact, one

[6] When the present author was working in the Research Department of the Steel Company of Canada, some consideration was given to increasing the entry speed of the transfer bar into the first stand of the finishing mill. The project was abandoned when the possibility of the steel becoming airborne was realized.

[7] The lack of pyrometers along the finishing train often causes difficulties when the mechanical and the thermal events at the mill stands are modelled.

of the major contributors to it. It is possible to measure roll surface temperatures by thermocouples embedded in the roll, their tips positioned close to the surface[8]. A mathematical model would then be necessary to extrapolate the temperatures to the surface. There are usually scale breakers before the first stand of the finishing train, consisting of one or two sets of pinch rolls, exerting only enough pressure on the strip to break off the scale. The strip exits from the finishing train at a thickness of 1–4 mm thickness. The Hylsa steel mill in Monterrey, Mexico, produces hot rolled strip of 0.91 mm thickness. Bobig and Stella (2004) describe the semi-endless rolling and ferritic rolling processes. These, introduced in the thin slab rolling plant EZZ Flat Steel in Egypt, produce 0.8 mm thick coils. The ferritic rolling process leads to reduced scale growth and lower roll wear.

During the last decade, the materials used for the rolls on the hot strip mill were changed from chill cast to tool steels, reducing roll wear in a most significant manner[9]. There have also been reports of significant changes of the coefficient of friction in the roll gap after the switch of roll materials. Tool steels rolls, once implemented correctly, do provide benefits that offset their higher costs. The impact of lubricant interactions with these new roll chemistries have not been fully explored (Nelson, 2006, Private communication, R&D Department, Dofasco Inc.).

1.2.5 Cooling

After exiting the finishing mill, the strip, at a temperature of 800–980°C, is cooled further under controlled conditions, by a water curtain on the run-out table. The run-out table may be as long as 150–200 m. Cooling water is sprayed on the top of the steel at a flow rate of 20 000–50 000 gpm; and on the bottom surface at 5000–20 000 gpm (1 gallon/min = 4.55 l/min). The purpose of cooling is, of course, to reduce the temperature for coiling and transportation but also to allow faster cooling of the finished product, resulting in higher strength. The cooling process also plays a major role in the thermal–mechanical schedule, designed to affect the microstructure.

1.2.6 Coiling

At the exit of the run-out table, the temperature of the strip is measured, and the strip is coiled by the coiler. After further cooling, the steel coils are ready for shipping.

---

[8] This would not, of course, be permitted in a production mill. Results on the rise of the surface temperature of the roll, obtained using eight thermocouples embedded in the work roll of a laboratory mill are presented in Chapter 5, Tribology.

[9] Roll wear will be discussed in Chapter 5, Tribology.

Figure 1.4 The seven-stand finishing mill of Dofasco Inc. (courtesy Dofasco Inc.) (See Plate 3).

1.2.7 The hot strip mill

A photograph of a hot strip mill of Dofasco Inc. is shown in Figure 1.4. A pair of work rolls is visible, stored in the foreground of the figure and ready to be placed in the stands[10].

1.2.8 The Steckel mill

The Steckel mill usually consists of a four-high reversing mill with two coilers placed within heated furnaces, on either side of the mill stand. In the rolling process the slab is rolled first to a thickness of about 12–13 mm on the reversing mill. This is then followed by employing the furnace coilers, and the strip is then coiled, uncoiled and rolled several times until the desired final thickness is reached. The roll wear in the Steckel mill is severe, higher than in conventional mills. Roberts (1983) writes that only a limited number of Steckel mills are in operation. Thaller et al. (2005) describe a modern Steckel mill at VOEST-ALPINE Industrieanalgenbau, giving its capabilities, which include rolling of plates as well as strips.

[10] The rolls are changed at regular intervals in the hot strip mill. The change takes place very fast, such that the mill need not be shut down.

1.3 CONTINUOUS CASTING

Irwing (1993) describes the history of the development of continuous casting and identifies Mannesmann AG where a production plant went into operation in 1950. A continuous casting plant was installed at Barrow Steel, in Great Britain in 1951. The essential idea of the process is simple: molten steel is poured into a water-cooled, oscillating mould. The cooled copper wall of the mould solidifies the outer layer of the steel and as the steel is moving vertically downward, the solidified skin thickens. As the steel leaves the mould, it is cooled further by water sprays. The solidifying steel is supported by rollers, which prevent outward bulging.

The continuous casting process replaced the ingot casting quite some time ago and succeeded in increasing productivity. The complete continuous casting process is shown in Figure 1.5, reproduced from Groover (2002). The figure shows the ladle into which the molten steel is poured. From the ladle the steel is metered into the tundish and from there it enters the water-cooled, oscillating mould. As the steel strand exits the mould, it solidifies further; an indication of the solidification front is also shown in Figure 1.5. Using the withdrawal rolls and the bending rolls, the now solid but still very hot strand is straightened and cut to pre-determined sizes by the cut-off torch.

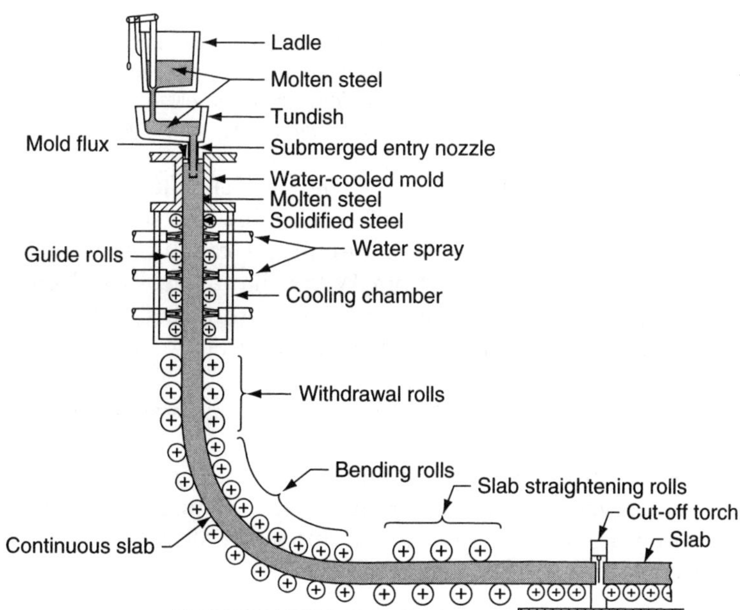

Figure 1.5 Continuous slab casting (Groover, 2002; reproduced with permission).

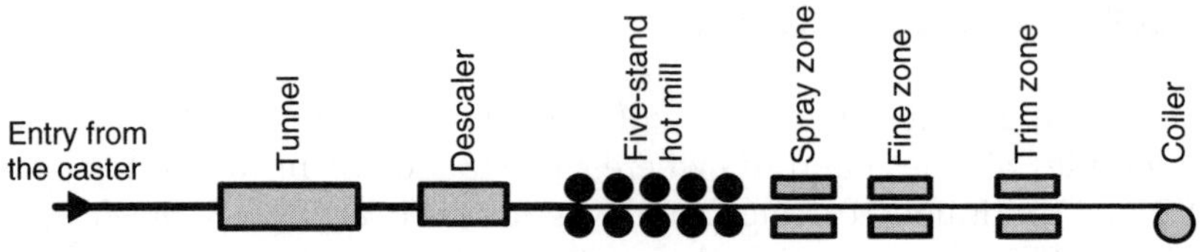

Figure 1.6 Continuous casting and direct rolling (following Pleschiutschnigg et al., 2004).

There are two possible subsequent activities at this point. The slabs may be allowed to cool and are then stored in the slab-yard, retrieved as needed by customers and reheated in the reheat furnaces and rolled, in the hot strip mill, as depicted in Figure 1.3 above. Alternatively, they may be rolled directly, as shown in Figure 1.6 (after Pleschiutschnigg et al. 2004)

1.4 MINI-MILLS

The American Iron and Steel Institute's website gives the following definition for mini-mills:

> *Normally defined as steel mills that melt scrap metal to produce commodity products. Although the mini-mills are subject to the same steel processing requirements after the caster as the integrated steel companies, they differ greatly in regard to their minimum efficient size, labour relations, product markets, and management style.*

Currently in the United States 52% of the steel is rolled by 20 integrated steel mills and 48% by more than 100 mini-mills. The integrated mills roll approximately 400 tons/h while the mini-mills are capable of 100 tons/h.

Information is also available from Wikipedia, a web-based encyclopaedia. It identifies mini-mills as secondary steel producers. Also, it mentions NUCOR, as one of the world's largest steel producers, which uses mini-mills exclusively.

A very impressive number (79%) of mini-mill customers expressed satisfaction with their suppliers[11].

The website "http://www.environmentaldefense.org" gives information regarding the recycling activities of mini-mills, stating that they conserve 1.25 tons of iron ore, 0.5 tons of coal and 40 lbs of limestone for every ton of steel recycled.

---

[11] 2001 Customer Satisfaction Report, Jacobson & Associates.

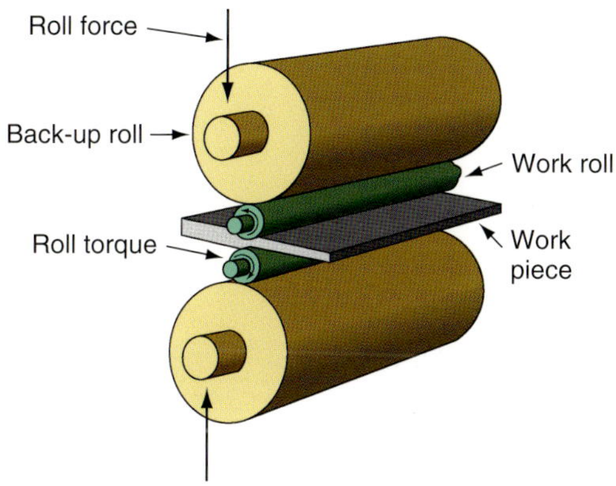

Plate 1 A schematic diagram of a single-stand, four-high set of rolls (See Figure 1.1, p. 2).

Plate 2 The coil box (courtesy The Steel Company of Canada) (See Figure 1.3, p. 5).

Plate 3 The seven-stand finishing mill of Dofasco Inc. (courtesy Dofasco Inc.) (See Figure 1.4, p. 8).

1.5 THE COLD ROLLING PROCESS

The layers of scales are removed from the surfaces of the strips by pickling, usually in hydrochloric acid. This is followed by further reduction of the thickness, produced by cold rolling. Essentially, there are three major objectives in this step: to reduce the thickness further, to increase the rolled metals' strength by strain hardening and to improve the dimensional consistency of the product. An additional objective may be to remove the yield point extension by temper rolling, in which only a small reduction, typically 0.5–5%, is used[12].

1.5.1 Cold rolling mill configurations

A large variation of configurations is possible in this process. An example of a modern cold rolling mill, for aluminum, is shown in Figure 1.7. The mill is six-high; having two small diameter work rolls of 470 mm diameter and two sets of back-up rolls. The diameter of the intermediate back-up roll is 510 mm, and the third back-up roll is of 1300 mm diameter. The mill is capable of producing strips of 0.08 mm thickness at speeds up to 1800 m/min.

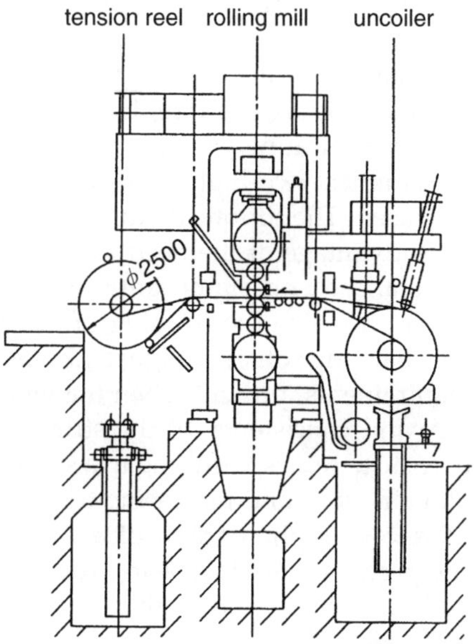

Figure 1.7 A schematic diagram of a modern cold rolling mill for aluminum (Hishikawa et al., 1990).

[12] Temper Rolling is discussed in Chapter 7.

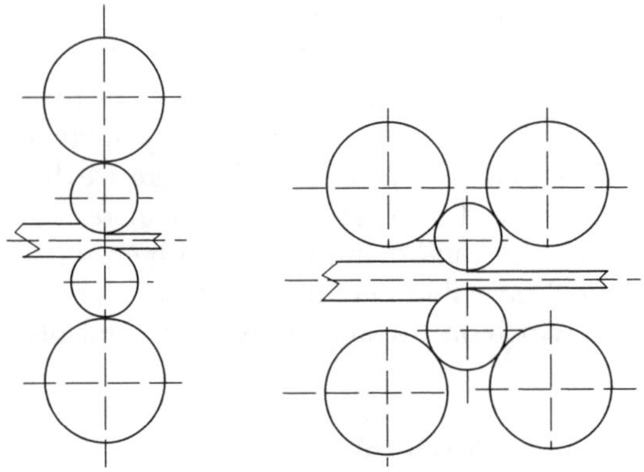

Figure 1.8 A two-high and a six-high mill.

Mill types, their design details and configurations are so numerous that it is impossible to list them all in a brief set of notes. Mill frames, bearings and chucks, screw-down arrangements, loopers, control systems, number of stands, drive systems, spindles, lubricant or emulsion delivery, roll cooling, roll bending devices, shears and coilers may have practically infinite variations in design. Roll materials may also vary, and as the recent literature indicates, the chill cast or high chrome rolls are being replaced by tool steel rolls. In what follows, only a set of figures indicating various roll arrangements is presented.

Figure 1.8 shows the simplest two-high version in which two work rolls of fairly large diameters are used. The simplicity–the low number of components–is outweighed by the disadvantage of the need for massive rolls to minimize roll bending. A more advanced and significantly more rigid arrangement is the six-high configuration, in which the bending of the work rolls is reduced substantially by the large back-up rolls, see Figure 1.7. Also, advantage is taken of the lower energy requirements, present when the work rolls are of smaller diameters. The accuracy and consistency of the strip dimensions increase as the number of back-up rolls increases, resulting in a significant reduction of the deflections of the small work rolls and increasing the stiffness of the complete rolling mill. Figure 1.9 shows a photograph of a twenty-high mill, built by SUNDWIG GmbH. The progressively increasing roll diameters, starting with the very small work rolls, are clearly observable.

Bill and Scriven (1979) describe the details of the Sendzimir mill – which is used for both hot and cold rolling – and show various designs and configurations. They describe the advantages and the disadvantages of using small diameter work rolls, and the history of how engineers attempted to maximize the former and minimize the latter. Tadeusz Sendzimir, a Polish engineer and inventor, designed the cluster mill, which, named after him, was built as an

Figure 1.9 A twenty-high mill, for rolling copper and copper alloys built by SUNDWIG GmbH.

experimental rolling mill in 1931 in Düsseldorf, Germany. In one of the designs, a type 1-2-3-4 arrangement shown in Figure 1.10, similar to the twenty-high mill illustrated above, the work rolls are driven through friction contact. The mill, as well as the other versions of it, is capable of producing very high reduction in one pass and can roll a strip to a very low thickness. Backofen (1972) writes that the work roll may well have a diameter under 1″ (25.4 mm) and the exit thickness may be as low as one-thousandth of an inch (0.025 mm).

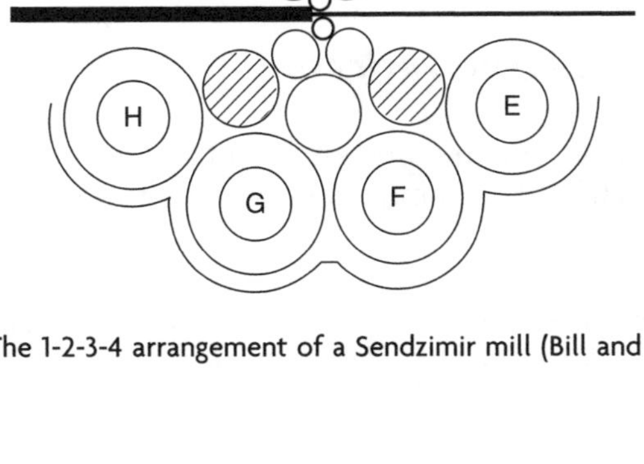

Figure 1.10 The 1-2-3-4 arrangement of a Sendzimir mill (Bill and Scriven, 1979).

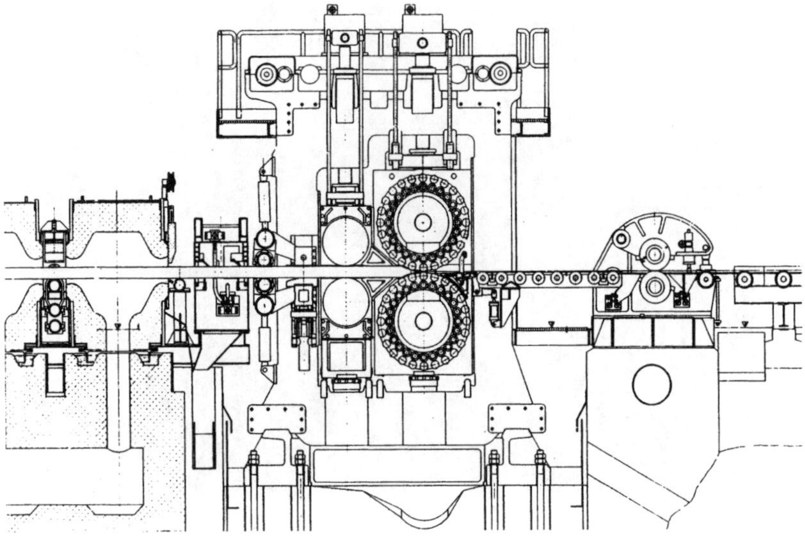

Figure 1.11 The Platzer planetary mill (Fink and Buch, 1979).

Further, since the small work rolls flatten less, they can continue to roll metal even after significant strain hardening with no need for intermediate annealing. The work rolls are often made of tungsten carbide, resulting in much longer roll life and producing a mirror finish on the rolled surfaces.

The ridges, sometimes created by the many small work rolls, are smoothed by subsequent operations.

The Platzer planetary mill, shown in Figure 1.11, is also capable of very high reductions. In some of the versions, the mill has two back-up beams, which are stationary. Around these are the intermediate and the work rolls. Feed rolls force the strip into the roll gap. The work roll diameters range from a low of 75 mm to 225 mm, depending on the width, much larger than in the Sendzimir mill of Figure 1.10. Fink and Buch (1979) indicate that 98% reductions are achievable on the Platzer mill, in one pass. It is interesting to note that the small work rolls rotate in a direction, opposite the rolling direction. The number of roll contacts may be as high as 40–60/s.

1.6 THE WARM-ROLLING PROCESS

The temperature range for this process is not defined very closely; it starts somewhat below half of the homologous temperature[13] and ends somewhat

[13] The homologous temperature range is defined such that one of the end points is absolute zero while the other is the melting temperature of the particular metal.

above that. In the process both the strain and the rate of strain affect the mechanical and metallurgical attributes of the rolled metal and in process design these need to be accounted for carefully. The energy requirements are, of course, higher than those for hot rolling but lower than for cold rolling. The strength of the resulting product is higher than what can be achieved by hot rolling. While there is an accumulation of scales on the surfaces, the amount is significantly less than in the hot rolling process. The ferrite rolling, mentioned above, may be considered to be a warm-rolling process, albeit this statement may be somewhat controversial.

1.7 NEW EQUIPMENT

New mill concepts such as continuous rolling, with transfer bar welding and high speed shears may increase productivity. Thin slab casting, tunnel furnaces, pair cross or continuous profile shape control, high quench units and strip casting have all led to new alternatives that can be considered in new mill designs and retrofits to existing mills. Combinations of these developments lead to extended capability to produce new steels and to produce existing steels better (Nelson, 2006, Private communication, R&D Department, Dofasco Inc.).

1.8 FURTHER READING

A large number of books, dealing with the rolling process, are available in the literature. Among these the excellent books of Roberts "Cold Rolling of Steel", "Hot Rolling of Steel" and "Flat Processing of Steel" (Roberts, 1978, 1983 and 1988) stand out. These are eminently readable, giving the history of the processes, detailed description of the equipment and the mathematical treatment. Rolling of shapes as well as flats is considered. Rolling of metals is considered exclusively by Underwood (1950), Starling (1962), Larke (1965), Tarnovskii et al. (1965), Tselikov (1967), Wusatowski (1969), Pietrzyk and Lenard (1991), Ginzburg (1993) and Lenard et al. (1999). Books dealing with the theory of plasticity or metal forming usually include chapters devoted to the rolling of metals. These include the books of Hill (1950), Hoffman and Sachs (1953), Johnson and Mellor (1962), Avitzur (1968), Backofen (1972), Rowe (1977), Lubliner (1990), Mielnik (1991), Hosford and Cadell (1983) and Wagoner and Chenot (1996).

The list of technical publications dealing with various aspects of the rolling process is prohibitively long to be included here.

In order to appreciate some of the discussions, it may be necessary to review the background to plastic forming of metals. The reader is referred to textbooks dealing with the mathematical theory of plasticity, theory of elasticity as well as continuum mechanics. Perusing books dealing with the metallurgical phenomena of hot and cold metal forming may also be useful.

1.9 CONCLUSIONS

The concern of the present book, strips and thin plates, were defined according to their geometry, such that the ratio of their width to thickness is much larger than unity. The flat rolling process, capable of producing strips and plates, was described in general terms. The integrated steel mill and hot strip mill, including its components, were described in some detail. Hot, warm and cold rolling were mentioned, and the temperature ranges for each were given. A brief presentation of some mill configurations was also given, including two-high, four-high, six-high and twenty-high arrangements. The Steckel mill, the Sendzimir mill and the planetary mill were discussed, accompanied by several illustrations. Mini-mills were presented and some comparisons of their capabilities to integrated steel mills were demonstrated. Material for further reading was also included, classified into two sections. In one, texts dealing with a general treatment plastic deformation of metals are listed. These include the necessary theory of plasticity in addition to the application of the theories to the analyses of bulk and sheet metal-forming problems. The second category includes specialist books, dealing with the process of rolling.

CHAPTER 2

Flat Rolling – A General Discussion

Abstract

A general discussion of the flat rolling process is presented. The components of a metal rolling system are defined. Reference is made to the rolling mill, designed by Leonardo da Vinci and the scale-model, built following his drawings. A description of the physical and the metallurgical phenomena during the rolling process is given, including the events as the strip to be rolled is ready to enter the roll gap, as it is partially reduced and as the process becomes one of steady-state. The independent variables of the system – connected with the mill, the rolled metal and their interface – are listed. The minimum value of the coefficient of friction, necessary to commence the rolling process is given. Some of the simplifying assumptions that are usually made in mathematical models of the process of flat rolling are critically discussed: these include the ideas of "plane-strain plastic flow" and "homogeneous compression of the strip". Microstructures of a fully recrystallized Nb steel, an AISI 1008 steel and a cold-rolled low-carbon steel are presented.

2.1 THE FLAT ROLLING PROCESS

The essential concept of the flat rolling process is simple and it has been in use for centuries to produce sheets and strips, or in other words, flat products. Leonardo da Vinci employed it to roll lead, utilizing a hand-cranked mill, depicted in Roberts' book, "Cold Rolling of Steel" (1978). The Leonardo museum, located in the medieval Castello Guidi, built in the 11–12th century in the city of Vinci – about 30 km from Firenze – contains some interesting examples of Leonardo's plans for a rolling mill, shown in the web-site of the museum (http://www.leonet.it/comuni/vinci/). Two figures, reproduced from that web-site, are given below. Figure 2.1 is a page of Leonardo's plans on which the handwriting is, unfortunately, indecipherable. The scale model, built according to these plans and shown in Figure 2.2, is able to roll a sheet of tin 30 cm wide.

The basic idea for the production of flat pieces of materials by rolling has not changed since the process has been invented. Dimensions, materials,

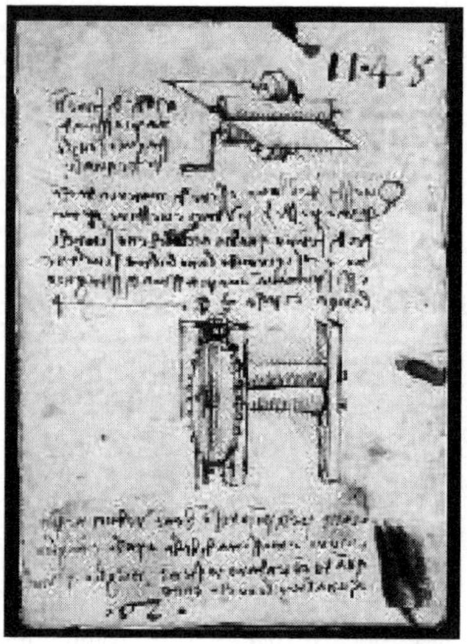

Figure 2.1 Leonardo's plans for a simple rolling mill.

Figure 2.2 The scale model of Leonardo's mill.

precision, speed, the mechanical and metallurgical quality of the product and most importantly, the mathematical analysis and the control of the process have evolved, however, and as a result, the flat rolling process may truly be considered one of the most successful "high-tech" processes, since for modern, efficient and productive applications the theories and practice of metallurgy, mechanics, mechatronics, surface engineering, automatic control, continuum mechanics, mathematical modelling, heat transfer, fluid mechanics, chemical engineering and chemistry, tribology and encompassing all, computer science are absolutely necessary.

2.1.1 Hot, cold and warm rolling

The rolling process may, of course, be performed at low and high temperatures, in the cold rolling mill or in the hot strip mill, respectively, as already mentioned in Chapter 1, Introduction. The formal distinction between what is low and what is high temperature, and in consequence, what are the cold and hot rolling processes, is made by considering the homologous temperature, in which the low end is at absolute zero and the high end is at the melting point of the metal to be rolled, T_m. When the process is performed at a temperature below $0.5T_m$, it is usually termed cold rolling while above that limit, hot rolling occurs. In addition to the above strict definitions of hot and cold rolling, there is the warm rolling process, as well. The temperature range for this phase is not defined very precisely but it starts somewhat below $0.5T_m$ and changes to hot rolling at some temperature above that. Each of these processes has advantages and disadvantages, of course. At high temperatures, at which hot rolling is performed, the metal is softer so less power may be needed for a particular reduction. Further, understanding the effects of the process parameters of the rolling process on the mechanical and metallurgical attributes allows the development of metals with specific, engineered properties; the process is termed thermal–mechanical treatment. The disadvantage of rolling at high temperatures concerns the development of a layer of scale on the surface and its effect on the process and on the quality of the resulting product. All of these need to be clearly understood, such that they may be controlled with confidence. Cold rolling follows after the pickling process in which the layer of scale is removed. Here the control of dimensional consistency and surface quality is the most important objective. Strict thickness and width tolerances must be maintained for the product to be commercially acceptable. During warm rolling, some of the disadvantages of the hot rolling process are minimized as scale formation is less intense. The energy requirements increase, however, as the metal's resistance to deformation is now higher.

2.1.2 Mathematical modelling

Use of sophisticated – on-line and off-line – mathematical models allows these activities to proceed. A number of these models have been developed, some simple and some making use of the availability of the finite-element method. Among the latter, the model, developed by the American Iron and

Steel Institute (Hot Strip Mill Model, HSMM) stands out. A quotation from the AISI web-site is given below:

> *"HSMM is one of several commercially licensed technologies developed under AISI's advanced process control program, a collaborative effort among steelmakers and the U.S. Department of Energy to create breakout steel technologies. HSMM simulates the steel hot-rolling process for a variety of steel grades and products, and forecasts final microstructure and properties, allowing the user to achieve a deeper insight into operations while optimizing product properties. Prior to its commercial release, it was used for several years by steel companies that helped develop the technology".*

In each of these models, the ideas of equilibrium, material behaviour and tribology are used to describe the physical phenomena. The first of these three is based on Newton's laws. The latter two require experimentation and the translation of the experimental data into mathematical expressions, for use in the models.

The traditional and simplest approach when mathematical analyses of metal forming processes are considered is to allow the material's resistance to deformation to be exclusively strain-dependent in the cold forming regime, and to be exclusively strain-rate-dependent in the hot range. In the warm forming range, the metal's strength is usually both strain- and strain-rate-dependent. It is acknowledged, of course, that these are much too oversimplified, and in Chapter 4, dealing with the attributes of the metals, other, more inclusive and more sophisticated material models are presented.

In advanced mathematical treatment, by finite difference or finite-element techniques for example, the metals' constitutive relation should be described in terms of several independent variables, including, at the very least, the strain, strain rate, temperature and metallurgical parameters, one of which is the Zener-Hollomon parameter, used extensively (see Chapter 4 on material attributes for the definition). In some instances the metal's chemical composition is also included in the equations.

The tribological events at the contact of the work roll and the rolled metal are also to be described in terms of parameters and variables.

2.1.3 The independent and dependent variables

The discussion above leads to the consideration of both the dependent and independent variables of the flat rolling process. It is first necessary to identify the boundaries of the domain under consideration. For the arguments presented in what follows, these define a single stand of a two-high or four-high mill, such that are used in most laboratories during the development of the data on forces, torques, energy requirements and the resulting microstructures. Under industrial conditions, a full-size mill stand, regardless of it being the roughing mill or one of the stands in the finishing train, is being considered. It is a metal rolling system that is, in fact, being defined and in the same sense as any metal forming system, this is divided into three, essentially independent but interconnected components: the rolling mill, the rolled metal and their interface.

The significant dependent variables depend on what the objectives of the industry, of the engineers operating the rolling process, of the researcher developing a mathematical model or of the customers happen to be. There may be three separate but interdependent objectives. One of these is the design of the rolling mill. Another objective is the design of the rolled strip while the third potential objective is the design of the interface between the rolls and the rolled strip.

The essential components of the rolling mill are the work rolls and the back-up rolls, the bearings, the mill frame, the drive spindles, the inter-stand tensioning devices, the heating and cooling equipment, the lubricant delivery apparatus and the driving motor. Their attributes, all which affect the rolling process, include their dimensions in addition to roll material, crowning, roll surface hardness, roll roughness and its orientation, mill frame load-carrying area, mill stiffness and hence, mill stretch. In choosing the driving motor, its power and speed are to be determined. Mill dynamics is also a significant contributor to mill performance and is a function of all of the above in addition to the process variables, such as the reduction, the speed and the dimensional consistency of the as-received metal. All of the above affect the quality of the rolled metal. All may be considered as independent variables.

The variables associated with the rolled metal include its mechanical, surface, metallurgical and tribological attributes: yield strength, tensile strength, strain and strain-rate sensitivity, bulk and surface hardness, ductility and formability, fatigue resistance, chemical composition, weldability, grain size and distribution, precipitates, surface roughness and roughness orientation.

Since the transfer of thermal and mechanical energy is accomplished at the interface of the work roll and the rolled strip, and the efficiency of that transfer is one of the most critical parameters of the process, the attributes of the contact are, arguably, the most important when the quality – dimensional accuracy, consistency and uniformity of the surface parameters – of the product is designed. The surface roughness of the rolls and their directions have been mentioned above. The attributes of the lubricant are to be considered here and among these are the viscosity and its temperature and pressure sensitivity, density, chemical composition and droplet dimensions if an emulsion is used. While environmental friendliness of the lubricant and the manner of its disposition after use are other, most important, considerations, they do not affect the quality of the product.

While all of these variables should be considered when the rolling process is designed and/or analysed, they rarely, if ever, are. Engineers simplify the task and consider only what is absolutely necessary. The rolling mill, its capabilities, its driving system and its peculiarities are, of course, given and will be changed only when forced by industrial competition or the development of processes that lead to increased productivity and/or reduced costs. The metal to be rolled must, in usual circumstances, be one in the product mix offered by the particular company. In unusual circumstances, if the customer requires a chemical composition or mechanical and metallurgical attributes different from those available, either the request would be refused and a metal, among the company's products and similar to the customer's prescription, would be offered or the costs of the development of the new metal would form a major part of the complete process costs. This second possibility would, of course, be prohibitively expensive. The choice of the lubricant, its volume flow and the lubrication system are often considered to be no more than a maintenance issue and in the opinion of the present writer, that is an uninformed and highly mistaken view. The most significant independent variables usually considered in the analyses of the flat rolling process are listed below, classified according to the three components of the metal rolling system. These will be dealt with in what follows, in some considerable detail.

The rolling mill

- Roll diameter and length
- Roll material
- Roll surface roughness
- Roll toughness and hardness

The metal

- Chemical composition
- Prior history – grain size, precipitates
- Constitutive relation
- Initial surface roughness

The interface

- Lubricant attributes – chemical compositions, viscosity, temperature and pressure sensitivity, density; droplet size if an emulsion is used.

The rolling process

- Reduction
- Speed
- Temperature

2.2 THE PHYSICAL EVENTS BEFORE, DURING AND AFTER THE PASS

These events have been mentioned by Lenard et al. (1999). In the discussion that follows, the earlier presentation has been enlarged and some, more important, ideas have been included. The first consideration is that in the presentation, the shapes of the work roll and the rolled strip are considered to be highly idealized. The work roll's cross-section is taken to be a perfect circle at the start and that makes the roll a perfect cylinder. The line connecting the roll centres is taken to be perpendicular to the direction of rolling and it remains so during the pass. The distance between the roll centres is considered to remain unchanged in most analyses, ignoring mill stretch[1]. The rotational speed of the work rolls is and remains constant, even after the loads on it increase and the inevitable slow-down under high torque loads is ignored. The strip to be rolled is straight, its sides make right angles to one-another, its thickness and its width are uniform as is the surface roughness.

The following events occur when the process of rolling a flat piece of metal starts, continues and is completed. As mentioned above, the work rolls are considered to be rotating at a constant angular velocity. The strip or the slab is moved towards the entry and is made to contact the rotating work rolls, either by vertical edge rollers or by a conveyor system or both. When contact is made, leading edge of the strip enters the deformation zone because of the friction forces exerted by the work rolls on it and it is not difficult to show that the minimum coefficient of friction, necessary for successful entry is given by

$$\mu_{\min} = \tan \phi_1 \tag{2.1}$$

where ϕ_1 is the bite angle.

In most analytical accounts of the entry, the first contact is assumed to be along a straight line, across the whole length of contact with the work roll. In reality, there must be a significant amount of deformation of both the edge of the strip and the roll, suggested by common sense as well as by the loud noise always heard on entry to the roll gap of an industrial mill. The leading edge of the strip must thicken somewhat on contact and the roll must flatten. Surprisingly, research concerning the geometrical changes at the instant of entry is not extensive. In one of the early attempts, Kobasa and Schultz (1968) used high-speed photography which allowed some, albeit limited, visualization of the entry conditions and the length of contact in the rolling process. The published photographs do not allow a clear, close look at the deformation at the initial contact, however.

The stresses in the metal increase and the limit of elasticity of the strip is reached soon after entry, followed by permanent, plastic deformation. One usually assumes that the flat strip rolling process can be described in terms of one independent variable, taken in the direction of rolling, and that the stresses do not vary across the strip thickness. This assumption in turn is based on the usual homogeneous compression assumption and the often-employed statement that "planes remain planes". With these assumptions, the elastic-plastic boundary becomes a plane perpendicular to the direction of rolling and on that plane the criterion of yielding is satisfied first. When the process is treated as a two-dimensional (2D) problem, as in many finite-element analyses, the elastic–plastic boundary may be quite different from that described above.

---

[1] Ignoring the mill stretch in the control software would, of course, lead to significant errors; mill stretch is accounted for in the set-up programs used for rolling mills. See Section 3.12.2 for a discussion of the mill stretch.

When strip rolling is considered and the roll diameter/strip thickness ratio is large – in industrial settings this ratio varies from about 25–30 at the first stand of the finishing train of the hot strip mill to as high as 400–1000 in the last stand – the one-dimensional (1D) treatment is perfectly adequate and yielding and the beginning of plastic flow are taken to occur on a plane, parallel to the line connecting the roll centres, as just described. The permanent deformation regime then remains in existence through most of the roll gap region, followed by the elastic unloading regime which starts when the converging channel of the roll gap begins to diverge. Often, the yielding process is simplified further, and rigid-plastic material behaviour is assumed to exist, ignoring the elastic deformation completely. With this assumption the rolled metal is taken to satisfy the yield criterion and to begin full plastic flow as soon as it enters the roll gap. The assumption of rigid-plastic behaviour is usually acceptable when hot rolling is analysed. It has often been shown that treating the cold rolling process requires the use of elastic-plastic material models.

These events are illustrated in Figures 2.3(a)–(c) which show a schematic diagram of a two-high mill and a strip ready to be rolled, rolled partway through and rolled continuously, in a steady-state condition. As well, the free

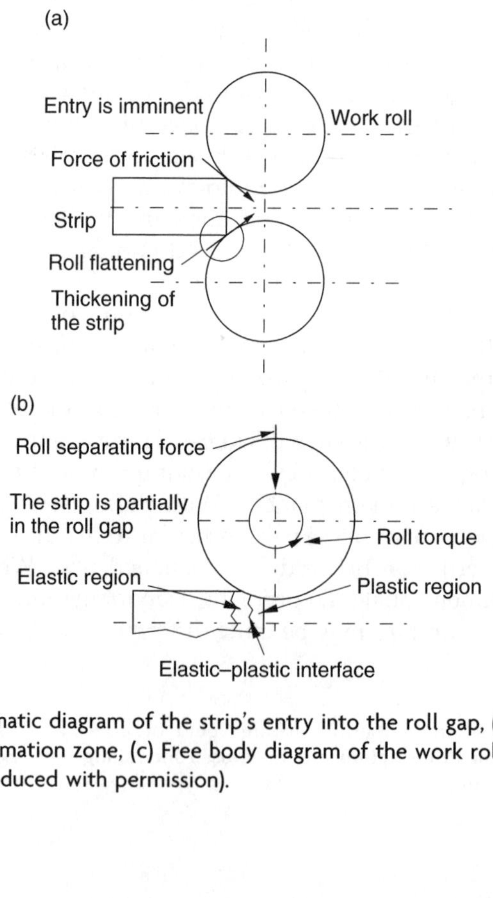

Figure 2.3 (a) Schematic diagram of the strip's entry into the roll gap, (b) The strip is partially in the deformation zone, (c) Free body diagram of the work roll (Lenard, Pietrzyk and Cser, 1999, reproduced with permission).

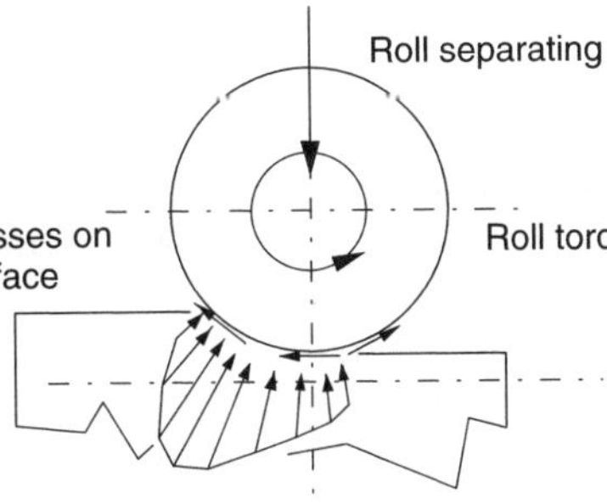

Figure 2.3 (Continued)

body diagram of the roll is indicated, showing the pressures, forces and torques acting on it. The conditions shown describe either a laboratory situation where no front and back tensions exist, or a single stand, reversing mill, such as a roughing mill.

Three stages of the rolling process are shown in Figures 2.3(a)–(c) respectively. In part (a), the strip is about to make contact. The strip velocity at this point is dependent on the edge rollers or the conveyor system but it is usually significantly less than the surface velocity of the work rolls. If the coefficient of friction is larger than the tangent of the bite angle, as indicated by eq. 2.1, a relationship that is often used to determine the minimum friction necessary to start the rolling process, the strip enters the deformation zone. In a laboratory mill, the usual practice is to carefully and lightly push the strip[2], placed on the delivery table, towards the work rolls and allow the friction forces to cause entry; under certain circumstances, when viscous lubricants are used or the roll speed is high, it is necessary to mill a shallow taper on the leading edge of the strip to facilitate the bite. Care must be exercised in this case, concerning the placement of the lubricant. In an experiment in the writer's laboratory, a highly viscous lubricant was used and to ensure bite, the tapered leading edge was left dry. Entry was achieved but when the rolls encountered the portion covered with the oil, the strip tore in two, in a violent fashion. It is concluded that sudden changes of the tribological conditions along the length of the strip to be rolled should be avoided.

In a hot strip mill, vertical edge rolls force the strip into the roll gap in the first stand. Entry creates some longitudinal compression of the strip and there will be some initial thickening as well, more than in cold rolling. This

[2] Caution is highly recommended here. The strip should never be pushed by hand but by a long piece of wood or another strip; this practice will eliminate the danger of rolling a few fingers of the operator.

is accompanied by local, elastic deformation of the work rolls, indicating that the usual simplification about the entry point – located where the perfectly straight edge of the strip encounters the undeformed, perfectly cylindrical roll – does not represent reality very well. In a cold mill, the strip is often threaded through the stationary mill and attached to the coiler on the exit side. The rolls then close, gradually reach the pre-determined roll gap and the mill is started, so no bite needs to be considered.

Part (b) of the figure shows the strip about half-way through the deformation zone. As previously mentioned, the unloaded metal first experiences elastic deformation, and when and where the yield criterion is first satisfied, plastic flow is observed. These two regimes are separated by an elastic–plastic boundary, the shape and the location of which should be determined by the mathematical analysis of the rolling process. In the elastic region, the theory of elasticity governs the deformation of the metal. In the permanent deformation region, the criterion of yielding, the appropriate associated flow rule, the condition of incompressibility and the appropriate compatibility conditions describe the situation in a satisfactory manner.

The rolls are further deformed – they bend and flatten. The magnitude of the roll stresses should not exceed the yield strength of the roll material. The theory of elasticity is to be used to determine the roll distortion and the corresponding changes of the length of contact.

In part (c), the leading edge of the rolled metal has exited and the rolling process is continuing, essentially as a steady-state event. The figure shows the pressures, the forces and the torques acting on the roll and on the strip. These include the roll pressure distribution and the interfacial shear stress, the integrals of which over the contact length lead to the roll separating force and the roll torque, respectively. These are the dependent variables the mathematical models are designed to determine. If front and back tensions are present, as would be the case under industrial conditions in the finishing mill stands, their effect on the longitudinal stresses at the entry and exit should be included in the definitions of the boundary conditions. The roll separating force and the roll torque may be used to study the metal flow in the roll gap. As well, they may be used to design the rolling mill itself. Knowledge of the magnitude of the roll separating force is needed to size the mill frame, the roll neck bearings and the roll dimensions, including roll crowning and roll flattening. The roll torque is necessary to establish the dimensions of the spindles, the couplings and the power required for the driving motor.

The surface velocities of the roll and the strip should also be considered. It may be assumed that the driving motor is of the constant torque variety and that the rolls rotate at a constant angular velocity, even though there may be some slow down under high loads. The strip usually enters the roll gap at a surface velocity less than that of the roll. The friction force always points in the direction of the relative motion, and on the entering strip it acts to aid its movement. As the compression of the strip proceeds, its velocity increases[3] and it approaches that of the roll's surface. When the two velocities are equal, the no-slip region is reached, often referred to as the neutral point[4]. At that location, the strip and the roll move together, and their relative velocity vanishes. If the neutral point is between the entry and the exit, the strip experiences further compression beyond it and its surface velocity surpasses that of the roll. Several researchers suggest that reference should be made to a *neutral region* instead of a *neutral point*, hypothesizing that the no-slip condition extends over some distance. In the region, between the neutral point and the exit, the friction force on the strip has changed direction and is now retarding its motion. The site between the entry and the neutral point is often referred to as the region of backward slip. The location between the neutral point and the exit is called the region of forward slip.

The forces shown in Figure 2.3c are the external loads acting on the work rolls. The rolls are in equilibrium, of course, and the surface forces at the contact must be balanced by other, also external forces. These originate at the bearings that exert the forces on the rolls to keep them in equilibrium, in a relatively stationary position. There are two types of loads at the roll bearings. One is a vertical force, minimizing the possibility of the roll moving upward, called the roll separating force. The other is a turning moment, originating from the drive spindle, referred to as the roll torque. In a two-high mill, these are the loads acting on the work roll, balancing the effects of the loads originating at the interface: the pressure of the strip on the roll and the interfacial frictional forces. If a four-high configuration is studied the forces – normal and shear – at the back-up roll and the work roll contact need to be included as part of the free-body diagram.

The picture changes somewhat when front and back tensions are also considered, as would have to be done to account for the effects of the preceding and the subsequent mill stands and the effects of the loopers – these are devices in between mill stands that keep some tensile forces in the strips. These forces act in the direction of rolling, of course, and would have an effect on the magnitudes of both the roll separating forces and the roll torques. It is possible and simple to include the effect of inter-stand tensions in the mathematical models of the process.

The knowledge of the roll separating force and torque is necessary for three possible purposes:

1. to design the mill – its frame, bearings, drive systems, lubricants and their delivery;
2. to determine the dimensions and the properties of the rolled metal;
3. to allow the development of control systems for on-line control of the process.

---

[3] Recall that the assumption of plane-strain flow implies no width changes. Incompressibility implies that the sum of thickness and the length strain should vanish.

[4] The ideas of the "neutral point" and the "neutral region" will be discussed in more detail in Chapter 7, dealing with temper rolling.

2.2.1 Some assumptions and simplifications

In dealing with the process of flat rolling, it is advantageous to consider two assumptions frequently made when mathematical models are developed. The first is to acknowledge the almost true fact that the width of the flat product is practically unchanged, the *plane-strain flow phenomenon*, and the second, again almost true, which allows the use of ordinary differential equations in the models, the *planes remain planes* simplification.

2.2.1.1 Plane-strain flow

The flat rolling process is usually taken to be essentially two-dimensional, in the sense that the width of the product doesn't change by much during the pass when compared to thickness and length changes and this makes the assumption of plane-strain plastic flow[5] quite realistic. When one is to study roll bending and the attendant changes of the shape of the rolled metal in addition to the changes of the field variables across the width – stress, strain, strain rate, temperature, grain distributions – a change from the two-dimensional mathematical formulation to 3D is unavoidable.

Of course, the width changes in the flat rolling process and this and its effect on the resulting product have been considered in numerous publications. As long as the width to thickness ratio is over 10, however, this change is not taken to be very significant. In rolling experiments, using strips of about 1 mm thickness and 10–25 mm width, the strain in the width direction is rarely over 2–3%.

Some further consideration of the term "the width doesn't change by much" is necessary here, in light of a recent publication by Sheppard and Duan (2002) who used FORGE3® V3, a three-dimensional, implicit, thermomechanically coupled, commercially available finite-element program to analyse spread during hot rolling of aluminum slabs. While the authors' predictions correspond to experimental and industrial data very well, the slabs they examined cannot be considered to behave according to the plane-strain assumption. In their study, the slabs measure 25 mm width and 25 mm entry thickness, rolled using a roll of 250 mm diameter. In the industrial example, the work piece's measurements are: 1129 mm width and 228 mm entry thickness. The roll diameter is 678 mm. In both cases, lateral spread – measured and calculated – is shown to be considerable.

2.2.1.2 Homogeneous compression

A discussion of the homogeneous compression assumption is also necessary here. This phenomenon has been studied experimentally by visio-plasticity methods in addition to observing the deformation of pins, inserted into the

[5] The deformation is deemed "plane-strain" when the strains in two directions are very much larger than in the third direction.

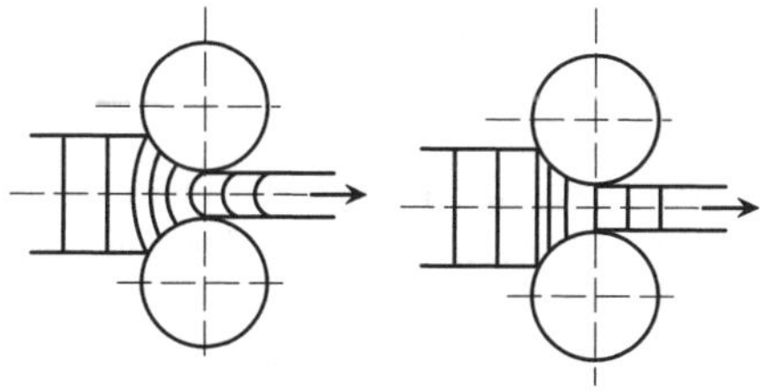

Figure 2.4 (a) Non-homogeneous compression. (b) Homogeneous compression.

rolled metal. Figure 2.4 above shows, in part (a) that the originally straight lines bend, while in (b) they don't and the original planes remain planes.

In the second case, the compression of the strip during the rolling pass is referred to as "homogeneous compression". Schey (2000) differentiates between the two possibilities, depending on the magnitude of the ratio of the average strip thickness in the pass, $h_{ave} = 0.5\left(h_{entry} + h_{exit}\right)$, and the length of the contact, $L = \sqrt{R'\Delta h}$, where R' is the radius of the flattened but still circular work roll (this idea will be discussed later, in Chapter 3, dealing with mathematical modelling of the process) and $\Delta h = h_{entry} - h_{exit}$, of course. When h_{ave}/L is larger than unity, the deformation in inhomogeneous and the originally straight planes bend, as shown in Figure 2.4(a). When the ratio is under unity, the effects of friction on the rolling forces and torques are significant and the homogeneous compression assumption may be made with confidence.

When strip rolling is discussed, whether hot or cold, the "planes remain planes" assumption is very close to reality, with one possible exception. This concerns metal flow in the first few passes of the slab through the roughing train of a hot strip mill where the strip thickness is in the order of 200–300 mm and the work rolls may be 1 m or more in diameter, leading to a roll diameter/strip thickness ratio in the order of 3–5. In the finishing train, this ratio increases by at least an order of magnitude and the plane-strain assumption becomes acceptable. Venter and Adb-Rabbo (1980) examined the effect of Orowan's (1943) inhomogeneity parameter on the stress distribution in the rolled metal. They concluded that the effect is more significant when sticking friction is considered to exist, compared to sliding friction[6]. The distributions of the roll pressure, with or without the inhomogeneity parameter differed by about 10%.

When one considers strip rolling however, in which case the roll diameter to entry thickness ratio is large in comparison to unity in addition to

[6] While sticking friction has been assumed to exist in hot rolling in past analyses, recent studies indicate that it rarely occurs in the flat rolling process; use of lubricants reduces the coefficient of friction.

the width/thickness ratio also being large, homogeneous compression, that is, planes remaining planes during the pass, as well as the assumption of plane-strain flow are quite close to the actual events. In what follows, both assumptions will be made without any further reference. Further simplifications and assumptions will be detailed and discussed in Chapter 3, dealing the details of mathematical modelling of the flat rolling process.

2.3 THE METALLURGICAL EVENTS BEFORE AND AFTER THE ROLLING PROCESS

The rolling process begins by continuous casting[7], or if an older, not modernized steel plant is considered, by ingot casting. In the most modern mills, continuous casting is followed directly by hot rolling, see Figure 1.6. In all of these cases the pre-rolling structure consists of dendrites which are subsequently removed in the reheat furnaces in which most of the alloying elements enter into solid solution. It may be assumed then that at the start of the rough rolling process, the sample is in the austenite range and that it has been fully annealed and recrystallized before entry into the roughing mill stands[8]. The structure is made up of strain free, equiaxed grains. The steel is reduced in the rougher, in several steps, all performed at relatively high temperatures and not excessive rates of strain and it then passes on to the finishing train. The grain structure at this stage depends on the pass schedule in the rougher, but, as has been mentioned above, the influence of the metallurgical structure prior to entry into the finishing train has little influence on the final attributes. Two typical examples of the steel's structure are shown in Figure 2.5(a) and (b), reproduced from the publication of Cuddy (1981). The figures show the microstructures, obtained by subjecting the samples to several, sequential plane-strain compression tests[9]. The chemical composition of the microalloyed steel was also given; it contained 0.057% C, 1.44% Mn and 0.112% Nb. The steel was reheated to 1200°C and deformed by 55% in five passes. Figure 2.5(a) shows a fully recrystallized structure, obtained at a deformation temperature of 1100–1070°C. The test shown Figure 2.5(b), which was conducted at a lower deformation temperature of 1000–960°C, indicates flattened grains and, as a result, some strain hardening.

Following rough rolling the transfer bar enters the finishing train where the microstructure undergoes further changes, again depending on the draft

---

[7] See Figure 1.5, Chapter 1.

[8] It is difficult to prove the validity of this assumption as it is impossible to interrupt the rolling process to remove a piece of the hot steel for metallography. Some of the micrographs that are shown have been obtained from various laboratory simulations.

[9] Use of sequential, multi-stage hot compression tests in simulating the multi-pass rolling process will be discussed in Chapter 4, Material Attributes.

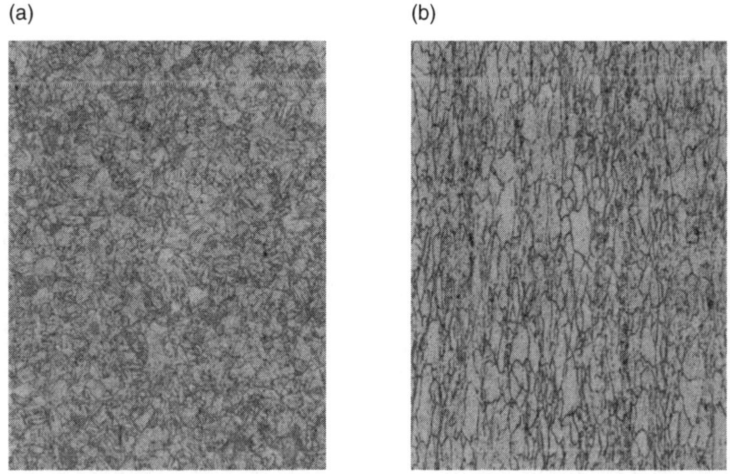

Figure 2.5 (a) The microstructure of a Nb carrying steel fully recrystallized after 55% deformation in five passes, at 1100–1070°C; The magnification is 100x, (b) The same steel, subjected to the same deformation pattern but at a lower temperature of 1000–960°C, shows significant grain elongation. The magnification is 100x (Cuddy, 1981, reproduced with permission).

schedule, which is usually prepared off-line, using mathematical models that are able to predict the expected metallurgical and mechanical attributes. There are prohibitively many possibilities to consider in one book so only a typical structure is shown in Figure 2.6, reproduced from the ASM Handbook (1985).

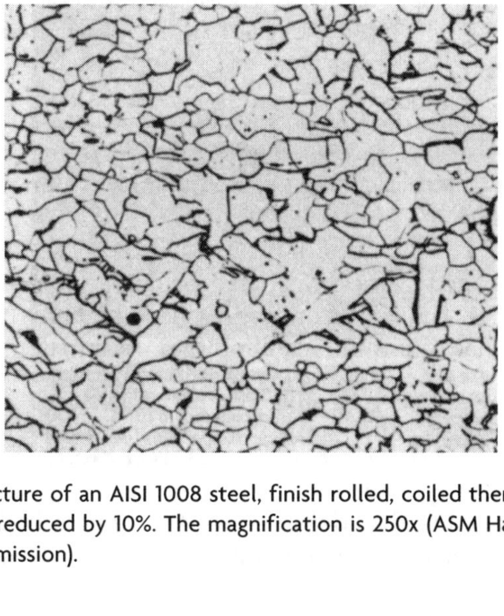

Figure 2.6 The structure of an AISI 1008 steel, finish rolled, coiled then hot rolled from a thickness of 3 mm, reduced by 10%. The magnification is 250x (ASM Handbook, 1985, reproduced with permission).

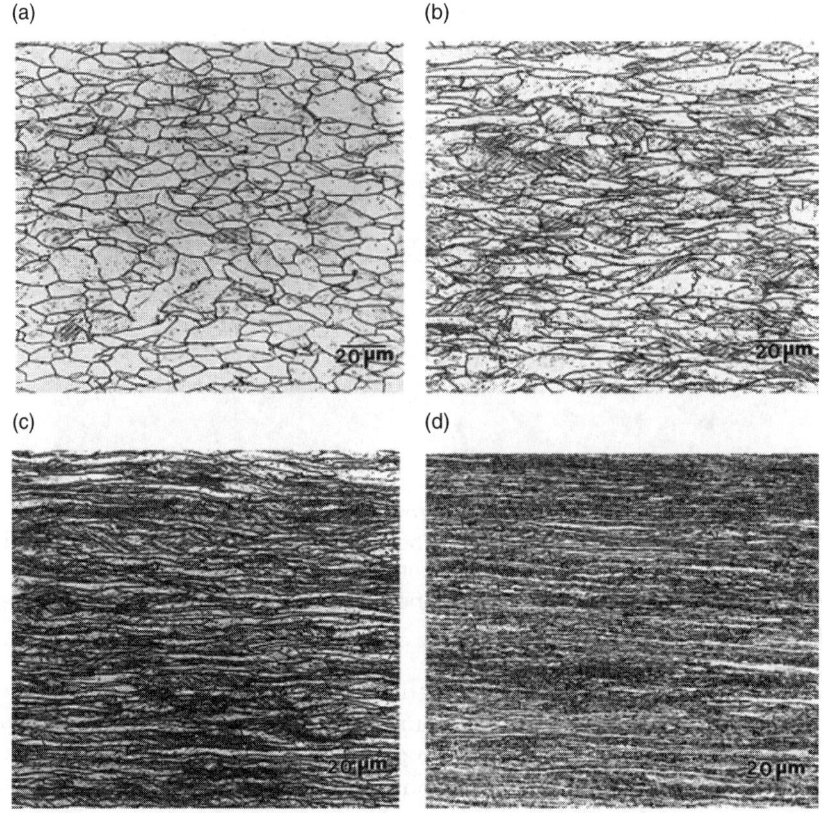

Figure 2.7 Microstructure of a cold-rolled, low carbon steel sheet showing ferrite grains at (a) 30%; (b) 50%; (c) 70% and (d) 90% cold reduction. The magnification is 500× (Benscoter and Bramfitt, 2004, reproduced with permission).

The structure of a capped, AISI 1008 steel is shown at a magnification of 250. The steel was finish rolled, coiled then hot rolled from a thickness of 3 mm, reduced by 10%. The steel was then cooled in air, resulting in the fully ferritic microstructure.

The next step that follows is the cold rolling process after the hot rolled, scaled surface is cleaned by pickling in hydrochloric acid. Several passes reduce the thickness further. The effects of progressively higher reductions are shown in Figure 2.7, demonstrating the resulting grain elongation.

2.4 LIMITATIONS OF THE FLAT ROLLING PROCESS

There are several limits that designers of the draft schedule of flat rolling must be aware of. One of these, the minimum coefficient of friction necessary to initiate the process, has been mentioned above, see equation 2.1. Other limitations of the process include the minimum rollable thickness, alligatoring and edge cracking. The first of these appears to be caused by the creation of a hydrostatic state of stress in the deformation zone. The latter two are also the consequence of the stress distribution; specifically the tensile stresses associated with the elongation of the rolled samples.

2.4.1 The minimum rollable thickness

This phenomenon[10] is observed to occur when a thin, hard strip is to be reduced in a single rolling pass, using large diameter rolls. In order to increase the reduction, the work rolls are progressively brought closer and closer in an attempt to reduce the roll gap. As the reduction is increased, the compression on the strip is also increasing and the work rolls deform more and more. After a certain gap dimension is reached, no further reductions of the thickness of the strip are possible; the minimum rollable thickness has been reached. A hydrostatic state of stress is supposed to have been built up within the strip in the deformation zone. Recalling that the material undergoing permanent plastic deformation retains its volume, no further change of the dimensions of the metal is possible. If the work rolls are forced to close still further, they flatten more, the mill frame stretches further and the minimum rollable thickness cannot be reduced any more. Further attempts are likely to cause damage to the mill. This thickness is a function of the material attributes of the metal as well as the elastic attributes of the work roll and of the mill frame.

Early researchers estimated the magnitude of the minimum obtainable thickness in a rolling pass. Stone (1953) presented the formula

$$h_{\min} = \frac{3.58\, D \mu \sigma_{\mathrm{fm}}}{E} \tag{2.2}$$

where the roll diameter is D, its elastic modulus is E, and σ_{fm} is the mean resistance of the rolled material to reduction (see eq. 3.2). Tong and Sachs (1957) also predict that the minimum rollable thickness is proportional to the same parameters, as in eq. 2.2. Johnson and Bentall (1969) hypothesize that the minimum rollable thickness does not actually exist in practice. Domanti et al. (1994) write that rolled thickness, beyond those predicted was achieved in foil rolling mills.

Nevertheless, the minimum rollable thickness is a real, actual limitation of the industrial rolling process and its existence has been demonstrated in several instances.

Researchers, using small-scale laboratory rolling mills are cautioned against attempting to demonstrate the existence of the minimum thickness. It is possible to force the work rolls together more and more, of course, but the chances

[10] The minimum thickness problem will be mentioned again in Chapter 7, Temper Rolling.

Figure 2.8 Edge cracking of an aluminum alloy, hot rolled at 505°C to a strain of 0.6 (Duly et al., 1998).

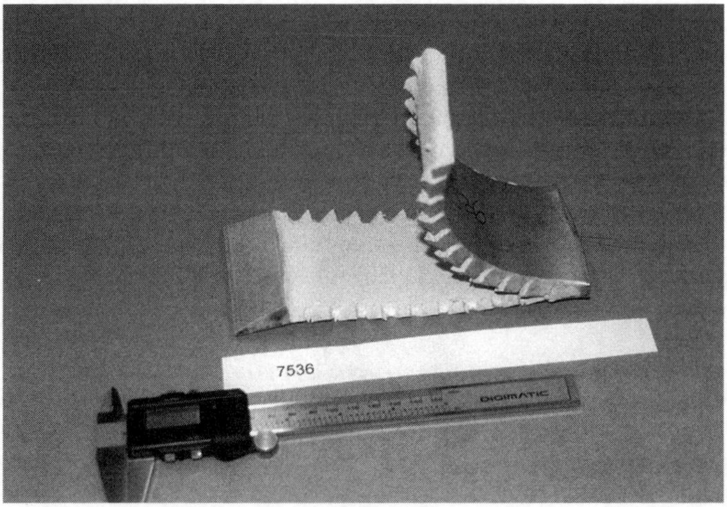

Figure 2.9 Alligatoring and edge-cracking of an aluminum alloy, hot rolled at 497°C to a strain of 0.56.

of creating permanent damage to the mill and the attendant costs of replacing the cracked rolls are both usually prohibitively high.

2.4.2 Alligatoring and edge-cracking

The rolled strip's length grows while it is being reduced and the tensile strains in the direction of rolling often limit the reductions possible in a single pass. The stress distribution in the deformation zone may cause either alligatoring or edge-cracking or both. These were purposefully created while hot rolling aluminum strips with tapered edges (Duly et al. 1998), in order to examine the workability of the alloys. In each pass, the work rolls were covered with a light coating of mineral seal oil. Severe edge cracking of the sample is shown in Figure 2.8, rolled at a temperature of 505°C. Edge cracking and alligatoring are demonstrated in Figure 2.9.

Workability and the limits of the process during hot rolling of steel and aluminum were considered in some detail by Lenard (2003).

2.5 CONCLUSIONS

A brief, general presentation of the flat rolling process was given. Two assumptions – the "planes remain planes" and "homogeneous compression" – necessary for the understanding of the flat rolling process, were critically examined. The physical and the metallurgical events experienced by the steel were discussed. These included the examination of the free-body-diagram of the work roll, in three conditions: the strip is ready to enter the roll gap; it is partially through and steady-state rolling has been reached. As far as the metallurgical phenomena are concerned, several micrographs were presented, each showing the microstructure of the rolled strips, undergoing various rolling schedules. The limits of the process were presented.

CHAPTER 3

Mathematical and Physical Modelling of the Flat Rolling Process

Abstract

There are two interrelated concepts covered in this chapter. First, the modelling of the mechanical events during the flat rolling process is considered, including the ideas of static and dynamic equilibrium of the rolled strips and plates, the elastic and plastic response of the materials to loading, interfacial friction and temperature effects. These are followed by a discussion and modelling of the metallurgical phenomena, as a result of the treatment the strip receives during its passage through the rolling mill, including the hardening and the restoration mechanisms. In each component, mathematical models of the processes are developed or presented with the observations and ideas described in terms of the laws of nature, empirical relations, physical simulation and assumptions.

The essential, basic ideas in mathematical modelling of the flat rolling process are presented first. Empirical and one-dimensional (1D) models, applicable for strip rolling, are described and their predictive capabilities are demonstrated. Extremum principles – specifically the upper bound theorem – are shown. The use of artificial intelligence (AI) in predicting the rolling variables is discussed. The need whether to include the effect of inertia forces in (1D) models is considered. A model, employing the friction factor instead of the coefficient of friction, is derived and its predictive abilities are examined in some detail. The development of the microstructure – as a result of the restoration and hardening phenomena – during hot rolling and its effect on the resulting mechanical attributes are given. Thermal–mechanical treatment is briefly discussed, and the physical simulation of the flat rolling process is also included. In the last section, several phenomena often ignored in the traditional mathematical models of the process, are given. These include the forward slip, mill stretch, roll bending, the lever arm and the effects of cumulative strain hardening. An approach to modelling, which considers the difficulties associated with determining the relevant values of the coefficient of friction and the metals' resistance to deformation, is suggested.

3.1 A DISCUSSION OF MATHEMATICAL MODELLING

Mathematical models of the flat rolling process are numerous and are easily available in the technical literature. The publications date from the early days of the 20th century to the present. Their complexity, mathematical rigour, predictive ability and ease of use vary broadly. In what follows, models, applicable to strip and plate rolling only will be presented, such that the large roll diameter to strip thickness ratios allow the application of the "planes remain planes" assumption, implying that homogeneous compression is present in the deforming metal. This step and the additional assumption of the plane-strain plastic flow condition[1] ensure that there will be only one independent variable in the equations, the distance along the direction of rolling or the angular variable around the roll. Thus, ordinary differential equations will be obtained, the integration of which is considerably less difficult than that of partial differential equations that would be obtained without the two assumptions.

The available models can be listed according to the objectives their authors have while devising them. They are applicable equally well to hot, warm or cold rolling. These objectives may include the following:

- A simple, fast calculation of the roll separating forces;
- In addition to the roll separating force, the roll torque, the temperature rise and the required power are to be calculated;
- In further addition to the above, the determination of the metallurgical parameters and the material attributes as a result of the hot and cold rolling are to be determined.

A more extensive list of the use of mathematical models of the rolling process is given by Hodgson et al. (1993). The authors add setup and online control of the rolling mills and the rolling process to the use of the models, in addition to the following:

- Minimize mill trials for product and process development;
- Evaluate the impact of different mill configurations and new hardware on the process and the work piece;
- Predict variables which cannot be easily measured (e.g. bulk temperature, temperature distribution, austenite grain size, post-cooling mechanical and metallurgical attributes);
- Perform sensitivity analyses to determine which process variables should be measured and controlled to achieve the required quality, or final properties, of the product;
- Aid hardware design;
- Further understand the physical process.

Another comment needs to be mentioned in the context of using the models for predictions of rolling loads, etc. The present author has been involved in the study of one-dimensional (1D) models of the flat rolling process for quite some time. The studies involved experimentation as well as modelling, and the

[1] These two assumptions have been discussed in Chapter 2.

predictive abilities of several 1D models were investigated. When the research studies began to appear in the technical literature, using finite element models to investigate the flat rolling process, the following suggestion was made to several authors: experimental data would be provided and let us all compare our predictions. None took up the challenge. One comment was received: "Our analyses are performed to get insight into the mechanics of the process, not for predictions".

In what follows, some of the basic, classical (1D) models[2] are reviewed in addition to some of the more recent efforts. While the following list is not complete, it gives the most popular and well-known formulations. A model which includes an account of the variation of frictional effects along the roll/strip contact is also described, employing the friction factor instead of the coefficient of friction. Upper bound analysis of the process is discussed. The use of neural networks for the prediction of the variables in the rolling process is demonstrated. The development of the metallurgical structure of the rolled strips is then reviewed and empirical relations, allowing the calculation of these parameters, are listed. As well, relations that predict the attributes of the material after the rolling process are given. Further, the predictive abilities of the models are presented and compared to each other and to experimental data.

Each of these models will be developed in more detail in subsequent sections, classified as follows:

- *The empirical models*. An example of these, which can be used with considerable ease, is presented by Schey (2000). Manual calculations, spread sheets or simple computer programs are sufficient while calculating the roll separating force. The major objective of the models is just that: a simple and fast but reasonably accurate prediction of the roll separating force. The roll torque, the power and the temperature rise may also be obtained but their accuracy is usually not quite as good as that of the force, no doubt because of the assumptions made in their determination;
- *The one-dimensional models*. These are capable of predicting the roll separating forces as well as the roll torques quite well. The traditional models of these types are based on the classical Orowan (1943) approach, including the idea of the "friction hill" and its simplifications. For cold rolling the Bland and Ford (1948) technique and for hot rolling the Sims' model (1954) are often used in the steel industry, usually as a first approximation, often followed by adjusting the predictions to data taken on a particular rolling mill. Alternatively, the Cook and McCrum (1958) tables, based on the 1D Sims' model may be employed. The predictive ability here is enhanced by accounting for the flattening of the work roll under the action of the roll

[2] It is recognized that these models have been published quite some time ago, yet they often form the bases of existing online models.

pressure. The well-known Hitchcock formula (1935) is used in these models to estimate the magnitude of the radius of the flattened but still circular work roll while in a more refined 1D version (Roychoudhury and Lenard, 1984) the elastic deformation of the roll is analysed, using the two-dimensional (2D) theory of elasticity. Interfacial frictional phenomena are modeled in two ways: mostly using the coefficient of friction and sometimes the friction factor[3]. The objectives here are similar to those above: that of the calculation of the roll separating force and the roll torque. The models can also be used to estimate the dimensions of some components of the rolling mill, such as the cross-sectional areas of the load-carrying columns of the mill frame, the dimensions of the bearings, the drive spindles and possibly the power of the driving motor[4]. In addition to the above, the empirical relations, describing the evolution of the metallurgical structure (the amount of static, dynamic and metadynamic recrystallization, recovery, precipitation, retained strain, volume fraction of ferrite, as well as the mechanical and metallurgical attributes after hot forming and cooling) during and after hot rolling, may be added to the 1D models. These equations are based on the studies of Sellars (1979, 1990), Roberts (1983), Laasraoui and Jonas (1991a,b), Choquet et al. (1990), Hodgson and Gibbs (1992), Yada (1987), Beynon and Sellars (1992), Sakai (1993), Kuziak et al. (1997) and Devadas et al. (1991). The predictive abilities of the above relations, while combined with 2D finite-element mechanical models, were reviewed by Lenard et al. (1999). A discussion of modelling the rolling process has been given in Chapter 16 of the Handbook of Workability and Process Design (2003);
- *Extremum theorems.* A model based on the extremum principles which, using the upper bound theorem, gives a conservative estimate of the power necessary for the rolling process;
- *Artificial neural networks (ANN).* The models based on AI approaches are also included in this chapter. For the derivation of the mathematical bases of AI the reader is referred to specialized texts. In this section, AI methods are mentioned only briefly, and the focus is on the predictive ability of the models, performed after the necessary training.

In addition to the care in the formulation of the models, the predictive ability of all of them depends, in a very significant manner, on the appropriate mathematical description of the rolled metal's resistance to deformation and on the way the frictional resistance at the contact surfaces is expressed. Both of these phenomena are considered in detail in subsequent chapters. Material and metallurgical attributes are the topic of Chapter 4 while Tribology is treated in Chapter 5.

---

[3] See Chapter 5 for the definitions.

[4] Note that the driving motor must have sufficient power to overcome friction losses and to compensate for the significantly less than 100% efficiency of the drive train. See eqs 3.8–3.11.

Caution is necessary when the choice of the most appropriate model in a particular set of circumstances is made. Often there is the tendency among researchers to select an advanced model and expect superior predictive capabilities; this step usually results in disappointment. The guiding principle should always be to make the complexity of the model match the complexity of the objectives. Further, the mathematical rigour of all components, such as the material and the friction models, should match the rigour of the mechanical and the metallurgical formulation.

A few comments, concerning the propensity of researchers to comment on the predictive capabilities of their models, are appropriate here. The usual tendency, when the predictions are compared to a few experimental results and the numbers compare well, is to proclaim that the model represents physical events very well. There are two concepts to consider, however, before good predictive ability is to be claimed. These are *accuracy* and *consistency*. A model that is accurate only sometimes and an error analysis has not been considered[5] in the study is essentially useless. A model, whose predictions may not be most accurate but are consistent, as demonstrated by the low standard deviation of the difference between calculated and measured data, is always useful as it can always be adjusted, a practice often followed in industry.

While mathematical models of the flat rolling process have been regularly published in the technical literature, a complete list of all of them is much too large to be included in the present volume. Several conferences have been held in the recent past, entitled "Modelling of Metal Rolling Processes" and the issues involved with all aspects of the rolling process have been discussed at these gatherings.

An interesting review of online and off-line mathematical models for flat rolling was recently published by Yuen (2003). He examined the models that account for the flattening of the work rolls as well as those that exclude it. He also discussed the models available for foil and temper rolling. He concluded that more sophisticated models are expected to be adopted for online applications in the future and added that there is an

> *urgent need for robust algorithms in order to implement these superior models.*

Mathematical models of the rolling process are now also available commercially. One of the outstanding ones, mentioned already in Chapter 2, has been made available by the American Iron and Steel Institute, the details of which can be obtained from the website http://www.integpg.com/Products/HSMM.asp.

A large collection of software for simulation and process control can also be found at the website http://www.mefos.se/simulati-vb.htm.

---

[5] It is indeed rare to see the analysis of the magnitudes of possible errors in the mathematical models.

3.2 A SIMPLE MODEL

A simple model, fast enough for online calculations of the roll separating force, has been presented by Schey in his text "Introduction to Manufacturing Processes", 3rd edition (2000). The model expresses the roll separating force per unit width (P_r) in terms of the average flow strength of the rolled metal in the pass, the projected contact length, $L = \sqrt{R'\Delta h}$, a multiplier, identified by Schey as the pressure intensification factor, Q_p, to account for the shape factor and friction and a correction for the plane-strain flow, $(2/\sqrt{3}) \approx 1.15$, in the roll gap. The radius of the work roll, flattened by the loads on it, is designated by R'(see below for Hitchcock's equation, eq. 3.3). For the case when homogeneous compression of the strip may be assumed and frictional effects are significant, that is $L/h_{ave} > 1$, h_{ave} being the average strip thickness, the model is written:

$$P_r = 1.15 Q_p \sigma_{fm} L \tag{3.1}$$

where the mean flow strength of the metal, σ_{fm}, is obtained by integrating the stress–strain relation over the total strain, experienced by the rolled strip:

$$\sigma_{fm} = \frac{1}{\varepsilon_{max}} \int_0^{\varepsilon_{max}} \sigma\,(\varepsilon)\; d\varepsilon \tag{3.2}$$

The radius of the flattened roll, R', is obtainable from Hitchcock's relation (1935)

$$R' = R\left[1 + \frac{16\left(1-\nu^2\right)}{\pi E\left(h_{entry} - h_{exit}\right)} P_r\right] \tag{3.3}$$

Hitchcock's equation, and the assumptions on which it is based, has been controversial ever since it was published quite some time ago. While the critiques are valid – the roll does not remain circular in the contact zone and the roll pressure distribution is not elliptical – equation 3.3 still enjoys widespread use.

Roberts (1978) examines the validity of Hitchcock's equation[6] and concludes that

> *The generally accepted Hitchcock equation, even considering elastic strip flattening, is not adequate to predict the length of the arc of contact between roll and strip.*

Schey (2000) also presents an approach to deal with the rolling of thick plates where the assumption of homogeneous compression is not valid any

---

[6] See also Section 3.4.1 where roll deformation is discussed in more detail.

longer. In these cases the shape factor is less than unity, $L/h_{ave} \leq 1$. The plastic deformation is affected less by friction and a different pressure intensification factor is to be used[7].

The simplicity of the model of eq. 3.1 is evident when one considers its relation to simple compression, akin to open die forging, expressing the force needed by the product of the mean flow strength and the projected contact length. This is then adapted to that of flat rolling by the application of the pressure intensification factor and the correction for plane-strain flow.

In a rolling pass, the total true strain is $\varepsilon_{max} = \ln(h_{entry}/h_{exit})$ and in cold rolling the stress–strain relation is usually taken as $\sigma(\varepsilon) = K\varepsilon^n$ or $\sigma(\varepsilon) = Y(1+B\varepsilon)^{n_1}$ where K, n, Y, B and n_1 are material constants. Other formulations for the metal's resistance to deformation are also possible of course, and some of these will be presented in Chapter 4. When hot rolling is analysed the mean flow strength is expressed in terms of the average rate of strain, $\dot{\varepsilon}_{ave}$:

$$\sigma_{fm} = C(\dot{\varepsilon}_{ave})^m \tag{3.4}$$

where the average strain rate is given, in terms of the roll surface velocity, v_r, the projected contact length and the strain by:

$$\dot{\varepsilon}_{ave} = \frac{v_r}{L} \ln \frac{h_{entry}}{h_{exit}} \tag{3.5}$$

The material parameters C, m, K, n may be taken from the data of Altan and Boulger (1973) for a large number of steels and non-ferrous metals or may be determined in a testing program. Choosing more complex constitutive relations requires the use of non-linear regression analysis to determine the material constants, such as Y, B and n_1.

Care is to be taken when previously published material models are considered for use. Karagiozis and Lenard (1987) compared the predictive capabilities of several published constitutive relations, all claimed by their authors to be valid for low carbon steels. On occasion, the predictions varied by a factor of two (see Figure 4.13). The recommendation therefore is as follows: if there is any doubt about the applicability or the accuracy of the material model that describes the resistance to deformation of the metal to be rolled, independent testing for the strength is necessary[8].

Finally, the multiplier Q_p, the pressure intensification factor, is obtained from Figure 3.1 in terms of the coefficient of friction and the shape factor L/h_{ave} where h_{ave} is the average of the entry and exit thickness. The torque to drive

[7] This problem is not dealt with in the present manuscript. If interested, please refer to the original reference, Schey (2000).

[8] Testing techniques are described in Chapter 4.

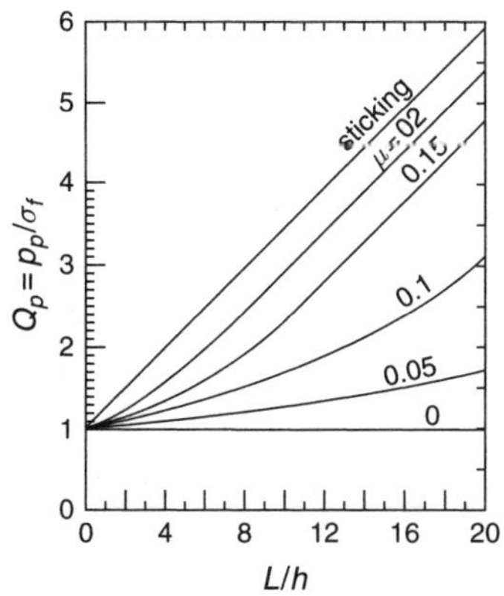

Figure 3.1 The pressure intensification factor Q_p, (Schey, 2000; reproduced with permission).

both rolls per unit width is then expressed, assuming that the roll force acts halfway between the entry and the exit:

$$M = P_r L \tag{3.6}$$

The lever arm, the distance by which the roll separating force is to be multiplied to determine the roll torque, is thus defined as the projected contact length[9].

The power to drive the mill is determined using the torque, from eq. 3.6, and the roll velocity. The relation gives the power in watts, provided the contact length is in m, the velocity in m/s, the roll radius in m and the units of the width, w, match those of the roll separating force/unit width:

$$P = P_r w L \frac{v_r}{R'} \tag{3.7}$$

Note that eq. 3.7 gives only the power for plastic forming of the strip and thus, it is not to be confused with the power needed to drive the rolling mill, which is significantly larger. In order to develop the specifications for the power of the driving motor of the mill, friction losses and the efficiency of all drive-train components need to be considered. Rowe (1977) defines the overall power requirement in terms of these parameters, in the form:

$$P_{\text{total}} = \frac{1}{\eta_m \eta_t}(2P + 4P_n) \tag{3.8}$$

where η_m is the efficiency of the driving motor and η_t is the efficiency of the transmission, including all of its components. Equation 3.7 supplies the power

[9] See Section 3.12.5 where the lever arm is discussed in some more detail.

required for plastic deformation, and the friction losses in the four roll-neck bearings (P_n) are given by $\mu_n P_r w d\dot{\phi}$, where μ_n is the coefficient of friction in the bearings, d is the bearing diameter and $\dot{\phi}$ is the angular velocity of the roll. Roberts (1978) includes the resistance of the material in an equation that can also be used to predict the mill power:

$$P_{\text{total}} = \frac{1}{\eta_m}\left\{h_{\text{entry}} w v_{\text{r}}\left[\sigma_{\text{fm}} r + \frac{(\sigma_{\text{entry}} - \sigma_{\text{exit}})(1-r)}{1-0.5r}\right] + P_n\right\} \tag{3.9}$$

where $\sigma_{\text{fm}}$is the dynamic, constrained yield strength of the rolled strip in N/m^2, w is the width in m, v_{r} is the velocity in m/s and r is the reduction in fractions; the power then will be obtained in watts. The yield strength at room temperature, in lb/in^2, of the "softer" strip is given in terms of the reduction as[10]

$$\sigma = 40000 + 1773r - 29.2r^2 + 0.195r^3 \tag{3.10}$$

where the reduction, r, is given again as a fraction. The constrained yield strength is then taken to depend on the rate of strain:

$$\sigma_{\text{fm}} = 1.155\left(\sigma + 4460 \log_{10} 1000\dot{\varepsilon}\right) \tag{3.11}$$

Roberts (1978) presents a set of calculations of the overall power required to cold roll hard and soft low carbon steels through a 6-stand rolling mill. The predicted and measured input motor powers varied by at most 20% and were under 10% in most cases.

In the calculations the efficiency of the driving motor was taken from a low of 76.7 to 88.7%, fairly reasonable values. In light of the successful predictive ability, the use of eq. 3.9 is recommended. As an alternative, the upper bound method may be used to calculate the power necessary to roll the strip; as is well known, the upper bound method gives a conservative estimate of the power for plastic deformation[11].

In an unpublished study, a simple experiment, to estimate the power losses due to friction and drive-train inefficiency, was conducted in the present writer's laboratory. The work rolls were compressed to a certain magnitude of the roll separating force with no strip in between. The mill was turned on and only the torque to drive the mill was measured. The power thus obtained was in the order of 30% of the power when a strip was reduced in a similar fashion.

[10] The conversion to SI units is: 1 lb/in^2 = 6.89476×10^{-3} MPa.

[11] The upper bound approach is treated later in this Chapter, see Section 3.9.1.

The rise of the temperature of the strip in the pass due to plastic work may be estimated by

$$\Delta T_{\text{gain}} = \frac{P}{\text{mass flow} \times \text{specific heat}} \tag{3.12}$$

where the power is to be expressed in J/s, the mass flow is to be in kg/s and the specific heat of the metal (c_p) is to be in J/kg °C. Equation 3.12 may be written in terms of the roll force, the geometry and the density in the form:

$$\Delta T_{\text{gain}} = \frac{P_r L/R'}{\rho c_p h_{\text{ave}}} \tag{3.13}$$

Note that while neither the speed of rolling nor the width of the strip appears in the equation, the strain rate would increase with increasing speeds and that would affect the magnitude of the roll separating force, and hence, the power. Note further that an error was knowingly committed in estimating the mass flow: the roll surface speed and the average strip thickness were used instead the thickness at the no-slip point which should have been used. While the location of the neutral point may be estimated, it is not known precisely, so the small error, no more than 10%, may be forgivable. Further, care must be exercised in the use of units. The roll separating force is to be in N/m; the contact length is to be in m; the roll radius to be in m; the density is to be in kg/m^3, the specific heat is to be in J/kgC and the average strip thickness is to be in m.

Roberts (1983) also gives a useful expression to estimate the temperature rise of the strip in the pass, which simply takes the work done/unit volume and assumes that all of the work done is converted to heat. The temperature increase, due to reduction r may then be calculated by

$$\Delta T_{\text{gain}} = \frac{\sigma_{\text{fm}}}{\rho c_p} \ln \frac{1}{1-r} \tag{3.14}$$

A numerical experiment illustrates the magnitude of the predicted rise of the temperature of a hot rolled steel strip. For example, consider a 30% reduction of an initially 10 mm thick strip, using 500 mm radius work rolls which rotate at 50 rpm. Assume that the coefficient of friction is 0.2, a reasonable magnitude when some lubrication is used. Let the density be 7570 kg/m^3, take the specific heat of the steel to equal 650 J/kg K and let the average flow strength in the pass be 150 MPa. The temperature rise is now predicted to be 20°C by eq. 3.12. The process and material parameters change drastically in the last stand of the finishing mill. Let the entry thickness be 2 mm, the roll's speed be 150 rpm and the reduction to be 50%, so the final strip thickness will be 1 mm. The temperature of the strip is lower now, so the average flow strength is 250 MPa; this number includes the effect of the strain rate, caused by the increased rolling speed. Let $\mu = 0.15$. The temperature rise is now much higher, calculated to be 131°C.

The temperature loss in the pass, due to conduction only is obtained as suggested by Seredynski (1973) in terms of the pass parameters, the heat transfer coefficient[12] and the density and specific heat of the rolled steel. Seredynski's formula is

$$\Delta T_{\text{loss}} = 60\alpha \sqrt{\frac{r}{h_{\text{entry}} R}} \left(T_{\text{strip}} - T_{\text{roll}}\right) \left[(1 - r)\, \pi \rho\, c_{\text{p}} N\right]^{-1} \tag{3.15}$$

where α is the heat transfer coefficient at the roll/strip interface (Seredynski gives its value as 44 kW/m^{2}K); r is the reduction in fractions, h_{entry} is the entry thickness, R is the original, undeformed roll radius, T_{strip} and T_{roll} are the temperatures of the strip and the roll, respectively, ρ is the density (7570 kg/m^3), c_{p} is the specific heat (650 J/kgK) and N is the roll rpm[13]. The temperature loss in the above two examples may be estimated now, using eq. 3.15. Most of the numbers are known except one: the temperature of the roll. Roberts (1983) shows the experimental results of Stevens et al. (1971) who used thermocouples embedded in a full-scale work roll to monitor the rise of the temperature of the surface. The results indicate that the roll surface temperature may rise by as much as 500°C[14]. With these numbers, the strip entering the roll gap at 1000°C may cool by as much as 19°C. In the second example the loss of temperature is estimated to be 25°C, affected by the shorter contact time, the larger reduction and the thinner strip. The final temperature of the strip after rolling will be the algebraic sum of these two values[15].

Roberts (1983) also presents the analysis of Stevens et al. (1971) to estimate the rise of the surface temperature of the roll, in terms of its bulk temperature, the time of contact and the thermal properties of the roll material: it's thermal conductivity, thermal diffusivity and the conductance. The calculations presented show that the rise of the surface temperature of the roll is somewhat less than those of the experiments of Stevens et al. (1971).

The rise of the roll's surface temperature may be estimated by the relation, developed by Stevens et al. (1971). The equation relates the roll's surface temperature (T_{roll}), the roll's temperature some distance below the surface (T_0), the strip's temperature at the entry (T_{strip}) to the time of contact (t), the density

---

[12] The heat transfer coefficient will be discussed in detail in Chapter 5, Tribology.

[13] The numbers are taken from Roberts (1983).

[14] The experiments of Tiley and Lenard (2003) on an experimental mill indicate that the roll's surface temperature may rise by as much as 200°C.

[15] Note that in the example only two phenomena were considered: temperature rise due to plastic work done and temperature loss due to conduction. A more advanced thermal treatment needs to consider the temperature changes associated with radiation, convection and the metallurgical events.

and to several thermal parameters of the roll material. The formula is written in the form (Roberts, 1983):

$$\frac{T_{\text{roll}} - T_0}{T_{\text{strip}} - T_0} = \alpha\sqrt{\frac{t}{k\rho c_{\text{p}}}} \tag{3.16}$$

where k is the thermal conductivity of the roll material in W/m K. Roberts (1983) writes that the magnitude of T_0 used in the calculations is not a critical variable. Typical calculations may be performed to appreciate the validity of the assumptions made above concerning the rise of the temperature of the roll's surface. The thermal conductivity, in W/m K, is dependent on the temperature, as indicated by Pietrzyk and Lenard (1991):

$$k = 23.16 + 51.96 \exp\left(\frac{-2.025\,T}{1000}\right) \tag{3.17}$$

where T is the temperature of the strip in Kelvin. When the time of contact is 0.01 s, and the heat transfer coefficient, the specific heat and the density are as in the example above, the conductivity calculated to be 28 W/m K by eq. 3.17, the strip is at a temperature of 900°C, the roll's bulk temperature is 100°C, the roll's surface is predicted to rise by 400°C, close to the measurements of Stevens et al. (1971).

3.3 ONE-DIMENSIONAL MODELS

3.3.1 The Classical Orowan model

Most of the 1D models are based on the equilibrium method in which a slab of the deforming material is isolated and a balance of all external forces acting on it is used to develop a differential equation of equilibrium[16]. Since the original treatment, published by Orowan (1943), is often considered to be the industry standard and other models' predictions are usually compared to its calculations, it is worthwhile to review it in some detail. A detailed review and a thorough critical discussion of the method have also been given by Alexander (1972)[17] who also published a computer program, in FORTRAN, to analyse the flat rolling process. The model is based on the static equilibrium of the forces in a slab of metal undergoing plastic deformation between the rolls (see Figure 3.2).

[16] If inertia forces are expected to be significant contributors to the stresses, equations of motion need to be developed, equating the sum of all forces to the product of the mass and the acceleration. This concept is dealt with in Section 3.5 of this Chapter.

[17] Note that Alexander indicated the existence of compressive stresses in the direction of rolling, acting on the isolated slab. In Figure 3.2 these stresses are shown as tensile and the boundary conditions are expected to determine if they are tensile or compressive.

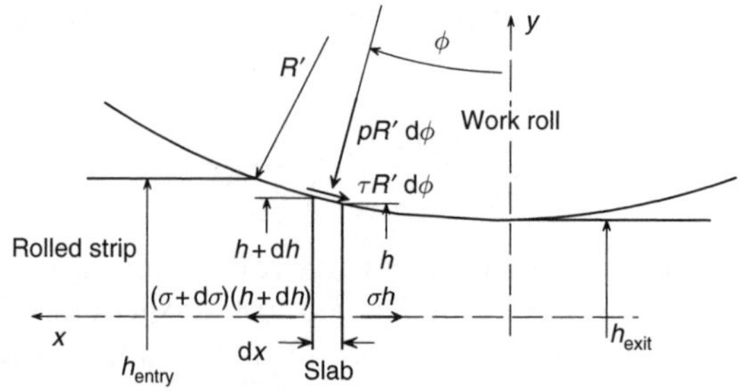

Figure 3.2 The schematic diagram of the rolled strip and the roll showing the forces acting on a slab of the deforming material.

The forces due to the roll pressure, distributed along the contact arc, the interfacial shear stress and the stresses in the longitudinal and the transverse directions form the force system, the equilibrium of which in the direction of rolling leads to the basic equation of balance. Assuming that planes remain planes allows this relation to be a 1D, ordinary differential equation of equilibrium in terms of the dependent variables: the roll pressure p, the strip thickness h, the radius of the deformed roll R', the interfacial shear stress τ, the stress in the direction of rolling σ_x and the independent variable x, indicating the distance in the direction of rolling, measured from the line connecting the roll centres:

$$\frac{\mathrm{d}(\sigma_x h)}{\mathrm{d}x} + p\frac{\mathrm{d}h}{\mathrm{d}x} \mp 2\mu p = 0 \tag{3.18}$$

where the $\mp$ sign indicates that the above equation describes the conditions of equilibrium between the neutral point and the entry (when using the negative sign), as well as between the neutral point and the exit (when using the positive sign). In fact, Eq. 3.18 is comprised of two independent ordinary, first order differential equations, containing four dependent variables: σ_x, p and h, all of which depend on R' in turn, in addition to the coefficient of friction. The interfacial shear stress has already been replaced by the product of the coefficient of friction and the normal pressure in eq. 3.18, as suggested by the Coulomb–Amonton formulation.

The necessary additional independent equations are obtained from the theory of plasticity and the geometry of the deformation zone. These include the Huber–Mises criterion of plastic flow, relating the stress components in the direction of rolling and perpendicular to it to the metal's flow strength. With the assumption of plane-strain plastic flow, the criterion becomes

$$\sigma_x + p = 2k \tag{3.19}$$

where k designates the metal's flow strength in pure shear. The other variable, the strip thickness, can be obtained from geometry:

$$h = h_{\text{exit}} + 2R'(1 - \cos\phi) \approx h_{\text{exit}} + \frac{x^2}{R'} \tag{3.20}$$

The approximate formula is valid as long as the angles are much smaller than unity, true in the case of thin plate and strip rolling. The radius of the flattened roll is obtained using the original Hitchcock equation (see eq. 3.3).

In order to integrate eq. 3.18, the metal's resistance to deformation is to be described and the interfacial shear stress needs to be given, usually as a function of the coefficient of friction and the roll pressure

$$\tau = \mu p \tag{3.21}$$

as was done already in eq. 3.8. Substituting eq. 3.19 into eq. 3.18 leads to

$$\frac{\mathrm{d}p}{\mathrm{d}x} \pm 2\mu\frac{p}{h} = \frac{2k}{h}\frac{\mathrm{d}h}{\mathrm{d}x} + \frac{\mathrm{d}(2k)}{\mathrm{d}x} \tag{3.22}$$

which, with the use of eq. 3.20 and an expression for $2k$ – see eqs 3.23 and 3.24 – is ready to be integrated.

The computation to determine the roll separating force and the roll torque begins with the integration of the equilibrium equations for the roll pressure. Starting at entry, using the appropriate boundary conditions [$p_{\text{entry}} = \sigma_{\text{entry}} - 2k_{\text{entry}} - \tau \tan\phi_1$], where ϕ_1 is the roll gap angle, and the – sign of the coefficient of the friction term, integration leads to a curve for the roll pressure. The next step is integration from the exit, and again using the appropriate boundary condition there [$p_{\text{exit}} = 2k_{\text{exit}} - \sigma_{\text{exit}}$] and now the + sign, leads to another curve for the pressure distribution. Two curves thus produced give the pressures exerted by the rolled strip on the roll, referred to as the friction hill. (Note that the subscripts "entry" and "exit" in the parentheses refer to the values of the designated parameters at those locations. The terms σ_{entry} and σ_{exit} indicate the front and the back tensions, respectively. In most laboratory mills or a single-stand roughing mill, these are not applied.)

The location of the intersection of the curves is defined as that of the neutral point, at which the roll surface velocity and that of the strip are equal, and no relative movement between them takes place. Further integration of the roll pressure distribution over the contact, from entry to the exit, leads to the roll separating force. Integration of the product of the roll radius and the shear forces from the entry to the exit leads to the roll torque. The necessity of accounting for the flattening of the work roll makes an iterative solution unavoidable. In the first set of calculations, rigid rolls are assumed to exist, that is $R = R'$. In the second iteration, the roll separating force, that has just been determined, is used to calculate the flattening of the roll employing Hitchcock's relation, see eq. 3.3, and using the radius of the flattened roll, a new roll force is obtained. The iteration is stopped when a pre-determined tolerance level on the roll force is satisfied. Corrections for the contribution of the elastic entry and exit regions can also be included in Orowan's model; for the details see Alexander (1972).

Equation 3.18 may be used to analyse either the cold, warm or hot flat rolling processes, the difference being the manner of the description of the term $2k$, the metal's resistance to deformation. If cold rolling is considered, one may follow Alexander (1972) and use the relation

$$2k = \frac{2}{\sqrt{3}} Y \left(1 + \frac{2}{\sqrt{3}} B \ln \frac{h_{\text{entry}}}{h}\right)^{n_1} \tag{3.23}$$

where the $2/\sqrt{3}$ multiplier corrects the stress–strain relation, obtained in a uniaxial tension or compression test, to be applicable for the analysis of the plane-strain flow problem of flat rolling. If hot rolling is to be studied, the resistance to deformation needs to be expressed in terms of the strain rate, at the very least. A form, often used, is

$$2k = \frac{2}{\sqrt{3}} C \dot{\varepsilon}^m \tag{3.24}$$

where C and m need to be determined in independent tests. In a more advanced approach the equation should include several more parameters. These will be discussed further in Chapter 4, Material Attributes.

3.3.2 Sims' model

Sims (1954) takes advantage of the fact that the angles in the roll gap are small when compared to unity, leading to the approximations $\sin\phi \approx \tan\phi \approx \phi$ and $1 - \cos\phi \approx \phi^2/2$. He also assumes that the product of the interfacial shear stress and the angular variable is negligible when compared to other terms and that sticking friction, that is $\tau = k$ is present in the contact between the roll and the strip[18]. These simplifications, in addition to assuming that the material of the rolled metal is characterized as rigid-ideally plastic, allow for a closed form integration of the equation of equilibrium, and the roll separating force per unit width is obtained as:

$$P_r = 2k\,L\,Q_p \tag{3.25}$$

---

[18] Note that the sticking friction assumption is not appropriate, even in hot rolling. The coefficient of friction, as a function of the temperature, will be discussed in Chapter 5, Tribology.

an equation that is similar to that of Schey, see eq. 3.1. In eq. 3.25, the term $2k$ stands for the yield strength of the metal, obtained in plane-strain compression. The contact length is as given above, $L = \sqrt{R'\Delta h}$, and the multiplier Q_p is dependent on the ratio of the radius of the flattened roll, the exit thickness of the rolled strip and the thickness of the strip at the neutral point, h_{np}:

$$Q_p = \left[\frac{\pi}{2}\sqrt{\left(\frac{1-r}{r}\right)}\tan^{-1}\sqrt{\left(\frac{r}{1-r}\right)} - \frac{\pi}{4} - \sqrt{\left(\frac{1-r}{r}\right)\left(\frac{R'}{h_{\mathrm{exit}}}\right)} \right.$$

$$\left. \times \ln\frac{h_{\mathrm{np}}}{h_{\mathrm{exit}}} + \frac{1}{2}\sqrt{\left(\frac{1-r}{r}\right)\left(\frac{R'}{h_{\mathrm{exit}}}\right)}\ln\frac{1}{1-r}\right] \quad (3.26)$$

where the thickness of the strip at the neutral point is found by equating the magnitudes of the roll pressures there. The location of the neutral point, ϕ_n, is obtained from:

$$\frac{\pi}{4}\ln(1-r) = 2\sqrt{\frac{R'}{h_{\mathrm{exit}}}}\tan^{-1}\sqrt{\frac{R'}{h_{\mathrm{exit}}}}\phi_n - \sqrt{\frac{R'}{h_{\mathrm{exit}}}}\tan^{-1}\sqrt{\frac{r}{1-r}} \quad (3.27)$$

In the above relations, r stands for the reduction in the pass, expressed as a fraction. Because of its simplicity, Sims' model often forms the basis of online roll force models for hot rolling in the steel rolling industry, albeit it is adapted for the particular mill on which it is used. The mathematical model of Caglayan and Beynon (1993), called SLIMMER, makes use of Sims' approach and combines it with several relationships that describe the microstructural evolution of the rolled metal. The model developed by Svietlichnyy and Pietrzyk (1999) for online control of hot plate rolling also uses Sims' model to calculate the roll separating forces.

3.3.3 Bland and Ford's model

In addition to the small angle assumption, Bland and Ford (1948) assume that the roll pressure equals the stress in the vertical direction and since the difference between them is a function of the cosine of very small angles, the error is not large, especially in cold rolling where roll diameters are usually much larger than the thickness of the strip. As with the Sims' model, this allows a closed form solution to be obtained. The roll force is then expressed:

$$P_{\mathrm{r}} = 2kR'\left\{\int_0^{\phi_n}\frac{h}{h_{\mathrm{exit}}}\exp(\mu H)\,\mathrm{d}\phi + \int_{\phi_n}^{\phi_1}\frac{h}{h_{\mathrm{entry}}}\exp\left[\mu\left(H_{\mathrm{entry}} - H\right)\right]\mathrm{d}\phi\right\} \quad (3.28)$$

where H is given by:

$$H = 2\sqrt{\frac{R'}{h_{exit}}}\tan^{-1}\left(\sqrt{\frac{R'}{h_{exit}}}\phi\right) \tag{3.29}$$

The location of the neutral point is calculated by :

$$\phi_n = \sqrt{\frac{h_{exit}}{R'}}\tan\frac{H_n}{2}\sqrt{\frac{h_{exit}}{R'}} \tag{3.30}$$

and the term H_n is determined by the formula

$$H_n = \frac{H_{entry}}{2} - \frac{1}{2\mu}\ln\frac{h_{entry}}{h_{exit}} \tag{3.31}$$

where the subscripts indicate the conditions at the entry or at the exit. The angular distance at the entry, the roll bite, is obtained by $\phi_1 = \sqrt{(h_{entry} - h_{exit})/R'}$. The Bland and Ford model is often used in the rolling industry in the analysis and the control of the cold rolling process.

Puchi-Cabrera (2001) used the Bland and Ford approach in a different way. He considered cold rolling of an aluminum alloy, from a thickness of 6 mm to a final thickness of 0.012 mm. The industrial practice is to roll the alloy in three stages. In the first stage, the work piece is reduced from 6 to 0.68 mm in four passes. In the second stage, the alloy is annealed and in the third stage reduced, in several passes, to the 0.012 mm thickness. In each of the rolling passes the reduction and hence, the rolling load is reduced and the mill's full capability is not utilized. The author considered the effects of maintaining a constant load during each of the multistage reductions and concluded that this may create clear advantages in terms of productivity, product quality and roll life.

3.4 REFINEMENTS OF THE OROWAN MODEL

Introducing the equations of elasticity to analyse the elastic entry and exit regions, as well as the deformation of the work roll leads to a somewhat more fundamental model of the flat rolling process (Roychoudhury and Lenard, 1984). The model is still based on the equilibrium method – the Orowan approach – and it is applicable when the roll radius to strip thickness ratios are much larger than unity, allowing for the assumptions of homogeneous compression and plane-strain flow in the roll gap. The differences between this model and the Orowan approach are as follows:

- Hitchcock's equation is replaced by a 2D elastic analysis of roll deformation. The rolls are assumed to be solid cylinders initially, which deform under the action of non-symmetrical normal and shear stresses during the pass. Two-, four-or six-high roll arrangements can be treated by the analysis, depending on how the work rolls are kept in balance is described mathematically. The 2D theory of elasticity, coupled with the elastic–plastic 1D treatment of the rolled strip, is used to determine the contour of the deformed roll;

- The elastic loading and unloading regions in the rolled strip at the entry and exit, respectively, are analysed using the 1D theory of elasticity. The locations of the elastic/plastic interfaces at both locations then become parts of the unknowns and are determined during the solution process by using the Huber–Mises criterion of plastic flow;
- The equation of equilibrium is written using the variable in the direction of rolling as the independent variable. The roll gap is divided into a finite number of slabs, each of which is assumed to be either elastic or ideally plastic. As the metal is deformed and strain hardens during the rolling process, the flow strength of each slab is changed accordingly. As well, the roll pressure and the interfacial shear stresses are expressed in terms of Fourier series. A closed form solution for each slab is thus obtained. Assembling the slabs is accomplished by enforcing horizontal equilibrium, leading to the complete solution for the pressure distribution and hence, to the roll separating force and the roll torque;
- The roll pressures and the interfacial shear stress distributions thus obtained are then used to calculate the contour of the deformed roll, using the Fourier series and the biharmonic equation;
- The current shape of the flattened work roll is used to determine the roll pressure, interfacial shear stress, the roll separating force and the roll torque, which are employed to re-calculate the roll contour;
- As above, the iteration is continued until satisfactory convergence of the roll force is reached.

Since the details of the model have been published (Roychoudhury and Lenard, 1984), only a brief exposition is given below. The schematic diagram from which the equation of equilibrium is derived, shown in Figure 3.3, differs from the one used by Orowan (1943) and Alexander (1972) in that the roll contour is taken to be an unknown function $y = f(x)$, to be determined as part of the computations.

The balance of the forces on a slab of the rolled metal is now derived using the direction of rolling as the independent variable, leading to:

$$\frac{\mathrm{d}}{\mathrm{d}x}\left[h\left(p - 2k \mp \tau\frac{\mathrm{d}y}{\mathrm{d}x}\right)\right] = 2\left(p\frac{\mathrm{d}y}{\mathrm{d}x} \pm \tau\right) \tag{3.32}$$

and as before, the positive algebraic sign in front of the interfacial shear stress indicates the region between the neutral point and the exit and the lower sign designates the region between the neutral point and the entry. The thickness

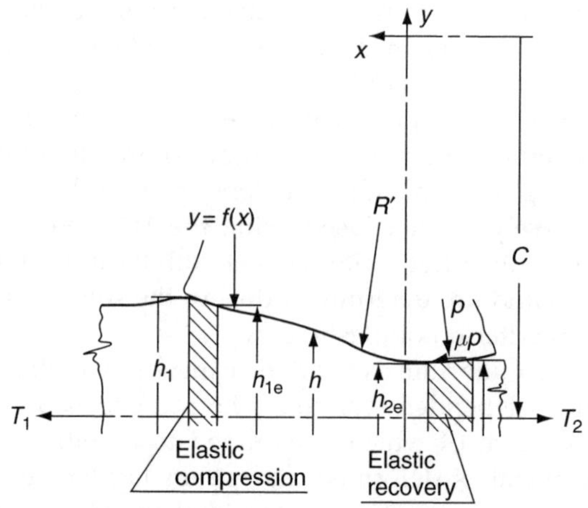

Figure 3.3 Schematic diagram of the rolled metal and the roll; the model of Roychoudhury and Lenard, (1984).

of the strip, using C to designate half the distance between the two centres of the two roll-neck bearings, is:

$$h = 2\ (y + C) \tag{3.33}$$

Using the coordinate system in Figure 3.3, the term y in eq. 3.33 becomes a negative number. Expressing the roll contour of one particular slab as a straight line, defined by

$$y = ax + b \tag{3.34}$$

and, as mentioned above, assuming that each slab is either elastic or is made of an ideally plastic metal simplifies eq. 3.32, and closed form integration, slab by slab, is now possible. The constants of integration are determined by assembling the slabs such that horizontal equilibrium is assured.

The elastic regions at the entry and exit are also accounted for in the model, explicitly. The equation of equilibrium, eq. 3.32 is valid in those regions as well. Combining them with the 1D plane-strain form of Hooke's law leads to the stress distributions in the elastic loading and recovery regions. Using the Huber–Mises yield criterion, and matching the elastic and the plastic stress distributions, the locations of the elastic–plastic boundaries are thus determined.

The analysis requires the explicit determination of the constants a and b, defining the roll contour at each slab. By expressing the roll pressures and the interfacial shear stress distributions in terms of Fourier series and analysing roll flattening following Michell's 2D elastic treatment (Michell, 1900), the roll separating forces are obtained as:

$$P_r = p\left[\int_{x_{entry}}^{x_n}\left(1+\mu\frac{dy}{dx}\right)dx+\int_{x_n}^{x_{exit}}\left(1+\mu\frac{dy}{dx}\right)dx+\int_{x_{entry}}^{x_n}\left(\mu-\frac{dy}{dx}\right)dx\right.$$
$$\left.+\int_{x_n}^{x_{exit}}\left(\mu+\frac{dy}{dx}\right)dx\right] \tag{3.35}$$

and the roll torques:

$$M/2=p\int_{x_{entry}}^{x_n}\left[x-y\frac{dy}{dx}+\mu\left(y+xy\frac{dy}{dx}\right)\right]dx-p\int_{x_n}^{x_{exit}}\left[x-y\frac{dy}{dx}-\mu\left(y+x\frac{dy}{dx}\right)\right]dx \tag{3.36}$$

3.4.1 The deformation of the work roll

The critique of Roberts (1978) concerning the use of Hitchcock's formula to calculate the radius of the flattened but still circular work rolls, mentioned above, is well accepted. The elastic flattening of the rolls has been treated by Jortner et al. (1960). The authors considered the effect of a force on the deflection of a solid cylinder and showed that the rolls do not in fact, remain circular in the deformation zone. Non-circular roll profiles have also been developed by Grimble (1976) and Grimble et al. (1978).

The problem of the deformation of the work roll is treated here by assuming that the work roll is a solid cylinder, subjected to non-symmetrical loads. The loading diagram is shown in Figure 3.4 (Roychoudhury and Lenard, 1984) where the roll pressure is designated by $p(\phi)$ and the interfacial shear stress by $\tau(\phi)$. The roll is kept in equilibrium in one of two ways. If there is a back-up roll, the pressure between the two rolls will keep the work roll in its place. If a two-high mill is considered, the roll centre is taken to be stationary, achieved by letting $2\xi = \pi$, where 2ξ is the extent of the pressure distribution of the now imaginary back-up roll.

The stress distribution in any problem of linear elasticity should satisfy the biharmonic equation, which in 2D cylindrical coordinates is

$$\left(\frac{\delta}{\delta r^2}+\frac{1}{r}\frac{\delta}{\delta r}+\frac{1}{r^2}\frac{\delta^2}{\delta\phi^2}\right)\left(\frac{\delta^2\varphi}{\delta r^2}+\frac{1}{r}\frac{\delta\varphi}{\delta r}+\frac{1}{r^2}\frac{\delta^2\varphi}{\delta\phi^2}\right)=0 \tag{3.37}$$

where the stress components are defined in terms of the Airy stress function, φ, as

$$\sigma_r = \frac{1}{r}\frac{\delta\varphi}{\delta r}+\frac{1}{r^2}\frac{\delta^2\varphi}{\delta\phi^2} \tag{3.38}$$

$$\sigma_\phi = \frac{\delta^2\varphi}{\delta r^2} \tag{3.39}$$

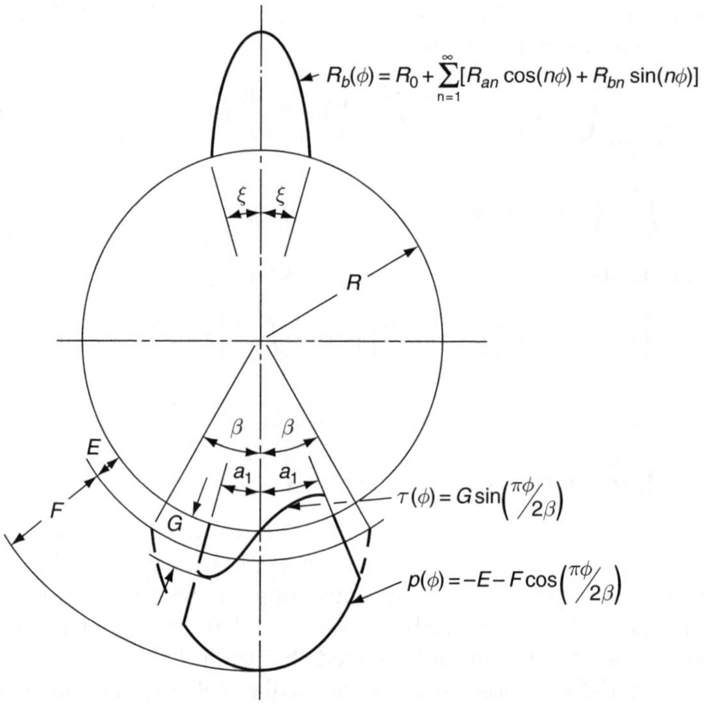

Figure 3.4 The loading diagram of the work roll showing the roll pressure and the interfacial shear stress distributions in addition to the forces that keep the roll in equilibrium (Roychoudhury and Lenard, 1984).

and

$$\tau_{r\phi} = -\frac{\delta}{\delta r}\left[\frac{1}{r}\left(\frac{\delta\varphi}{\delta\phi}\right)\right] \tag{3.40}$$

Following Michell (1900), the stress and the strain distributions may be calculated using biharmonic functions

$$\varphi = c_0 r^2 + d_1 r^3 \sin\phi + d_2 r^3 \cos\phi + \sum_{n=2}^{\infty}\left(a_{1n}r^n + b_{1n}r^{n+2}\right)\sin(n\phi) + \left(a_{2n}r^n + b_{2n}r^{n+2}\right)\cos(n\phi) \tag{3.41}$$

where the constants $a_{1n}, a_{2n}, b_{1n}, b_{2n}, c_0, d_1$ and d_2 need to be determined such that the stress boundary conditions at $r = R$ are satisfied:

$$\sigma_r = p(\phi) \quad \text{and} \quad \tau_{r\phi} = \tau(\phi) \tag{3.42}$$

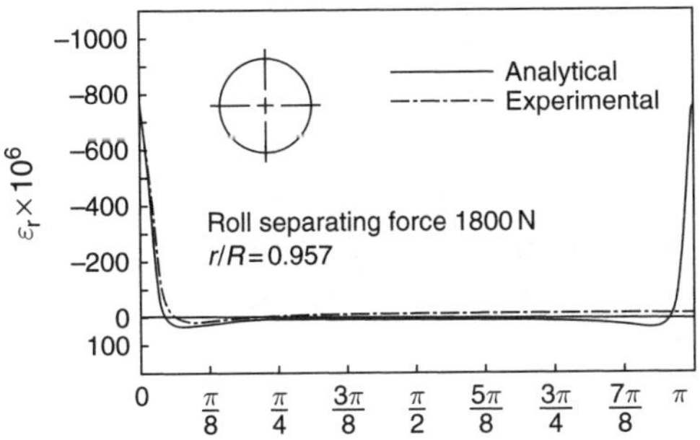

Figure 3.5 The calculated and measured radial strains of the work roll; the strain gauge is placed 95.7% of the roll radius from the centre (Roychoudhury and Lenard, 1984).

The coefficients in eq. 3.41 can be determined by representing the normal and the shear loading on the roll's surface in terms of Fourier series

$$p(\phi) = p_{a0} + \sum_{n=1}^{\infty} [p_{an} \cos(n\phi) + p_{bn} \sin(n\phi)] \tag{3.43}$$

and

$$\tau(\phi) = q_{a0} + \sum_{n=1}^{\infty} [q_{an} \cos(n\phi) + q_{bn} \sin(n\phi)] \tag{3.44}$$

where the coefficients may be determined by the Euler formulas[19].

The roll flattening, thus determined, was tested in a simple experiment. The side of the 125 mm radius work roll was fitted by a strain gauge, and the strains during rolling of commercially pure aluminum alloys were measured[20]. These were compared to the calculated strains. The results are shown in Figure 3.5, plotting the radial strains against the angular distance around the work roll, using the data from the strain gauge near the edge, at 119.6 mm from the roll's centre. It is observed that the predicted strains by the 2D elastic analysis compare well to the measurements.

3.5 THE EFFECT OF THE INERTIA FORCE

The metal to be rolled enters the roll gap at some velocity, which is usually lower than the surface velocity of the roll. As the thickness is reduced, the

[19] The detailed development of the 2D analysis of Michell (1900) is given by Pietrzyk and Lenard (1991).

[20] The 50 mm wide, 2 mm thick aluminum strips were reduced by 5% in the tests.

width remains unchanged and the length grows, the metal accelerates and exits from the roll gap at a velocity larger than that of the roll under most circumstances. Hence, there exists a force due to this acceleration and its effect on the rolling variables needs to be established. While 1D models usually ignore this contribution, the finite-element models usually include it in their analyses. In what follows, the validity of these approaches will be discussed and the potential effect of the mass × acceleration term on the roll force, etc. will be given in numerical terms.

3.5.1 The equations of motion

The effects of the inertia forces on the rolling process have rarely been analysed explicitly. In what follows, this effect will be considered in some detail. Equating the forces acting on a slab of the material in the roll gap to the product of the mass of the slab and its acceleration, and using the distance in the direction of rolling as the independent variable, leads to

$$\frac{\mathrm{d}(\sigma_x h)}{\mathrm{d}x} + p\frac{\mathrm{d}h}{\mathrm{d}x} \mp 2\mu p = \frac{ma}{\mathrm{d}x} \tag{3.45}$$

where the mass/unit width is given by $m = h\,\mathrm{d}x\rho$, the density is designated by ρ and the acceleration is a. The left side of eq. 3.45 is, of course, identical to that of eq. 3.18. In order to develop a relation for the acceleration of the slab, use is made of mass conservation which requires that $\mathrm{d}(vh) = 0$, leading to $a = \mathrm{d}v/\mathrm{d}t = -(v/h)\,(\mathrm{d}h/\mathrm{d}t)$. The time derivative of the strip's thickness may be obtained from the simplified version of eq. 3.20, $h = h_{\mathrm{exit}} + x^2/R'$, written in terms of x, the variable along the direction of rolling, in the form $\mathrm{d}h/\mathrm{d}t = 2xv/R'$, where the strip velocity, v, in terms of the roll's surface velocity and the thickness at the neutral point is given as $v = v_{\mathrm{r}} h_{\mathrm{np}}/h$. Substituting the above into eq. 3.45 along with the Huber–Mises criterion of plastic flow yields the differential equation of motion:

$$\frac{\mathrm{d}p}{\mathrm{d}x} \pm 2\mu\frac{p}{h} = \frac{2k}{h}\frac{\mathrm{d}h}{\mathrm{d}x} + \frac{\mathrm{d}(2k)}{\mathrm{d}x} + \frac{2x\rho}{R'}\frac{\left(v_{\mathrm{r}} h_{\mathrm{np}}\right)^2}{h^3} \tag{3.46}$$

It is now possible to estimate the orders of the magnitudes of the terms of eq. 3.46. The magnitude of the roll pressure is in the order of several hundred MPa. The magnitude of the last term of the equation, for any reasonable set of rolling parameters, is less than 1% of that.

3.5.2 A numerical approach

In another, simpler approach, the inertia force acting on the whole of the mass in the deformation zone can be determined. The acceleration is then given by

$a = (v_{exit} - v_{entry})/\Delta t$ and the time taken for a cross-sectional plane to travel from the entry to the exit is $\Delta t = \sqrt{R'\Delta h}/v_r$. The mass of the metal in the roll gap is $m = \rho w h_{ave}\sqrt{R'\Delta h}$ so the inertia force is

$$F_I = \rho w h_{ave} v_r (v_{exit} - v_{entry}) \tag{3.47}$$

For the inertia force to be a significant contributor in the analysis of permanent deformation, the stress it creates over the average cross-section of the rolled metal should be similar in magnitude to the yield strength. From eq. 3.47 equate the stress created by the inertia force to the yield strength:

$$\sigma_{yield} = \frac{F_I}{w h_{ave}} = \rho v_{roll} (v_{exit} - v_{entry}) \tag{3.48}$$

For any realistic set of numbers, the difference between the exit and entry velocities becomes unrealistically high, underscoring the conclusions drawn above: the contribution of the inertia force may be safely ignored. To get a numerical estimate, take a steel whose density is 7850 kg/m^3 and let a 1 m diameter roll have a rotational speed of 100 rpm, leading to a roll surface velocity of 5.24 m/s. Let the entry thickness be 5 mm, the exit thickness be 2 mm, and the entry velocity be 5 m/s. The exit velocity is then, from mass conservation, 12.5 m/s. Substituting these numbers in the right side of eq. 3.48 leads to a stress, due to inertia effects alone of 0.31 MPa, clearly negligible in comparison to the magnitudes of all other stress components.

3.6 THE PREDICTIVE ABILITY OF THE MATHEMATICAL MODELS

The decision to be made when choosing a mathematical model to analyse the flat rolling process is not an easy one. In what follows, it is assumed that the objective of the analysis is to predict only some of the rolling variables, namely the roll separating force. The predictive abilities of some of the models discussed above will be presented and critically discussed. The experimental data developed by McConnell and Lenard (2000) will be used. In that project low carbon steels were rolled at various rolling speeds and to various reductions, using low viscosity oils for lubrication. The roll separating forces and the roll torques were measured.

In the calculations that follow, the predictive abilities of three models – those given by Schey (2000), Bland and Ford (1948) and Roychoudhury and Lenard (1984) – are compared. The results of the comparison are shown in Figure 3.6 in terms of the ratios of the measured and the calculated roll forces for each of the methods of calculation, as functions of the rotational speed of

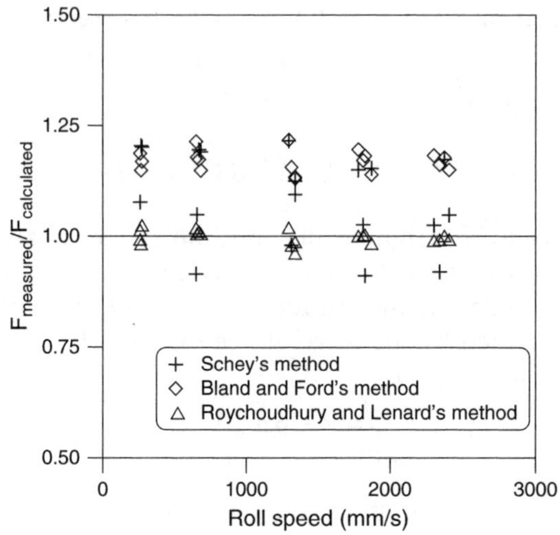

Figure 3.6 Comparison of the predictive capabilities of three simple models for cold rolling of low carbon steel strips.

the roll. The reductions vary from a low of 14% up to 50% in the rolling passes. The data given in the figure need to be discussed very carefully and in some detail.

The essential data necessary for modelling includes the material's resistance to deformation and the coefficient of friction. The former was given by McConnell and Lenard (2000) for the steel used here, obtained in traditional uniaxial tension tests, as $\sigma = 150\,(1+234\varepsilon)^{0.251}$ MPa. The latter was determined by inverse calculations, matching the measured separating forces to those calculated by the approach of Roychoudhury and Lenard (1984). This is evident in Figure 3.6, as the triangles of Roychoudhury and Lenard's predictions are always very close to unity, as expected, of course. The magnitude of the coefficient of friction, thus obtained, was then used in the other two methods of calculations. The diamonds of the Bland and Ford (1948) approach are approximately 20% over unity. The crosses of the Schey (2000) technique are not very consistent. It needs to be pointed out that when judging a model for its predictive ability, consistency is much more important than accuracy, since predictions with low standard deviation can always be adjusted by the use of carefully determined factors.

Both of these approaches – Bland and Ford's and Schey's – could have been used to determine the coefficient of friction, in an inverse manner, of course. Both would have yielded values for μ that would vary broadly and would be quite different from those used in Figure 3.6, indicating that the inverse method for the determination of the coefficient may not be the most suitable approach. Instead, independent experiments, to be discussed in Chapter 5, are recommended.

3.7 THE FRICTION FACTOR IN THE FLAT ROLLING PROCESS

The referee of a manuscript of the present author and his student (Lenard and Barbulovic-Nad, 2002) questioned the use of the coefficient of friction in bulk-forming processes, stating correctly that at high normal pressures the physical meaning of μ is lost. The rebuttal, that was accepted by the editor, is quoted below:

> *It is realized that the coefficient of friction obtained by inverse modelling, while it may be close to the actual value, is in fact an effective one. Further, while application of the traditional definition of the coefficient, as the ratio of the tangential to the normal forces, in metal forming operation has been questioned, it still remains a parameter in widespread use. As shown by Schey (2000), μ reaches a maximum as the normal stresses increase. This condition, while it may be reached during dry contact, is not likely to be observed when forming occurs in the boundary or mixed lubrication regimes. Azarkhin and Richmond (1992) also show that the friction factor will be less than unity, even when adhesion is the main cause of frictional resistance.*

Nevertheless, the comments of the referee were taken seriously and they gave the impetus to develop a model of flat rolling, using the friction factor, instead of the coefficient of friction.

Pashley et al. (1984) examined the three most significant factors that contribute to surface interactions involving adhesion: the area of real contact, the interfacial bond strength and the mechanical properties of the interface. They used a tungsten tip and a nickel flat, the tungsten being nearly ten times harder than the nickel. When the surfaces were clean, the junction failed at a stress level roughly equal to the yield strength of the metal. Li and Kobayashi (1982) included the effect of the relative velocity of the sliding surfaces in their formulation of the frictional model. A similar model is used in Elroll[21], a finite element software developed by Pietrzyk (1982) in which the coefficient of friction is defined in terms of a constant value, $\overline{\mu}$, the relative velocity of the roll and the strip, Δv, and a constant, a, which is chosen to be 10^{-3}, in the form

$$\frac{\mu}{\overline{\mu}} = \frac{2}{\pi} \tan^{-1} \frac{\Delta v}{a} \tag{3.49}$$

Most commercially available finite-element software packages allow the user to choose the manner in which friction is to be modeled. A random search on

[21] Elroll, a finite-element program that analyses the flat rolling process has been developed in the Department of Modelling and Information Technology AGH in Kraków, Poland. The distributor of the software my be reached by e-mail: miti@metal.agh.edu.pl.

the Internet yielded a 1994 newsletter from the MARC Corporation, giving an equation for the friction force:

$$f_t = \mu f_{\mathrm{n}} \frac{2}{\pi} \tan^{-1} \frac{\Delta v}{C} \tag{3.50}$$

where μ is the coefficient of friction, f_{n} is the normal force and C is a constant.

Another relationship for the coefficient of friction in terms of the relative velocity of the roll and the rolled strip was given by Gratacos et al. (1992), which interestingly combines both the coefficient and the friction factor:

$$|\tau| = m \frac{\sigma_{\mathrm{fm}}}{\sqrt{3}} \frac{\Delta v}{\sqrt{\Delta v^2 - K^2}} \tag{3.51}$$

where K is identified as the "regularization parameter for the friction law", given later as a very small number, in the order of 0.001. Another interesting expression was presented by Nadai (1939) for the interfacial shear stress as a function of the relative velocity of the strip and the roll, the lubricant's viscosity and the thickness of the oil film

$$\tau = \frac{\eta\,(v - v_{\mathrm{r}})}{h_{\mathrm{film,ave}}} \tag{3.52}$$

where the thickness of the oil film is to be the average over the rolling pass.

A third possibility in modelling friction is presented by Carter (1994) by relating the fractional shear strength of the contacting interface to the normal component of the deviatoric stress, through a "constant of proportionality", identified as "much like the coefficient of friction". Carter also states, unfortunately without referencing the information, that in simple compression "the fracture strength of the junction is close to the shear strength of the softer material."

Regardless of the manner in which friction is to be modeled, some difficulties, uncertainties and unknowns will always remain. In what follows, the friction factor is used in developing a 1D model of the flat rolling process, partially as the result of the comments of the reviewer, mentioned above. Three factors aid in making the decision to use the friction factor, in spite of the lack of precise knowledge of the magnitude of "k", the shear strength of the interface. One is consideration of the pressure sensitivity of lubricants, which, for an SAE 10W oil is given as 0.0229 MPa^{-1} by Booser (1984) who also gives the viscosity as 32.6 mm^2/s. If the roll pressure is 800 MPa, not an unreasonable magnitude when cold rolling steel, the Barus equation gives the viscosity at that pressure as 2.9×10^9 mm^2/s. This number, while possibly unrealistically high, indicates that shearing the lubricant at that pressure may require as much of an effort as shearing the metal. The other is the comment, referred to above, concerning the level of stress at which the tungsten–nickel junction failed (Pashley et al., 1984) and the third, also mentioned above, is the conclusion that ploughing was the major frictional mechanism (Lenard, 2004; Dick and Lenard, 2005) when a sand-blasted roll was used. Carter (1994) and Montmitonnet et al. (2000) reinforce this last factor by indicating that ploughing may be as important as adhesion in understanding frictional resistance. The objectives are then

- To develop a 1D model of the flat rolling process using the friction factor;
- To determine the dependence of the friction factor on speed and the reduction by using data, developed earlier on cold rolling of steel strips (McConnell and Lenard, 2000)
- To test the predictive capability of the model by comparing the predictions to experimental data and
- To develop a correlation of the coefficient of friction and the friction factor.

3.7.1 The mathematical model

The choice of the level of complexity in mathematical modelling of the flat rolling process depends on the objectives of the researchers. The comments in the last paragraph of the concluding chapter of Pietrzyk and Lenard (1991) are still valid: if the aim is to analyse mechanical events in strip rolling, where the roll diameter/strip thickness ratio is large, a 1D treatment is sufficient. This is followed in the present analysis in which the usual simplifying assumptions of previous workers are also employed. These include the assumptions of rigid rolls, homogeneous compression, a rigid-plastic material which remains isotropic and homogeneous as the rolling process continues. The angles are taken as small when compared to unity. As well, inertia forces are small in comparison to other forces, and are therefore, ignored.

The usual, 1D schematic diagram of the flat rolling process is used and the balance of forces in the direction of rolling on a slab of the rolled metal then leads to the well-known relation

$$\frac{\mathrm{d}(\sigma_x h)}{\mathrm{d}x} + p\frac{\mathrm{d}h}{\mathrm{d}x} - 2\tau = 0 \tag{3.53}$$

where σ_x is the stress in the direction of rolling and p the roll pressure.

In simplifying eq. 3.53, the Huber–Mises flow criterion, $\sigma_x + p = 2k$, is used, the strip thickness is taken in terms of the independent variable, $h \cong h_{\text{exit}} + x^2/R$, and the interfacial shear stress is defined by the friction factor, $\tau = mk$. The shear strength of the softer material, the rolled steel, is taken as k and as usual, $0 \leq m \leq 1$. A first order ordinary differential equation for the roll pressure is then obtained

$$\frac{\mathrm{d}p}{\mathrm{d}x} = \frac{2k}{h_{\text{exit}} R + x^2}(2x - mR) \tag{3.54}$$

where the friction factor, m, is to expressed in terms of the significant variables and parameters. Equations 3.49–3.52 above expressed the coefficient of friction in various forms, as a function of the relative velocity of the strip with respect to the roll, Δv, acknowledging the well-known speed dependence of the frictional resistance, the viscosity, the thickness of the oil film and several constants. Following these but also recognizing the dependence of friction on the normal stresses, the friction factor is now written as dependent on both load and speed:

$$m = \bar{a}\left(x^2 - x_{\text{np}}^2\right)p + \bar{b}\tan^{-1}(\Delta v/q) \tag{3.55}$$

where $\bar{a}$ and $\bar{b}$ are constants to be determined and q is a constant, taken arbitrarily to be 0.1. The relative velocity is given in terms of the location of the neutral point, x_{np}, and the surface velocity of the roll, v_{r}

$$\Delta v = v_{\text{r}}\frac{x_{\text{np}}^2 - x^2}{h_{\text{exit}}R + x^2} \tag{3.56}$$

allowing the friction factor to vary from the point of entry to exit. Since the numerator in eq. 3.56 changes algebraic sign as x varies, the friction factor also changes sign at the neutral point. At this stage of the calculations the constants $\bar{a}$ and $\bar{b}$ in eq. 3.55 and the location of the neutral point, x_{np}, are not known.

The computations start by integrating eq. 3.54, using a Runge–Kutta approach, for the roll pressure, starting at the entry with the appropriate boundary condition, and using assumed values for all three unknowns, $\bar{a}$, $\bar{b}$ and x_{np}. The boundary condition at the exit is satisfied by adjusting the location of the neutral point. Integral of the roll pressure distribution, thus obtained, over the contact length, is the roll separating force. By adjusting the constants $\bar{a}$ and $\bar{b}$ in eq. 3.55 for the friction factor, repeating the integration, the calculated and the measured roll separating forces are compared and when satisfactory convergence is reached, the constants $\bar{a}$ and $\bar{b}$ and the location of the neutral point are deemed to have been determined. At this point uniqueness of the predictions is not considered.

The friction factor, thus determined, varies from a negative value at the entry to the neutral point where it reaches zero. Beyond that the factor becomes positive. Its average value, m_{ave}, is indicative of frictional resistance.

The roll torque is determined using the power, P, required to roll the metal. The power is obtained as the sum of the power for internal deformation and friction:

$$P = \frac{2\sigma_{\text{fm}}}{\sqrt{3}}\int_V \dot{\varepsilon}\,\mathrm{d}V + 2\int_S \tau\,\Delta v\,\mathrm{d}S \tag{3.57}$$

where the friction stress is as given above and the torque is then, for both rolls

$$M = \frac{RP}{v_{\text{r}}} \tag{3.58}$$

The closeness of the calculated and measured roll torques indicates that both constants $\bar{a}$ and $\bar{b}$ have been determined correctly.

3.7.2 Calculations using the model

The predictive abilities of the model are tested in two instances. First, the roll separating forces are compared to those measured in earlier cold rolling experiments, followed by comparing the calculated and measured roll pressure distributions.

3.7.2.1 Cold rolling of steel

Selected portions of the data, obtained by McConnell and Lenard (2000), are used to determine the friction factor. In that publication low carbon steel strips, having a true stress–true strain relation of $\sigma = 150\,(1+234\,\varepsilon)^{0.251}$ MPa and measuring $1 \times 23 \times 300$ mm, have been cold rolled, using lubricants, containing various additives and having broadly varying viscosities. The rolls, made of D2 tool steel and hardened to Rc = 63, were of 249.8 mm diameter. Their surface roughness was Ra = 0.2 μm. The objective was to determine the coefficient of friction by inverse calculations and by the use of Hill's formula. The data, used in the present study, involves a lubricant with a kinematic viscosity of 19.83 mm^2/s and a density of 861.6 kg/m^3.

The dependence of the average friction factor on the surface velocity of the roll and on the reduction is shown in Figure 3.7. As expected from previous studies, the friction factor decreases with both increasing rolling speed and reduction, affected by the same mechanisms that affected the coefficient of friction. As has been pointed out in several instances, as the speed increases, more oil is drawn into the contact, leading to lower friction and as the loads increase, the viscosity increases, also leading to lower friction, at least in the boundary and in the mixed lubricating regimes.

Statistical modelling, using non-linear regression analysis, gives the dependence of the friction factor on the two process parameters, load and speed. The relation is shown in eq. 3.59

$$m_{\text{ave}} = -1.607\,(\text{red}) - 0.00013 v_{\text{r}} + 1.256 \tag{3.59}$$

where "*red*" is the reduction in decimals. The predictive ability of eq. 3.59 was tested on data, not used in its determination. The data points from McConnell and Lenard (2000) were taken using a different lubricant whose viscosity was similar to the one used in developing eq. 3.59, its value being 20.03 mm^2/s. The roll surface velocity was 2308 mm/s. The results are given in Table 3.1 below. In Column 1 of the table the roll forces, as measured, are given. In Column 2, the forces, as calculated by the model are shown while in Columns 3 and 4, the torques are indicated. Column 5 lists the average friction factors that resulted in the calculated forces and torques. The friction factors, as predicted

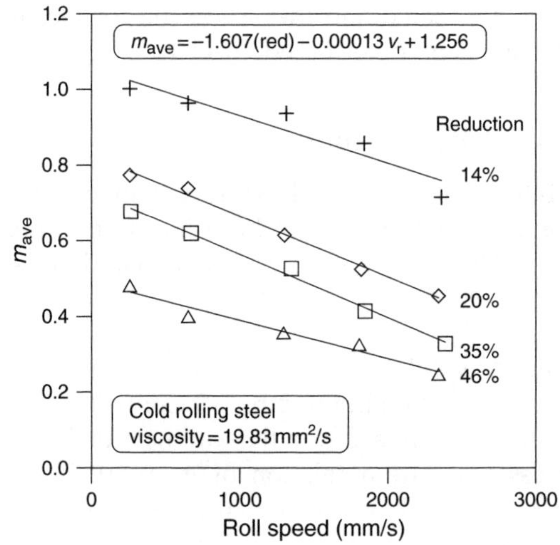

Figure 3.7 The friction factor as a function of the reduction and the roll surface velocity.

Table 3.1 A comparison of the predictions of the model and that of eq. 3.55

Roll force, measured (N/mm)	Roll force by the model	Roll torque, measured (Nm/mm)	Roll torque by the model	m_{ave}, by the model	m_{ave} from eq. 3.59
6086	5927	39.39	37.38	0.214	0.17
8225	8259	45.48	45.05	0.502	0.501
7782	7741	46.29	44.01	0.405	0.417
6871	6835	41.18	41.4	0.305	0.31

by eq. 3.59 are shown in Column 6, demonstrating that within the range of the process parameters of the experiments, the eq. 3.59 predicts the friction factor reasonably well.

As mentioned, the calculations proceed until the measured and calculated roll separating forces and roll torques are close, to within a pre-specified tolerance. The accuracy of the computations is shown in Figure 3.8, which gives the ratios of the measured and estimated loads on the mill against the number of tests. All speeds, from 261 to 2341 mm/s, and all reductions, from 14 to 46%, are included in the figure. In general, one may conclude that a reasonable accuracy has been reached. At lower reductions the differences between the experimental data and the calculations are larger, due to the deviation from homogeneous compression. The numbers fall to near unity as the loads increase.

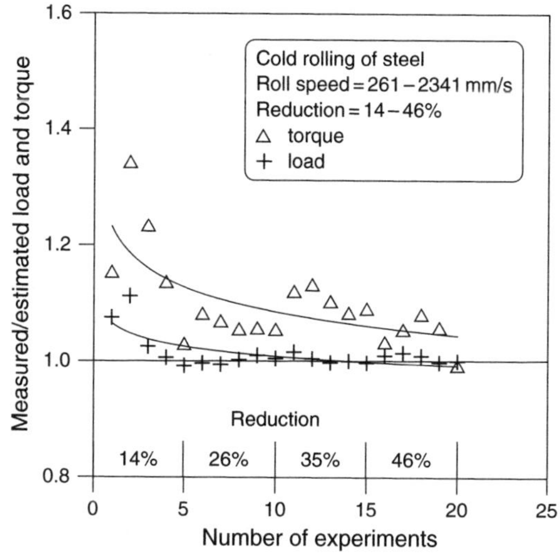

Figure 3.8 The accuracy of the computations: the ratio of the measured and calculated roll force and torque.

McConnell and Lenard (2000) determined the coefficient of friction using Hill's equation (see eq. 5.26, Chapter 5). These values are compared to the friction factor in Figure 3.9. A linear relationship is evident and the equation

$$m_{\mathrm{ave}} = 4.425\,\mu_{\mathrm{Hill}} + 0.01 \tag{3.60}$$

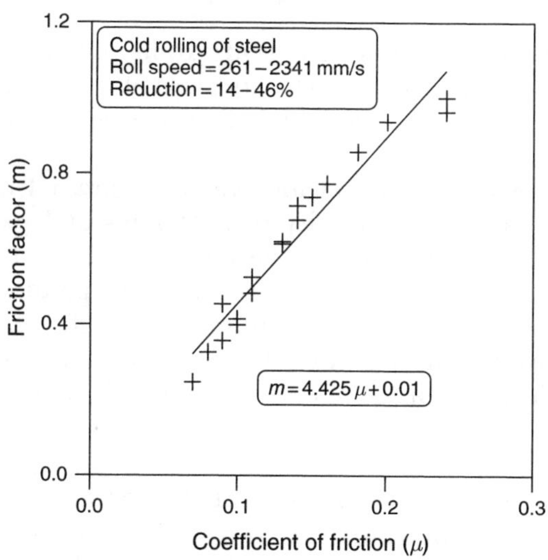

Figure 3.9 The friction factor versus Hill's coefficient of friction.

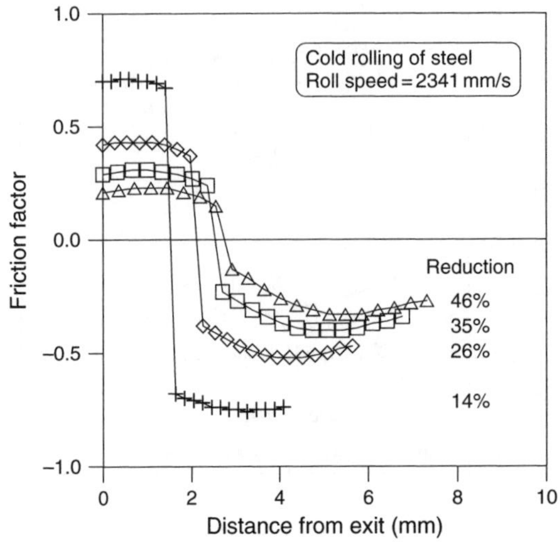

Figure 3.10 The variation of the friction factor along the roll/strip contact.

relates the two descriptions of frictional events in the roll gap. A relationship between m_{ave} and μ, for use in forging, has been suggested by Kudo (1960) in the form

$$\mu = \frac{m_{ave}/\sqrt{3}}{p_{ave}/\sigma_{fm}} \tag{3.61}$$

Using the data of the flat rolling tests leads to the conclusion that the values of m_{ave}, predicted by eqs 3.60 and 3.61 are close at low speeds, underscoring the importance of the speed in defining either the coefficient of friction or the friction factor.

The variation of the friction factor over the contact length is shown in Figure 3.10, at a roll surface velocity of 2341 mm/s and reductions ranging from 14 to 46%. It is observed that the positive values at the exit and the negative values at the entry are quite similar in magnitude, indicating that the surfaces in contact have been well lubricated. High values of the friction factor at the exit would imply that the lubricant has not been carried through the location of maximum pressure and the contact may have been starved of lubricant.

3.7.2.2 Distribution of the roll pressure at the contact

While integrating the friction hill over the contact gives realistic magnitudes of the roll separating forces, its shape has been shown to be unrealistic in several publications, starting with the work of Siebel and Lueg (1933). The friction hill is the result of the 1D models' traditional approach, in which the intersection of the curves, obtained by integrating from the entry and from the exit, is taken as the location of the maximum pressure and of the neutral point. In the present work the shooting approach is followed in which integration proceeds from the exit and satisfaction of the boundary condition at the entry indicates success. The roll pressure distribution, thus obtained, is different from the sharp cusp of the usual 1D models.

Comparisons of the predicted roll pressures to experimental data are given in Figures 3.11 and 3.12. In Figure 3.11, the measurements of Lu et al. (2002) are used. Employing pins and transducers in the work roll, the authors rolled low carbon steel strips at 1000°C and reported on the distribution of the roll pressures and the interfacial shear stresses over the contact zone. One of these experiments is used here and the measured and calculated interfacial stresses are shown in Figure 3.11. The test was conducted at 35 rpm, and the 20 mm thick slab was reduced by 20%. The roll separating force is read off Figure 3 of Lu et al. (2002) as 3000 N/mm and the roll torque as 66 Nm/mm. (Note that this figure is for a roll speed of 40 rpm, and no roll force data are given for a speed of 35 rpm.) The average flow strength of the steel, at the temperature and the strain rate used, is obtained by Shida's (1969) relations as 124.74 MPa. Figure 3.11 indicates that the predicted pressures and shear stresses match the measurements quite well, indicating that use of the friction factor, as a variable in the contact zone, is quite realistic.

Roll pressure and interfacial shear stress distributions, obtained during warm rolling of 1100-H14 aluminum alloy strips have been presented by

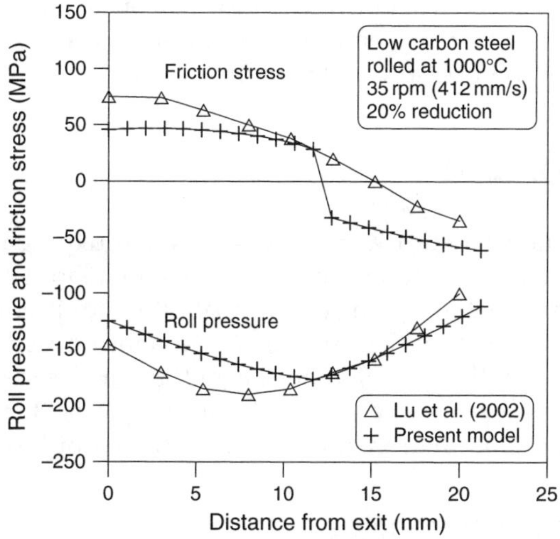

Figure 3.11 Comparison of the roll pressures, as measured by Lu et al. (2002) and calculated by the present model; hot rolling of steel.

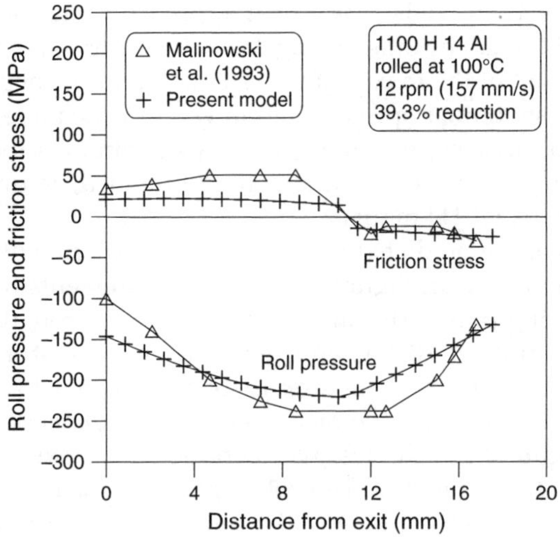

Figure 3.12 Comparison of the roll pressures, as measured by Malinowski et al. (1993) and calculated by the present model; warm rolling of aluminum.

Malinowski et al. (1993). A comparison of the predictions of the present model to the measurements by Malinowski et al. (1993) is shown in Figure 3.12. The 6.28-mm thick strip has been reduced by 39.3% at 100°C at a roll speed of 12 rpm. The average flow strength of the metal is taken as 163 MPa in the calculations. The roll separating force was measured to be 3240 N/mm. The present model calculated it to be 3293 N/mm.

Examination of the two figures leads to the conclusion that allowing the friction factor to vary from the entry to the exit in the roll gap leads to realistic calculations of the roll pressure distribution.

3.8 THE USE OF ANN

Because of the characteristics which can satisfy the requirements for on-line control, that is, short computing time, accuracy and adaptive learning, neural networks are most suitable, among other capabilities, for predictive type calculations. While the general background of ANN is introduced in the following sections, the mathematical development is not given here as it is considered to be beyond the scope of the book. The technique is then applied to the prediction of roll separating forces during cold rolling of low carbon steels.

3.8.1 Structure and terminology

ANN, simply called neural networks, take their name from the networks of nerve cells in the brain. Similar to biological neural networks of the brain, they are composed of a large number of nodes, divided into several layers. The input layer receives the external input from the environment and the output layer provides the final output to the environment. The middle layer, called hidden layer, is isolated from the outside environment. There is no limit to the number of hidden layers.

3.8.2 Interconnection

The nodes in the network are interconnected. There are three types of interconnections (Simpson, 1990) between nodes: intralayer, interlayer and recurrent connections. In an intralayer connection, also called lateral connection, a node is connected with nodes in the same layer. Interlayer connections are connections between nodes in different layers. Recurrent connections are those that come back to the same node. The connection in which nodes are interconnected to all of the nodes in neighbouring layers is called full connection.

3.8.3 Propagation of information

If the information is allowed to propagate only in one direction, it is called feedforward. Usually the arrows in the connections indicate the directions of the information flow. Feedback allows information to flow in either direction and/or recursively.

3.8.4 Functions of a node

Each interconnection between two nodes has an associated weight w_{ij}, which represents the connection strength from node i to node j. In addition, there may be an additional value x_0 which is modulated by the weight w_{0j}. This term is considered to be an internal threshold value that must be exceeded for activating that node. The node performs two simple processes: calculating a weighted sum of the inputs and utilizing the weighted sum as an argument to calculate the output through a threshold function.

3.8.5 Threshold function

Threshold functions are also referred to as activation functions, squashing functions or signal functions. There are four types of functions adopted as threshold functions: linear, ramp, step and sigmoidal functions (Rich and Knight, 1991). The logistic sigmoidal function is used often for threshold functions.

3.8.6 Learning

Learning, also referred to as training or encoding the knowledge, is an important procedure in teaching an ANN to acquire knowledge and must be done before an ANN can work. Depending on whether guidance is received from an outside agent, the learning scheme is divided into two major categories: supervised learning and unsupervised learning. Before learning, the associated weights of connections are given random values. In supervised learning the weights are adjusted based on the difference between the outputs produced by the ANN and the desired outputs (target). Unsupervised learning, also called self-organization, is a process that incorporates no teacher and relies only on local information of input data and internal control.

3.8.7 Characteristics of neural networks

Rich and Knight (1991) summarized the characteristics of ANN, as follows:

- A large number of very simple neuron-like processing nodes are used;
- A large number of weighted connections exist between nodes. The weights on the interconnections encode the knowledge of the network;
- Highly parallel, distributed control leads to ease of usage;
- An emphasis on automatic learning aids updating of the information.

Based on their characteristics, neural networks have been described as a powerful statistical mapping technology (White, 1989). By using the traditional statistical method, such as multiple non-linear regression, to search for the relations between input and output data, all possible formulae must be given to a computer manually before the optimal solution is obtained. This work takes time and human resources. Unlike the conventional statistical methods, a neural network can directly map input data to output data by learning, without knowing the relations between inputs and outputs *a priori*. When a neural network encounters new data, it can self-adjust by learning automatically. It also can eliminate useless data by learning. A well-trained neural network has the ability to respond to the inputs which are not included in the training pairs before. Because of the parallel and distributed structures, the network can provide correct outputs even if the inputs are incomplete or partially inadequate (Dayhoff, 1990).

3.8.8 Back-propagation neural networks

A back-propagation neural network (B-P network) has one or more hidden layers. The information is merely allowed to propagate in one direction, from input layer to output layer. The nodes in the B-P network are fully connected between neighbouring layers. Each interconnection has an associated weight, w_{ij}, which is given a random value initially. The targets are used as a guide during learning. The errors between the network outputs and targets are corrected by changing the associated weight of each interconnection. The correction mechanism starts from the nodes in the output layer and propagates backward through the hidden layers towards the nodes in the input layer. This is why this algorithm is named "back-propagation".

3.8.9 General Delta Rule

The concept of the back-propagation learning algorithm was presented by Werbos (1974) and Parker (1982). The potential and applications of the B-P networks were introduced by Rumelhart and McClelland in their book, *Parallel Distributed Processing* (Rumelhart and McClelland, 1986). The mathematical basis of the back-propagation learning algorithm is the General Delta Rule (Freeman and Skapure, 1992). Based on this rule the updated value of the associated weight of each connection is determined.

3.8.10 The learning algorithm

The algorithm of training a B-P network can be summarized as follows:

- Apply the input vector x_i to the input nodes;
- Calculate the weighted sum of each node in the hidden layer;
- Calculate the output of each node in the hidden layer;
- Calculate the weighted sum of each node in the output layer;
- Calculate the output of each node in the output layer;
- Calculate the error term of each node in the output layer;
- Calculate the error term of each node in the hidden layer;
- Update the weights in the output layer;
- Update the weights in the hidden layer;
- The steps are repeated for each set of learning data until the error reaches an acceptable value.

3.8.11 Drawbacks of B-P networks

Although the B-P network is powerful in mapping, there are two drawbacks with regard to convergence and learning rate. Several theorems (Coben and Grossberg, 1983; Kosko, 1988) indicate the stability of neural networks, but it cannot be guaranteed that the training process will converge to the global minimum. Hence, when training is finished, the B-P network must be tested for performance. The second drawback is that the learning rate of B-P network is very slow, and the time required to train the B-P network may not be known *a priori*. In order to improve these problems, several learning algorithms were developed to accelerate the learning rate of B-P networks, including the Second Order Algorithm (Parker, 1987; Ricotti et al., 1988), the Gradient Reuse Algorithm (Hush and Salas, 1988) and the Accelerated General Delta Rule (Dahi, 1987).

3.8.12 Application of neural networks to predict the roll forces in cold rolling of a low carbon steel

In what follows, the predictive ability of a simple B-P network is examined. Cold rolling experiments were conducted on a low carbon steel, containing

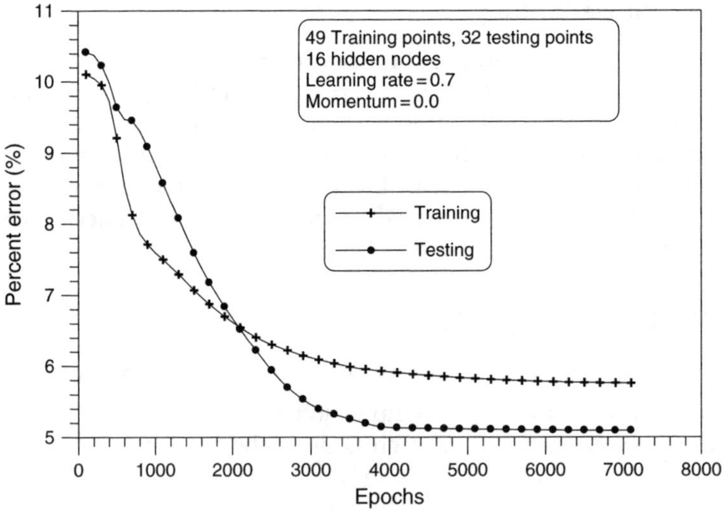

Figure 3.13 Roll separating force – neural network training and testing results.

0.04% C and 0.19% Mn. The true stress–true strain equation of the steel is $\sigma = 174.9\,(1 + 120.7\varepsilon)^{0.245}$ MPa, tested in uniaxial tension. The two-high rolling mill had rolls of 250 mm diameter, hardened to $R_c = 64$. The roll roughness was $R_a = 0.11\,\mu$m. An emulsion, consisting of tallow, sodium alkaryl sulfonate emulsifier and distilled water, was sprayed at the rolls, directly above and below the entry. The roll force and the roll torque were measured in each pass. The reduction/pass and the rolling speed were the independent variables. An ANN was trained and then tested for its predictive abilities. The results are given in Figure 3.13.

The figure shows how the per cent error drops as the repetitions increase. The network, with 16 hidden nodes, was trained first, using 49 data points and its predictive ability was tested on 32 extra measurements. After 4000 iterations the error reduced to about 5% and remained there.

3.9 EXTREMUM PRINCIPLES

Arguably, the most powerful of the approximate techniques available to analyse metal-forming processes are the extremum principles; specifically the upper bound and lower bound theories. Both theories are formulated to estimate the power required for plastic forming. The upper bound theorem can be shown to predict the power that is always more than necessary. The lower bound is designed to lead to a power that is less than needed. Hence, since the upper bound theorem is the more conservative and the more useful of the two, it will be described in some detail. In spite of the widespread use of these theories in the treatment of problems of metal forming, the upper bound approach has rarely been used to treat the process of flat rolling.

3.9.1 The upper bound theorem

The upper bound theorem is described by Avitzur (1968) as follows: Among all kinematically admissible strain rate fields the actual one minimizes the expression

$$J^* = \frac{2\sigma_{\mathrm{fm}}}{\sqrt{3}} \int_V \sqrt{\frac{1}{2}\dot{\varepsilon}_{ij}\dot{\varepsilon}_{ij}}\, \mathrm{d}V + \int_{S_\Gamma} \tau|\Delta v|\mathrm{d}S - \int_{S_\mathrm{i}} T_i v_i \,\mathrm{d}S \tag{3.62}$$

A strain rate field derived from a kinematically admissible velocity field is kinematically admissible. In eq. 3.62 J^* is the externally supplied power; the first integral represents the power for internal deformation over the volume of the body (V), the second evaluates power due to shearing over surfaces of velocity discontinuities (S_Γ) and the last term accounts for power supplied by body tractions over the surface, designated by S_i.

There are several concepts, mentioned above, that require careful definition. The term "kinematically admissible velocity field" implies the requirement that the velocity field must satisfy constancy of volume and all boundary conditions. The concept of velocity discontinuities is also mentioned above. As Avitzur (1968) explains, the velocity field within a deforming body need not be continuous. As shown in Figure 3.14, it is permissible to divide a body into several zones, in each of which a different set of velocities may exist. The boundary, at which the velocity may be discontinuous, is indicated in the figure; this boundary may be located at the die/metal contact or it may be within the deforming metal. When the flat rolling process is analysed, the roll/strip contact surface is considered one of these surfaces of discontinuity.

In Figure 3.14, two zones are identified, zone 1 and zone 2. The velocity of a material point in zone 1 is v_1; its component normal to the discontinuity is $v^N{}_1$ and the component parallel to the surface of discontinuity is $v^T{}_1$. In

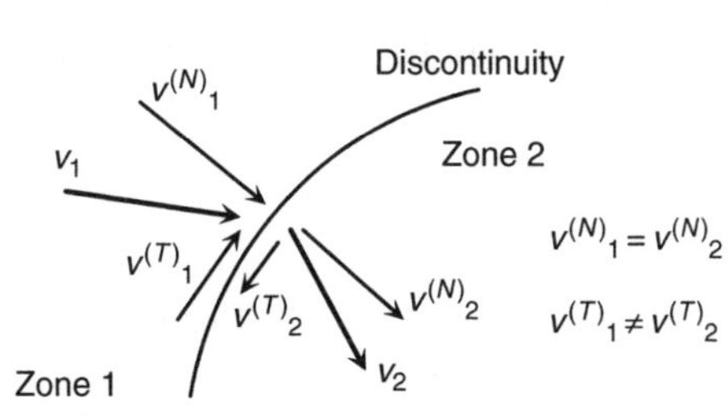

Figure 3.14 Velocity discontinuity within a metal forming system.

zone 2 the velocity is ν_2; its component normal to the discontinuity is $\nu^{N}{}_{2}$ and the component parallel to the surface of discontinuity is $\nu^{T}{}_{2}$. As shown in the figure, continuum mechanics requires only that the velocity components normal to the surface be continuous. The tangential components need not be equal on the two sides of the surface of discontinuity, giving rise to a region of high shearing stresses. Using these concepts and the assumption that the velocity of the rolled metal in the deformation zone moves towards the intersection of lines, tangential to the rigid rolls at the entry, the upper bound on the power per unit width, required for plastic deformation of the rolled metal, becomes (Avitzur, 1968):

$$J^* = \frac{2}{\sqrt{3}}\sigma_{\text{fm}}\nu_{\text{r}}\,h_{\text{exit}}\left[\ln\frac{h_{\text{entry}}}{h_{\text{exit}}} + \frac{1}{4}\sqrt{\frac{h_{\text{exit}}}{R}}\sqrt{\frac{h_{\text{entry}}}{h_{\text{exit}}}-1} + \frac{\sigma_{\text{entry}}-\sigma_{\text{exit}}}{(2/\sqrt{3})\sigma_{\text{fm}}}\right.$$

$$\left. + \frac{m}{\sqrt{h_{\text{exit}}/R}}\left(\sqrt{\frac{h_{\text{entry}}}{h_{\text{exit}}}-1} - \tan^{-1}\sqrt{\frac{h_{\text{entry}}}{h_{\text{exit}}}-1}\right)\right] \tag{3.63}$$

At the boundaries separating these zones only the normal velocity component must be continuous; the tangential component in one zone may be different than the corresponding component on the other side of the separating surface. This velocity discontinuity of magnitude

$$\Delta\nu = \nu_1^{(T)} - \nu_2^{(T)} \tag{3.64}$$

will create shearing stresses along the "surface of discontinuity" – S_Γ in eq. 3.62 – whose magnitude is given by

$$\tau = \frac{m\sigma_{\text{fm}}}{\sqrt{3}} \tag{3.65}$$

where the friction factor m is given by $0 \le m \le 1$.
The roll torque, for both rolls, may be obtained from the power as

$$M = \frac{R}{v_{\text{r}}}J^* \tag{3.66}$$

Tirosh et al. (1985) applied Avitzur's (1968) upper bound approach to analyse cold rolling of viscoplastic materials at high speeds. The authors focused their attention on the effect of the speed, the inertia and temperature dependence of the material's resistance to deformation on the roll separating forces and on the roll torques. The Bingham material model[22] was taken as the constitutive relation, defining the stress deviator tensor components in terms of the metals' viscosity as

$$S_{ij} = \frac{2\eta}{1 - {}^{k}\!\big/\!\sqrt{J_2}}\,\dot{\varepsilon}_{ij} \tag{3.67}$$

where the dynamic viscosity of the material is designated by η, k stands for the yield strength of the material in pure shear, $\dot{\varepsilon}_{ij}$ are the components of the strain rate tensor and J_2 is the second invariant of the of the stress deviator tensor, defined as

$$J_2 = \frac{1}{2} S_{ij} S_{ij} \tag{3.68}$$

For completeness, recall that the deviator stress tensor components are defined in terms of the stress components by the relation

$$S_{ij} = \sigma_{ij} - \frac{1}{3}\sigma_{kk}\delta_{ij} \tag{3.69}$$

where δ_{ij} is Kronecker's delta and σ_{kk} are the components of the spherical stress tensor.

In deriving the velocity field, the authors assumed that the arc of contact may be replaced by straight lines. The flow pattern then becomes a radially converging flow, leading to the statement that the resulting stress field is "unavoidably approximate in nature". In an interesting step, the coefficient of friction is taken to depend on the speed of the rolled strip at entry. They use the two relations given by Sims and Arthur (1952):

$$\mu = 0.08\exp(-0.54\nu_{\text{entry}}) \tag{3.70}$$

for $0 \le \nu_{\text{entry}} \le 0.25\,\text{m/s}$
and

$$\mu = 10^{-3}\exp\left(4\nu_{\text{entry}}^{-0.038}\right) \tag{3.71}$$

for $0.25 \le \nu_{\text{entry}} \le 1.5\,\text{m/s}$.

The predicted roll separating force and torque values compared very well to the experiments of Shida and Awazuhara (1973) on cold rolling of steel strips. Further, increasing speeds were found to cause increasing compressive loads on the rolls and increasing tensile stresses within the strip, both of which were most likely caused by the strain rate dependence of the rolled metal, as predicted by the Bingham material model.

---

[22] It is rare to see the Bingham model used in problems dealing with plastic forming of metals. One of the exceptions is the work of Haddow on the compression of a disk; see J.B. Haddow, "On the compression of a thin disk", Int. J. Mech. Sci., 1965, 7, 657–660.

3.10 COMPARISON OF THE PREDICTED POWERS

Several mathematical models have been introduced in this chapter: an empirical model, 1D and a model based on the upper bound theorem. Two formulas have also been given to estimate the power required to reduce a strip of metal in a rolling mill; see eqs 3.8 and 3.9. Their predictions can now be compared and discussed in light of their assumptions.

Consider the hot rolling of a low carbon steel strip in a single stand roughing mill. Let the entry thickness of the slab be 20 mm and its width to be 2000 mm. Its resistance to deformation is taken to be 150 MPa. It is reduced by 30% using work rolls of 800 mm diameter, rotating at 50 rpm. The coefficient of friction is taken to be 0.2. The estimated power for plastic deformation, by eq. 3.7 is obtained as 6115 kW. With four bearings of 400 mm diameter and a coefficient of friction of 0.01 in the bearings, the losses are estimated to be 500 kW. Assuming further that the efficiencies of the motor and the transmission are both 0.9, the total power to drive the mill is obtained, by eq. 3.8 as 8170 kW. Using the same numbers and eq. 3.9, the total power is estimated to be 5350 kW. The upper bound theorem, designed to give conservative estimates, also allows the prediction of the power required for plastic deformation of the strip. Equation 3.63, with the friction factor equal to 0.8, yields 7190 kW. Calculations using the refined 1D model (see Section 3.4) leads to a torque for both rolls of 524 Nm/mm, which when used to compute the power needed to reduce the strip, gives 5483 kW. Adding on the power losses in the four bearing and using 0.9 for the efficiency of the motor as well as the transmissions, one obtains 7262 kW. Based on these calculations[23], it is recommended to use either the refined 1D model or the empirical; a conservative number for the motor power is likely to result.

The second number appears to be closer to reality, based on Robert's calculations, as mentioned above.

3.11 THE DEVELOPMENT OF THE METALLURGICAL ATTRIBUTES OF THE ROLLED STRIP

A detailed exposition of state-of-the-art of the evolution of the microstructure and the resulting mechanical attributes after hot flat rolling has been presented by Lenard et al. (1999). Carbon and alloy steels were included in the discussion, and the predictions of the metallurgical model have been substantiated by

[23] It is to be noted that only one set of data was used in the exercise. Statistical analysis is necessary to prove the consistency of the predictions of any of the models.

comparing them to results obtained in the laboratory and in industry. In what follows, a revised and updated version, dealing with carbon steels is described. Numerical examples are also given.

The development of the draft schedule of the hot or the cold controlled rolling processes is usually performed offline, using sophisticated mathematical models, which are composed of mechanical, thermal and metallurgical parts. The objective of the process is the creation of steel with small, uniform ferrite grains and as the Hall–Petch equation demonstrates, this will increase the strength of the rolled metal. There are a limited number of parameters whose magnitudes may be chosen relatively freely, although several of them are connected through mass conservation. The parameters include

- the starting temperature;
- the strain per pass;
- the strain rate per pass (within strict limits);
- the inter-stand tensions and
- the inter-stand and the pre-coiler cooling rates.

The engineer must also consider the given parameters which cannot be altered: these involve the rolling mill and its capabilities. At this point the chemical composition of the steel is also given, and the designer of the draft schedule must keep that in mind. The thickness of the scale on the surface may be controlled to some extent by the scale breakers. The interfacial frictional forces may also be controlled, also only to some extent, by the careful design of lubricant and coolant delivery. Lenard and Pietrzyk (1993) showed in a numerical experiment that while the austenite grain size of a low carbon steel is not affected by the coefficient of friction, the higher the surface heat transfer coefficient, the lower the grain diameter near the surface of the rolled steel, as expected by the higher surface cooling rates.

There is some evidence that the metallurgical structure at the end of the rough rolling process does not affect the subsequent events in any significant manner, so the concern here is with the design of the passes on the finishing train and the cooling banks.

The metallurgical events that affect the eventual attributes of the rolled metal are the hardening and the restoration processes. The hardening processes include strain, strain rate and precipitation hardening; the restoration processes include recovery and recrystallization, static, dynamic or metadynamic. These, in turn, are affected by the mechanical and thermal events. The three critical temperatures need to be known:

- the precipitation start and stop temperature;
- the recrystallization start and stop temperature and
- the transformation start and stop temperature.

Sellars (1990) summarized the importance of modelling of the evolution of the microstructure:

- for a given composition of alloy, the high temperature flow stress is influenced to a large extent by the microstructure. Proper prediction of the rolling force is possible only if the relevant microstructure is known and
- the microstructure present at the end of the rolling and cooling operations controls the product properties.

Austenite, a face-centred-cubic (FCC) structure is formed after solidifying. It is designated by γ. On further cooling to the Ar_3 temperature, the ferrite (α) grains appear and the steel reaches the two-phase region. The structure of the ferrite is body-centred-cubic (BCC). As the temperature drops to the Ar_1 temperature, the transformation stops and the steel has become fully ferritic. Depending on the carbon content and cooling rates, other phases such as pearlite or bainite may appear, as well. The two temperatures, Ar_3 and Ar_1, are affected by the chemical composition, pre-strain, cooling rate and initial austenite grain size (see Hwu and Lenard, 1998).

The costs associated with industrial trials are prohibitively high. The trials are expensive, difficult to control and monitor and are necessarily constrained by the capabilities of existing plants. Laboratory simulation tests are unable to reproduce all conditions of industrial hot rolling[24]. Both the torsion and compression tests have limitations on the attainable strain rates, particularly in relation to strip or rod rolling. Further limitations are evident in torsion testing, in which the sample also develops different textures from those in flat rolling. On the other hand, the plane-strain and axisymmetrical compression tests cannot achieve the total strains of complete industrial rolling schedules. Hence, the use of off-line models – which, in spite of the critique just above, have been obtained from laboratory simulation tests – is very useful, especially if the consistency and the accuracy of their predictions can be demonstrated. It must be realized at this point that the predictive abilities of these models have been substantiated by comparing their predictions to a selected number of measurements. Statistical analysis of the predictions, while necessary, is not widespread.

3.11.1 Thermal–mechanical treatment

The two major objectives of the hot rolling process are to control the dimensions of the product and to affect the attributes the metal will possess on cooling.

For most commercial products in the steel industry, their external shapes are the result of hot deformation, such as hot rolling, while the necessary mechanical properties are obtained by alloying elements and heat treatment

[24] In spite of this limitation, laboratory simulation of the multistage hot rolling process yielded extremely useful results.

after deformation. However, metallurgical changes caused by hot deformation may result in additional beneficial effects on the mechanical properties of steels, and sometimes can eliminate heat treatment after deformation. Thermo-mechanical processing is a technique to combine shaping and heat treating of steel. Controlled rolling is a typical example of thermo-mechanical processing in which austenite is conditioned to produce a fine ferrite grain size. The development of controlled rolling approaches, used for carbon steels, is shown in Figure 3.15. There are four different techniques in the figure, in which "R" refers to rolling in the roughing mill and "F" indicates the finishing mill.

In the first method, both roughing and finishing are completed at a temperature at which the steel is fully recrystallized. The resulting product will emerge as soft and ductile. In the second, the finish rolling process is interrupted and the steel is allowed to cool for some time but the rolling process is completed in the full recrystallization region. The result is a steel that is somewhat harder than in the first process. In both the third and the fourth strategies, the processing temperatures are further decreased and rolling is completed such that the steel is only partially recrystallized or, in the last step, finish rolling is performed in the two-phase region.

Hodgson and Barnett (2000) review the practice of thermo-mechanical processing of steels. They list the processes in use in industry, classifying them as those carried out during the deformation process and those performed during the cooling phase, after deformation. These processes are

- Conventional controlled rolling to improve strength and toughness;
- Recrystallization controlled rolling to achieve fine grains by affecting austenite grain growth and higher strength by precipitation hardening;
- Accelerated cooling, direct quenching, quench and self-tempering to affect the transformation mechanisms;
- Warm forming to affect the ferrite phase and
- Intercritical rolling of the austenite–ferrite structure to increase the strength and toughness.

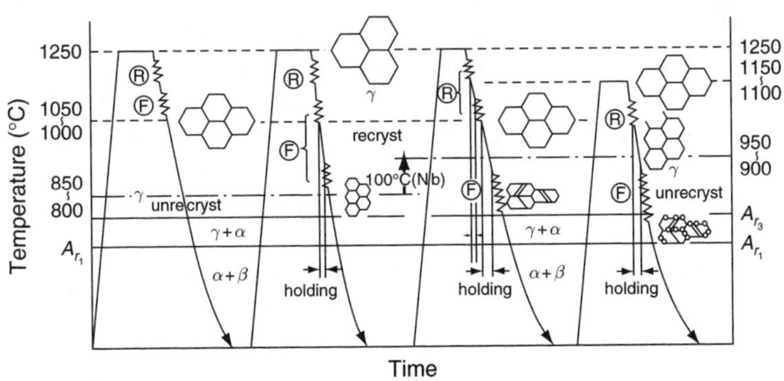

Figure 3.15 Controlled rolling strategies (Tamura et al., 1988).

The authors also write about how thermo-mechanical processes may develop in the future. They identify the processes that produce ultrafine grains by heavy plastic deformation (see Chapter 8, Severe Plastic Deformation) and by the application of magnetic fields. In another process (Hodgson et al. 1998) austenite grains are coarsened prior to deformation, followed by a small reduction and cooling. The result is a composite strip with ultrafine grains in the surface layers, possessing markedly increased strength.

3.11.1.1 Controlled rolling of C–Mn steels

Notch ductility and yield strength can both be improved by grain refinement. Among other techniques for grain refinement, European mills utilized controlled low temperature hot rolling in order to refine the grains and to increase the toughness. The following features were generally applied in this controlled rolling process:

- Interrupting the hot rolling operation when the slab has been reduced to the prescribed thickness, for example, 1.65 times the final thickness.
- Recommencing hot working when the slab has reached a prescribed temperature and finishing at temperatures in the austenite (γ) range, above the Ar_3 but lower than the conventional finishing temperatures, for example down to 800°C.

The low temperature finish rolling practise refines the γ grains, hence, the transformed α grains. A considerable additional grain refinement can be achieved by rolling in the non-recrystallized γ region, where deformation bands increase nucleation sites for α grains. However, the temperature range for non-recrystallized austenite in C–Mn steels is relatively narrow, and this mechanism for grain refinement cannot be effectively utilized, due to the risk of getting into deformation in the two-phase region.

3.11.1.2 Dynamic and metadynamic recrystallization-controlled rolling

In rod and bar rolling, using high strain rates (100–1000 s^{-1}), short interpass times (between few tens of milliseconds to few hundreds of milliseconds) and large strains per pass (0.4–0.6) dynamic recrystallization has been found to occur. It has been proposed that under appropriate conditions, dynamic recrystallization may also occur during strip rolling of niobium HSLA steels. The occurrence of dynamic recrystallization during simulated strip rolling of HSLA steels has been cited by several authors. The results of an analysis of the events during strip rolling also indicated that dynamic recrystallization is happening during rolling of Nb. Dynamic recrystallization affects rolling loads and is reported to produce considerably finer ferrite grains ($\sim 3\,\mu m$) than those transformed from pancaked austenite ($\sim 7\,\mu m$). However, there are concerns regarding the validity and applicability of the results obtained in all of the above studies to industrial practice, primarily due to the low strain rates employed in the simulation experiments.

Conventional controlled rolling relies on static recrystallization in the early stages of finish rolling to refine the austenite, followed by pancaking of the austenite grains in the last stages to enhance ferrite grain nucleation during transformation. In contrast, dynamic recrystallization favours higher reductions in the first few stands to exceed the critical strain for the onset of dynamic recrystallization. Dynamic recrystallization controlled rolling leads to greater ferrite grain refinement through austenite grain refinement. Another advantage of the initiation of dynamic recrystallization during rolling is a marked reduction in roll forces and torques, which in turn translates to savings in energy consumption and reduced roll wear. Also, the gauge accuracy will be enhanced due to the lower reductions required in the last stands.

3.11.1.3 Effects of recrystallization type on the grain size

Many different authors have attempted to develop models predicting grain sizes produced by static, dynamic and metadynamic recrystallization for different materials. The general observation, common in all these models, is that, statically recrystallized grain size is a function of the initial grain size, temperature and amount of strain, while dynamically and metadynamically recrystallized grain sizes are only functions of the Zener–Hollomon parameter, that is, the temperature and the strain rate, in an inverse power law form. This indicates that increasing strain rate and decreasing rolling temperature lead to more grain refinement, provided dynamic and metadynamic recrystallization are in place. Another common understanding is that rolling schedules with dynamic and metadynamic recrystallization produce finer final grain sizes compared to schedules with only static recrystallization. This idea is appealing to the steel manufacturers to achieve further grain refinement.

3.11.1.4 Controversies regarding the type of recrystallization in strip rolling

The occurrence of dynamic recrystallization by strain accumulation during industrial hot strip rolling schedules has been questioned. It has been argued that the kinetics of static recrystallization approaches those of dynamic recrystallization as the strain increases. In addition, interpass times are generally much greater than deformation times. Hence, softening of the material during strip rolling may be due to enhanced static recrystallization.

This controversy, in spite of its practical importance in terms of final mechanical properties and mill setup, still remains. The physical proof of the possibility of dynamic recrystallization during strip rolling is notoriously difficult, since it requires extremely fast quenching of the steel during deformation to freeze the structure and look for dynamically created grain nuclei. Most of the mill engineers do not believe in the possibility of dynamic recrystallization in any kind of steel during strip rolling. This belief has been reinforced by the fact that the possibility of dynamic recrystallization has not been taken into account in the conventional strip mill setup and in control modules developed by General Electric and Westinghouse. In these control modules, which are in use in North America, it is assumed that the steel repeatedly goes through only work hardening during deformation and static softening during interpass times. This assumption may lead to erroneous roll force prediction if the steel actually softens in one or more stands instead of hardening.

3.11.2 Conventional microstructure evolution models

Mathematical models of the evolution of the microstructure have been published in the technical literature. Sellars (1990), Roberts et al. (1983), Laasraoui and Jonas (1991a,b), Choquet et al. (1990), Hodgson and Gibbs (1992), Yada (1987), Beynon and Sellars (1992), Sakai (1995), Kuziak et al. (1997) and Devadas et al. (1991) presented various closed form equations, describing the processes of recrystallization and grain growth.

The restoration processes are time dependent and since in industrial hot rolling the strain rates are high, there is not enough time to trigger dynamic restoration of the work hardened material; note that the time available is determined by the ratio of the strain and the strain rate, $t = \varepsilon_{pass}/\dot{\varepsilon}_{pass}$. To demonstrate the validity of this statement, take a typical set of numbers in the first stand of a hot trip mill. Let the entry thickness be 15 mm and consider a fairly high, 50% reduction at a roll speed of 30 rpm. Take the diameter of the work roll as 500 mm. The average strain rate in the pass is then estimated to be $12\,s^{-1}$, the true strain is 0.69 and hence, the pass takes place in about 55 ms, indicating that concurrent static recovery, accompanied by static recrystallization, usually occurs after deformation. Both static recovery and recrystallization have been observed in austenite, although the extent of the former is rather limited. Some caution is introduced at this point. Biglou et al. (1995) considered industrial hot rolling schedules of Nb bearing microalloyed steels. Torsion testing was used to simulate the finish rolling schedules, and some softening, attributed to metadynamic recrystallization, was found in the third twist. As well, accumulating strains have been thought to contribute to dynamic recrystallization.

3.11.2.1 Static changes of the microstructure

The first step is to attempt to control the temperature at the entry to the first stand of the finishing mill. The success of this attempt is limited by the temperature of the just hot rolled strip, called the transfer bar, which is most likely waiting to exit the coil box. The temperature of the strip entering the coil box is controlled by the reheat furnace, held at 1200–1250°C and by the heat gains and losses in the roughing passes. The difference between the head and the tail temperatures is minimized while the steel is coiled up in the coil box within which the cooling rate is quite slow. The entry temperature therefore will depend on all of the above: the temperature in the reheat furnace, the gains and losses during rough rolling and the time spent in the coil box. The temperature above which recrystallization will occur is given by (Boratto et al., 1988):

$$T_{NRX} = 887 + 464\,[C] + (6445\,[Nb] - 644\sqrt{[Nb]}) + (732\,[V] - 230\sqrt{[V]}) + 890\,[Ti] + 363\,[Al] - 357\,[Si] \tag{3.72}$$

so a low carbon steel, containing 0.05% C, will recrystallize above 910°C. The higher the carbon content, the higher the temperature above which recrystallization will be present. The entry temperature into the first stand is usually higher than 900°C, unless ferrite rolling is contemplated.

The extent of static recovery, defined as a softening process in which the decrease of density and the change in the distribution of the dislocations after hot deformation or during annealing are the operating mechanisms, is rather limited in hot rolling processes. There is a general consensus that the maximum amount of softening during holding, attributable to recovery, is approximately 20%.

The hot deformation of austenite at strains typically encountered in plate or strip rolling processes leads to significant work hardening, which is usually not removed by either dynamic softening processes or by static recovery. This hardening creates a high driving force for static recrystallization. The mechanism of these processes is explained clearly by Hodgson et al. (1993). Some of his observations are presented briefly below:

- a minimum amount of deformation (critical strain) is necessary before static recrystallization can take place;
- the lower the degree of deformation, the higher the temperature required to initiate static recrystallization;
- the final grain size depends on the degree of deformation and to a lesser extent, on the annealing temperature and
- the larger the original grain size, the slower the rate of recrystallization.

During conventional pre-heating at high temperatures, incomplete recrystallization can take place at an early stage of the rolling process when small reductions are applied. The accumulation of strains then leads to full recrystallization in subsequent passes and, in consequence, the effect of the initial conditions on the downstream final microstructure is very small and is usually neglected.

The recrystallized volume fraction X is determined by the Johnson–Mehl–Avrami–Kolmogorov equation as a function of the holding time after deformation:

$$X = 1 - \exp\left[A\left(\frac{t}{t_X}\right)^k\right] \tag{3.73}$$

where t is the holding time, t_X is the time for a given volume fraction X to recrystallize, $A = \ln(X)$, and k is the Avrami exponent. The majority of microstructure evolution models has been developed for $X = 0.5$, indicating that t_X in eq. 3.72 represents the time for 50% recrystallization and the constant $A = -0.693$. The most commonly used form describing the time for 50% recrystallization ($t_{0.5X}$) is

$$t_{0.5X} = B\varepsilon^p D_\gamma^q Z^r \dot{\varepsilon}^s \exp\left(\frac{Q_{RX}}{RT}\right) \tag{3.74}$$

where ε is the strain, D_γ is the austenite grain size prior to deformation in μm, Z is the Zener–Hollomon parameter[25], $\dot{\varepsilon}$ is the strain rate, Q_{RX} is the apparent activation energy for recrystallization, R is the gas constant, and T is the absolute temperature. Sellars (1990) gives $B = 2.5 \times 10^{-19}$, $p = -4$, $q = 2$, $Q_{RX} = 300\,000$ J/mole and the Avrami exponent, k = 1.7. The exponents of the Zener–Hollomon parameter and the strain rate are indicated to equal zero. Equation 3.74 implies that the time for 50% recrystallization decreases with increasing strains and grows with the grain size. The time required for 50% recrystallization is given as a function of the temperature of the pass in Figure 3.16, for a set of realistic strains and pre-pass grain sizes. It is clear that the steel will recrystallize quite fast at higher strains and at higher temperatures.

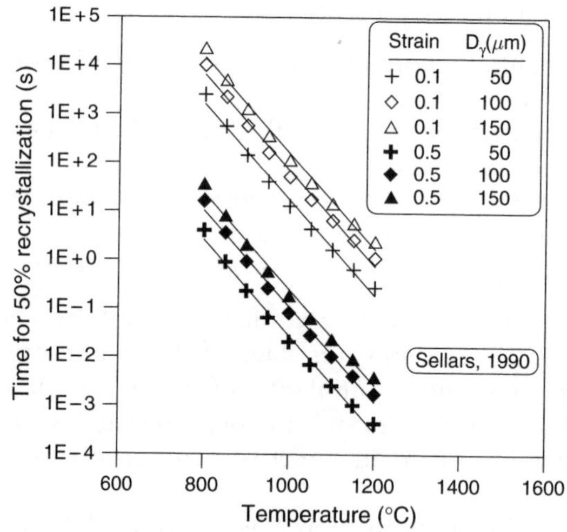

Figure 3.16 The time required for 50% recrystallization as a function of the temperature, the strain and the initial grain size.

[25] The Zener–Hollomon parameter is defined as $Z = \dot{\varepsilon} \exp(Q/RT)$.

The recrystallized grain size is reportedly sensitive to the temperature. The most commonly used form of the equation (Sellars, 1990) describes the dependence of the grain size after recrystallization (D_r) on the strain, the strain rate, the prior austenite grain size, the apparent activation energy and the temperature.

$$D_r = C_1 + C_2 \varepsilon^m \dot{\varepsilon}^n D_\gamma{}^l \exp \frac{-Q_d}{RT} \tag{3.75}$$

Sellars (1990) gives the magnitudes of the constants and the exponents in eq. (3.75) as follows: $C_1 = 0;\ C_2 = 0.5;\ m = -1;\ n = 0;\ l = 0.67;\ Q_d = 0$, indicating that the strain and the prior austenite grain size are the most significant variables. Roberts et al. (1983) provide somewhat different magnitudes. They give $C_1 = 6.2;\ C_2 = 55.7;\ m = -0.65;\ n = 0;\ l = 0.5;\ Q_d = 35\,000\,\mathrm{J/mole}$. Note that Sellars' equation excludes the dependence of the recrystallized grain size on the temperature but Roberts' accounts for it. Both researchers indicate that the grain size after recrystallization is independent of the rate of strain. Choquet et al. (1990) and Hodgson and Gibbs (1992) also gave various magnitudes for the coefficients and the exponents. Laasraoui and Jonas (1991a) offer another relation for the recrystallized grain size in a C–Mn steel in terms of the strain and the pre-deformation austenite grain size, similar to that of Sellars (1990) with a somewhat different exponent for the strain:

$$D_r = 0.5 D_\gamma{}^{0.67} \varepsilon^{-0.67} \tag{3.76}$$

The three equations predict different grain sizes for the same initial conditions. Assuming an initial grain size of 50 μm, a strain of 0.30 and at a temperature of 800°C, Sellars and Roberts predict a grain size of 23 μm, while Laasraoui and Jonas predict 15 μm. Increasing the initial size to 150 μm gives 48 μm by Sellars, 36 μm by Roberts and 32 μm by Laasraoui and Jonas. While it is difficult to recommend one of these relations for use without some more data, preferably analysed statistically, Sellars' relations have been shown to provide reasonable predictions.

The time for the completion of recrystallization is calculated from the Avrami equation for the recrystallized volume fraction, X, eq. 3.73. The constant A is taken to correspond to 50% recrystallization. In that case $X = 0.5$, both t and t_X equal $t_{0.5X}$ and $A = \ln(1 - 0.5) = -0.693$.

The time for X% of recrystallization is thus

$$t = \left[\frac{\ln(1 - X)}{A} \right]^{\frac{1}{k}} t_X \tag{3.77}$$

and the time for 95% recrystallization, when $t_X = t_{0.50}$ is

$$t_{0.95} = \left[\frac{\ln(0.05)}{\ln(0.5)} \right]^{\frac{1}{k}} t_{0.50} = 4.3219^{\frac{1}{k}} t_{0.50} \tag{3.78}$$

Situations when partial recrystallization takes place during interpass times are common in the industrial rolling processes. Beynon and Sellars (1992) present an equation to calculate the grain size at the entry to the next pass:

$$D_p = D_r X^{(4/3)} + D_\gamma (1 - X)^2 \tag{3.79}$$

where D_γ represents the grain size prior to deformation, D_r is the recrystallized grain size and X is the recrystallized volume fraction. Considering some of the grain sizes used above (800°C, $D_\gamma = 50\,\mu\text{m}$, $D_r = 23\,\mu\text{m}$ and 75% recrystallization) the average grain size of the rolled strip, entering the next stand is predicted to be nearly 19 μm.

3.11.2.2 Dynamic softening

All softening processes that take place during plastic deformation are referred to as dynamic ones. These include dynamic recovery and dynamic recrystallization.

The conventional models of dynamic recrystallization involve equations describing the critical strain, kinetics of dynamic recrystallization and the grain size after dynamic recrystallization. The critical strain at which dynamic recrystallization starts is given in terms of the Zener–Hollomon parameter, the grain size and several constants:

$$\varepsilon_c = AZ^p D_\gamma{}^q \tag{3.80}$$

Sellars (1990), considering a C–Mn steel, defines $A = 4.9 \times 10^{-4}$, $p = 0.15$, $q = 0.5$ and $Q_{\text{DRX}} = 312\,000$ J/mole. Laasraoui and Jonas (1991a) give $A = 6.82 \times 10^{-4}$, $p = 0.13$, $q = 0$ and $Q_{\text{DRX}} = 312000$ J/mole for a similar steel. A check of Sellars' predictions of the critical strain is possible by considering the true stress–true strain curve for a 0.05% C steel at 975°C at a strain rate of $1.4 \times 10^{-3}\,\text{s}^{-1}$, presented by Jonas and Sakai (1984). Reading the critical strain off the curve one obtains $\varepsilon_c \approx 0.14$. The grain size is taken as 65 μm and the predicted critical strain is then found to be 0.13.

The equation describing the dynamically recrystallized volume fraction is:

$$X_{\text{DRX}} = 1 - \exp\left[B\left(\frac{\varepsilon - \varepsilon_c}{\varepsilon_p}\right)^k\right] \tag{3.81}$$

where ε_p is the strain at the peak stress, usually calculated as $\varepsilon_p = \overline{C}\varepsilon_c$. Hodgson et al. (1993) give the coefficients in eq. 3.81 for C–Mn steels as $B = -0.8$, $k = 1.4$ and $\overline{C} = 1.23$. The strain for 50% recrystallization is calculated as:

$$\varepsilon_{0.5X} = 1.144 \times 10^{-3} D_\gamma{}^{0.28} \dot{\varepsilon}^{0.05} \exp\left(\frac{51880}{RT}\right) \tag{3.82}$$

The equation describing the grain size after dynamic recrystallization is:

$$D_{\mathrm{DRX}} = BZ^{r} \tag{3.83}$$

Sellars (1979) provides the coefficients in eq. 3.83 for C–Mn steels as $B = 1.8 \times 10^3$ and $r = -0.15$. The apparent activation energy is as given above, 312 000 J/mole.

3.11.2.3 Metadynamic recrystallization

When dynamic recrystallization starts during the deformation and the recrystallized nuclei continue to grow after the deformation ends, the phenomenon is identified as metadynamic recrystallization. The equation describing the time for 50% metadynamic recrystallization is:

$$t_{0.5} = A_1 Z^{s} \exp\left(\frac{Q}{RT}\right) \tag{3.84}$$

The constants in eq. 3.84 for carbon–manganese steels are given by Hodgson et al. (1993): $A_1 = 1.12$, $s = -0.8$, Q_d (the activation energy of deformation in the Zener–Hollomon parameter) = 300 000 J/mole and $Q = 230\,000$ J/mole. The metadynamic grain size is (Hodgson et al. 1993)

$$D_{\mathrm{MD}} = AZ^{u} \tag{3.85}$$

where $A = 2.6 \times 10^4$ and $u = -0.23$.
The grain size during metadynamic recrystallization is calculated as the weighted average of the contributing grains:

$$D(t) = D_{\mathrm{DRX}} + (D_{\mathrm{MD}} - D_{\mathrm{DRX}})\, X_{\mathrm{MD}} \tag{3.86}$$

In eq. 3.86, X_{MD} is the volume fraction after metadynamic recrystallization, calculated from the Avrami equation with $k = 1.5$.

3.11.2.4 Grain growth

Following complete static or metadynamic recrystallization, the equiaxed austenite microstructure coarsens by grain growth, assumed to be uniform. Nanba et al. (1992) and Hodgson and Gibbs (1992) presented an equation for C–Mn steels:

$$D(t)^{n} = D_{\mathrm{RX}}^{n} + k_g t \exp\left(\frac{-Q_g}{RT}\right) \tag{3.87}$$

where D_{RX} is the fully recrystallized grain size, t is the time after complete recrystallization, $Q_g = 66\,600$ J/mole, is the apparent activation energy for grain growth, and $n = 2$ and $k_g = 4.27 \times 10^{12}$ are constants.

3.11.3 Properties at room temperatures

Empirical relations, leading to the mechanical attributes of the rolled product have also been developed and in what follows, these are reviewed in some detail. At the Ar_3 temperature, given by

$$Ar_3 = 910 - 310\,[C] - 80\,[Mn] - 20\,[Cu] - 15\,[Cr] - 80\,[Mo] + 0.35(t-8) \quad (3.88)$$

the austenite grains begin their transformation to ferrite grains. Equation 3.88 was developed for plate rolling and t represents the thickness of the plate.

3.11.3.1 Ferrite grain size

The ferrite grain sizes may be estimated by the relation of Sellars and Beynon (1984):

$$D_\alpha = \left(1 - 0.45\varepsilon_r^{1/2}\right) \times \left\{1.4 + 5C_r^{-1/2} + 22\left[1 - \exp\left(-1.5 \times 10^{-2} D_\gamma\right)\right]\right\} \quad (3.89)$$

where D_α is the ferrite grain size in μm, C_r is the cooling rate in K/s, $D\gamma$ is the austenite grain size also in μm and ε_r is the accumulated strain. When the cooling rate is taken as 20 K/s, the accumulated strain as 0.4 and the grain size as 50 μm, eq. 3.89 predicts a ferrite grain size of 10 μm.

3.11.3.2 Lower yield stress

According to the Hall–Petch equation, the lower yield stress σ_y for a homogeneous microstructure is expressed as

$$\sigma_y = \sigma_0 + K_y D_\alpha^{-0.5} \quad (3.90)$$

where σ_0 is the lattice friction stress, K_y is the grain boundary unlocking term for high-angle grain boundaries, taken as 15.1–18.1 Nmm$^{-3/2}$, and D_α is the ferrite grain size.

Le Bon and Saint Martin (1976) present a simple equation for the lower yield stress of carbon steels, in terms of the ferrite grain size:

$$\sigma_y = 190 + 15.9\,(0.001 D_\alpha)^{-0.5} \quad (3.91)$$

The yield strength of the steel, just considered, is given as 349 MPa.

3.11.3.3 Tensile strength

Hodgson and Gibbs (1992) published a simple formula expressing the tensile strength of carbon steels with some alloying elements:

$$\sigma_u = 164.9 + 634.7\,[\mathrm{C}] + 53.6\,[\mathrm{Mn}] + 99.7\,[\mathrm{Si}] + 651.9\,[\mathrm{P}]$$
$$+ 472.6\,[\mathrm{Ni}] + 3339\,[\mathrm{N}] + 11\,(0.001 D_\alpha)^{-0.5} \quad (3.92)$$

The tensile strength of a carbon steel, containing 0.1% C and 0.6% Mn and 10 μm ferrite grains, is estimated as 370 MPa.

3.11.4 Physical simulation

In spite of the above comments regarding the limited abilities of physical simulation of thermal–mechanical treatment, useful and detailed information can be obtained about the hot response of steels. While the number of publications in the field is too numerous to be reviewed here, many of the equations, given above, have been obtained as a result of simulation experiments: multistage compression and torsion tests have been found to be very useful. Some of the publications have been reviewed by Lenard et al. (1999); in one of these, Majta et al. (1996) performed multistage hot compression of a high strength low alloy steel, measured the yield strength after cooling and compared it very successfully to the measurements of a large number of researchers (Morrison et al., 1993; Coldren et al., 1981; Irvine and Baker, 1984). Several meetings have been devoted to the subject; one of the outstanding conferences was held in Pittsburgh in 1981, entitled "Thermomechanical Processing of Microalloyed Austenite", edited by A.J. DeArdo, G.A. Ratz and P.J. Ray.

3.12 MISCELLANEOUS PARAMETERS AND RELATIONSHIPS IN THE FLAT ROLLING PROCESS

The mathematical models presented above take account of the contributions of the most significant variables and parameters. Several more phenomena are associated with the flat rolling process, however, and it is surprising that these are not usually included in the traditional analyses[26]. These are listed below, their definitions are given and simple formulae are presented for their evaluation.

3.12.1 The forward slip

The relative velocities of the strip and the roll have been identified as having an effect on the rate of straining, lubrication, friction, scaling and the interfacial forces. The forward slip, which is given in terms of the relative velocity, has, on occasion, been used to characterize tribological events. It is defined as

$$S_\nu = \frac{\nu_{\mathrm{exit}} - \nu_{\mathrm{r}}}{v_{\mathrm{r}}} \tag{3.93}$$

In determining the exit velocity of the strip, one may use a variety of approaches. Optical techniques, that monitor the roll and the strip velocities, offer the most accurate measurements. A method, often used, is to mark the work roll surface using equally spaced, parallel lines, the separation of which

[26] Online and off-line models used in the rolling industry often include these parameters.

is designated by l_r. These lines make their impressions on the surface of the rolled strips and their distance on the strip, l_s, may be measured using traveling microscopes. The forward slip can then be determined from these distances as

$$S_f = \frac{l_s - l_r}{l_r} \tag{3.94}$$

Using the idea of mass conservation or its equivalent, volume constancy, indicating that the volume of the rolled metal is constant, it can be shown that the two formulas for the forward slip are identical.

Researchers, studying the development of surfaces as a result of flat rolling may well object to marking the roll surface as the lines may affect the interactions of the surfaces and the lubricants.

The forward slip is often taken as a direct indication of frictional conditions in the roll gap. There are several formulas in the technical literature, connecting the coefficient of friction and the forward slip. These will be discussed in some more detail in Chapter 5, Tribology.

3.12.2 Mill stretch

When a certain exit thickness, h_{exit}, is required, and the roll gap is to be set such that it is to be achieved, it is necessary to account for the extension of the mill frame, as well. The formula expressing the thickness that will result when the roll gap is set to h'_1, is given below:

$$h_{exit} = h'_1 + \frac{P}{S} \tag{3.95}$$

where P is the roll separating force in N and S is the mill stiffness, measured in N/mm. A typical value for the mill stiffness is 5 MN/mm, however, this would have to be ascertained for each particular mill.

3.12.3 Roll bending

Rowe presents a simple formula to estimate the maximum deflection at the centre of the work roll (Rowe, 1977), treating the roll as a simply supported beam, loaded at its centre. The formula accounts for the deflections due to shear loading as well:

$$\Delta = \frac{PL_1{}^3}{EI} + 0.2\frac{PL_1}{AG} \tag{3.96}$$

where Δ is the maximum deflection of the roll at its centre in mm, P is the roll force in N and L_1 is the length of the roll, bearing centreline to bearing centreline, in mm. The elastic modulus is designated by E and taken as 200 000 MPa and the shear modulus by G, equal to 86 000 MPa. The cross-sectional area of the roll in mm^2 is A, and I is the moment of inertia[27] of the roll's cross-section in mm^4.

Roberts (1978) develops a more fundamental formula for the maximum deflection of the roll, based on the double integration method. The effects of both the normal and shear loads are included here as well

$$\Delta = \frac{P\overline{L}^2\,(5L + 24c)}{6\pi ED^4} + \frac{P\overline{L}}{2\pi GD^2} \tag{3.97}$$

where c is half of the bearing length and $\overline{L}$ is the barrel length.

The equation presented by Wusatowski (1969) includes several more geometrical parameters. He gives the roll deflection at the centre, including the effects of shear:

$$\Delta = P\left\{\frac{1}{18.8ED^4}\left[8L_1^3 - 4L_1b^2 + b^3 + 64c^3\left(D^4/d^4 - 1\right)\right]\right.$$
$$\left.+\frac{1}{GD^2\pi}\left[L_1 - 0.5b + 2c\left(D^2/d^2 - 1\right)\right]\right\} \tag{3.98}$$

where b is the width of the rolled strip.

Simple calculations indicate interesting magnitudes of the deflection of a work roll at its centre. In a laboratory experiment, using a 250 mm diameter steel roll, with the bearings 400 mm apart, and reducing a 25 mm wide low carbon steel strip by 50%, the roll separating force was measured to be 8000 N/mm. The roll deflection is then obtained as 0.337 mm by eq. 3.96. Similar calculation for an industrial case yields quite different numbers. Considering hot rolling of a 1500-mm wide low carbon steel strip and reducing it with 1000 mm diameter rolls, leads to roll separating forces of 24–34 MN, depending on the temperature and the reduction. Taking the bearing diameter to be 600 mm, the centre-to-centre distance to be 2500 mm, the barrel length to be 2100 mm, half the bearing length to be 200 mm and the roll force to be 24 MN, eq. 3.96 now indicates that the roll deflection will be 38.4 mm, a highly unrealistic magnitude. Since eq. 3.97 yields 0.52 mm and eq. 3.98 gives 0.91 mm, close enough, the recommendation is to use eq. 3.97, the simpler of the two relations. These magnitudes imply that crowning is necessary for consistent thickness to be produced.

3.12.4 Cumulative strain hardening

The cumulative effect of sequential straining on the resistance of the material to deformation is well understood. In what follows, a simple procedure to estimate this effect in multipass flat rolling is presented.

---

[27] A more suitable name for I is "second moment of the area".

In the example a strip of steel is to be rolled in two consecutive passes. In the first pass its thickness at the entry is h_{entry} and its thickness at the exit is $h_{exit,1}$ and the strain then becomes

$$\varepsilon_1 = \ln\left({}^{h_{entry}}/_{h_{exit,1}}\right) \tag{3.99}$$

and the average flow strength is obtained by integrating the true stress–true strain relation over the strain in the pass

$$\sigma_1 = \int_0^{\varepsilon_1} \sigma(\varepsilon)\, d\varepsilon \tag{3.100}$$

In the second pass the entry thickness is $h_{exit,1}$ and the exit thickness is $h_{exit,2}$ so the strain in the second pass is

$$\varepsilon_2 = \ln\left({}^{h_{exit,1}}/_{h_{exit,2}}\right) \tag{3.101}$$

and the total strain experienced by the strip so far

$$\varepsilon_{total} = \ln\left({}^{h_{entry}}/_{h_{exit,2}}\right) \tag{3.102}$$

The average flow strength in the *second pass* is then determined by integrating over the strain *in the second pass*:

$$\sigma_2 = \int_{\varepsilon_1}^{\varepsilon_{total}} \sigma(\varepsilon)\, d\varepsilon \tag{3.103}$$

The steps described above are illustrated by an example in which a low carbon steel strip is reduced, first by a strain of 0.1, followed by another pass creating the same magnitude of the strain. The true stress–true strain relation of the steel, in MPa, is $\sigma = 100\,(1 + 182.02\varepsilon)^{0.355}$ so the average flow strength in the first pass is obtained as 218 MPa and in the second, 326 MPa; Figure 3.17 shows the details.

3.12.5 The lever arm

In the empirical model of the flat rolling process (see Section 3.2) the roll torque was calculated by assuming that the roll separating force acts halfway between the entry and the exit, making the ratio of the torque for both rolls and the roll separating force – the lever arm – equal the projected length of the contact, L. As mentioned above, while the predictions of the roll separating forces by the empirical model are reasonably accurate and consistent, those of the roll torque are not quite so good. The reason is found in the assumption of the magnitude of the lever arm.

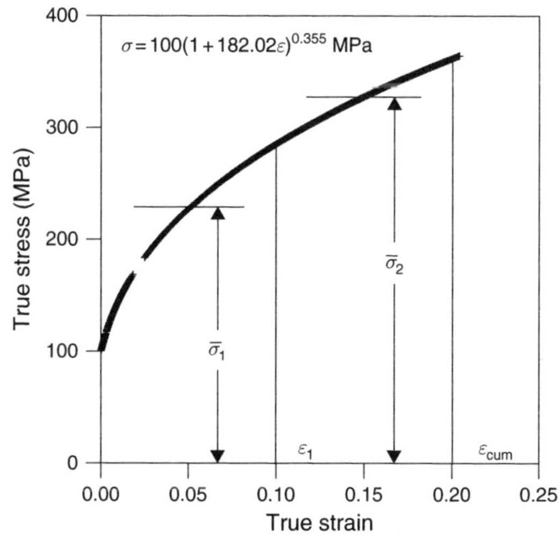

Figure 3.17 The true stress–true strain curve and the average flow strengths in the two passes.

In an effort to develop a better appreciation of the lever arm, the data of McConnell and Lenard (2000) are employed once again; these include approximately 250 experiments. The ratios of the measured roll torques and the roll separating forces are calculated, yielding the actual lever arm, l_a.

The results are shown in Figure 3.18, where the ratio L/l_a is plotted versus the roll separating force. All data are included, with roll speeds varying from a

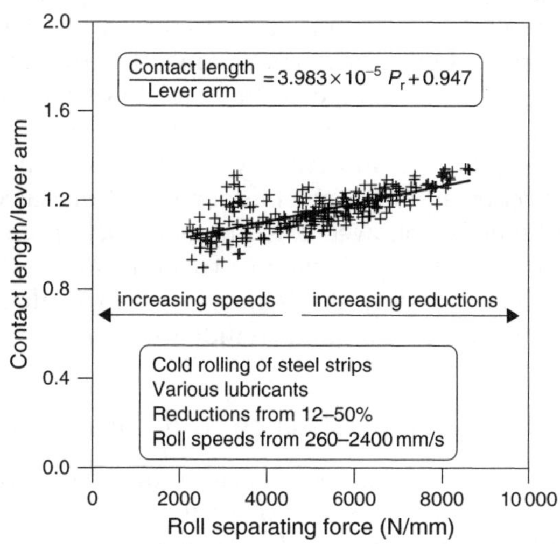

Figure 3.18 The dependence of the ratio of the contact length and the lever arm on the roll separating force.

low of 260 mm/s to a high of 2400 mm/s and reductions varying from a low of 12 to a high of 50% . The ratio is obtained as $1 \pm 30\%$. Lundberg and Gustaffson (1993) estimate the lever arm in edge rolling to be close to unity.

The ratio of the projected contact length and the lever arm are calculated, resulting in

$$\frac{L}{l_a} = 3.983 \times 10^{-5} P_r + 0.947 \tag{3.104}$$

3.13 HOW A MATHEMATICAL MODEL SHOULD BE USED

For successful predictions of the rolling variables while using any of the available mathematical models, knowledge of the accurate magnitudes of the coefficient of friction or the friction factor and the metal's resistance to deformation is absolutely necessary. While it is clear that without them the predictions become essentially useless, their determination may cause almost insurmountable difficulties in many instances. The following steps are then recommended.

3.13.1 Establish the magnitude of the coefficient of friction

Conduct a carefully controlled set of rolling tests and measure the roll separating force as a function of the rolling speeds and the reductions. If hot rolling is studied, the temperature also becomes one of the independent parameters and its effect also needs to be taken into account. Its measurement is not easy. Arguably, the best approach may be to embed thermocouples in the strip to be rolled, even though the stress concentration this causes may affect the magnitude of the reduction. Optical pyrometers may be used instead at both the entry and the exit and the average of their readings may give the average surface temperature quite closely. Once the data are collected, use one of the models of the rolling process and employing the inverse method, determine the coefficient of friction such that the measured and the calculated roll forces agree. This should then be followed by using non-linear regression analysis to develop a relationship of the coefficient of friction as a function of the speed and the reduction and possibly the temperature. In further modelling, this equation may then be used with good confidence.

3.13.2 Establish the metal's resistance to deformation

Use the plane-strain compression test to determine the material's resistance to deformation. If a plane-strain press is not available, a uniaxial tension or compression test will be acceptable; if both are possible, choose the compression test. The experimental difficulties increase if one deals with hot deformation. In an ideal case, isothermal tests should be conducted.

Whether hot or cold rolling is considered, the effects of friction and temperature rise should be removed from the data.

If experimental equipment, needed to determine the material's strength, is not available, one has no choice but to rely on published data, the perils of which have been pointed out elsewhere. For hot rolling, Shida's (1969) equations are recommended, but checking them against the data of Suzuki et al. (1968) would be helpful.

Another approach, that of using ANN for the prediction of rolling variables as well as for the control of the process, is also being accepted more and more in the industry. Having a large database, it is possible to train the network to predict the required values in the next rolling pass.

3.14 CONCLUSIONS

In this chapter, mathematical models that describe the mechanical and the metallurgical phenomena during flat rolling of metals were discussed. Modelling of the rolling process of strips and thin plates was examined exclusively so a 1D treatment was considered to be satisfactory. The potential objectives of modelling were listed first. The models presented were classified according to their level of sophistication. These started with an empirical model and were followed by several, well-known 1D models, including one, a 1D elastic–plastic model which takes careful account of the elastic entry and exit regions as well as the elastic flattening of the work roll. In another 1D model the coefficient of friction was replaced by the friction factor which was allowed to vary along the contact region from the entry to the exit. Based on past experience, the factor was taken to depend on the roll pressure, the relative velocity of the roll and the strip and the distance along the contact. The roll pressure distribution was calculated by using the shooting method: numerical integration of the equation of equilibrium was started at the exit and the location of the no-slip point was adjusted to meet the boundary condition at the entry.

The extremum theorem – the upper bound formulation – was also used to estimate the power needed to roll a strip. AI techniques, that is, ANN, were then discussed and their predictive abilities were given.

Following the mechanical models, the development of the microstructure during and after the rolling pass was described. Empirical relations that can be used to estimate the metallurgical parameters during rolling of low carbon steels were listed.

A few numerical examples indicated the predictions of the equations.

The chapter was concluded by presenting several parameters and relationships in the rolling process which are not usually included in mathematical models: the forward slip, mill stretch, roll bending, cumulative strain hardening and the lever arm.

The following recommendations regarding the choice of a model to analyse the flat rolling process may now be made.

- As long as the roll diameter to thickness ratio is much larger than unity, a criterion that is satisfied well in strip and thin plate rolling, the planes-remain-planes assumption is valid and 1D analyses are satisfactory;
- The choice of the model depends on the objectives of the user. The guiding principle should be to use the simplest model that satisfies the need;
- Regardless of the model chosen, the accurate knowledge of three parameters is necessary: the coefficient of friction, the heat transfer coefficient and the metal's resistance to deformation;
- If the roll separating force only is needed, the empirical model of Section 3.2 is adequate;
- If the force and the torque are needed, a 1D model should be used. For somewhat more confidence in the predictions, the refined 1D model should be employed;
- If, in addition to the above, the temperature changes, roll flattening, required power and the metallurgical events are to be determined, the refined 1D model is recommended, see Section 3.4;
- If the distributions of the normal pressures and the interfacial shear stresses on the work roll are wanted, the coefficient of friction or the friction factor should be expressed as a variable from the entry to the exit of the roll gap. The shape of the coefficient of friction distribution may be based on existing data;
- If rolling of thicker plates is to be analysed, the recommendation is to use the finite-element technique.

CHAPTER 4

Material Attributes

Abstract

The need for understanding the behaviour of metals, when subjected to plastic deformation under various process conditions, is emphasized. Newly developed steels are mentioned. The mechanical and metallurgical attributes are reviewed. Attention is paid to the available, traditional testing processes that may be used to establish the metal's resistance to deformation. Their advantages and disadvantages are listed. Corrections for frictional effects and for isothermal conditions are given. The mathematical models that describe the resistance to deformation are listed and compared. Their suitability for the analysis of the flat rolling process is presented, compared and discussed.

4.1 INTRODUCTION

The metal's resistance to deformation is often referred to by several names. It may be called the flow stress, the flow strength or the constitutive relation; the bottom line is that a relationship of the metal's strength to other, independent variables is being considered. The best identification is the term "resistance to deformation" since it describes a material property and its meaning is clear: it indicates how the material reacts when it is loaded and deformed by external forces. The need for understanding the intricacies of the material's resistance to deformation has been indicated in Chapters 2 and 3. This includes two ideas: the appreciation of the physical and metallurgical capabilities and the response of the materials while in service as well as the development of mathematical models of the metals' resistance to deformation. The former is necessary to provide insight and to aid in the design and the planning of the metal-forming processes. The latter is critical for the success of the control and the predictive models of the flat rolling process. Several steps need to be completed in order to reach these objectives. These are listed below.

- Determine the independent variables that are expected to affect the metal's resistance to deformation;

- Determine the metal's resistance to deformation in an appropriate test, one that allows the variation of the independent variables over an appropriate range of magnitudes and
- Develop, through mathematical modelling (non-linear regression analysis, artificial intelligence or storing data in a multidimensional matrix of data, for example) a true stress–true strain, strain rate, temperature, etc. relation.

In what follows, selected attributes of some of the steel and aluminum alloys used in the metal-forming industry will be briefly reviewed and will be compared, with special attention paid to the automotive industry. Recently developed alloys will also be briefly introduced. Traditional testing techniques to determine the metals' attributes will be discussed next; their advantages and disadvantages are given. This will be followed by a presentation of the mathematical description of some of the attributes. Metallurgical events will also be discussed, and the grain structures accompanying hot or cold deformation processes will be presented. Most of the comments will concern steels, reflecting the experience of the writer.

4.2 RECENTLY DEVELOPED STEELS

The traditional metals used in the metal-forming industry include the alloys of steel, aluminum, copper and titanium. New alloys have been developed in the last several decades, mostly driven by the need of the automotive industry to reduce weights, gasoline consumption and thus, reduce air pollution. This need created one of the most important current objectives of the carmakers, which is to develop the technology to produce lightweight components. The materials used in this regard must have attributes that include high strength and high ductility. The new ferrous metals being introduced include the interstitial free steels, bake-hardenable steels, transformation-induced plasticity (TRIP) steels, the high strength low alloy (HSLA) steels, dual-phase (DP) steels, martensitic and manganese–boron steels, having yield strengths that vary from a low of 200 up to 1250 MPa. The elongation of these steels decreases as the strength increases, from a high of nearly 40% to a low of 4–5%, affecting the design of the subsequent applications. A recent review (Ehrhardt et al., 2004) lists many of these steels and indicates that light construction steels with induced plasticity possess tensile strength in the order of 1000 MPa and remarkably high total elongation of 60–70%.

The website of the American Iron and Steel Industry also includes up-to-date information concerning the description and the processing of recently developed steel alloys. As given in the site, advanced high strength steels (AHSS) in use in the automotive industry include the dual phase (DP) steels – microstructure of which includes ferrite and up to 20 and 70% volume fraction of martensite. While the use of bainite helps to enhance the capability to resist stretching on a blanked edge, the ferrite phase leads to high ductility and creates high work hardening rates, which give the DP steels higher tensile strength than conventional steels. Further, TRIP steels are also used, the microstructure of which consists of a ferrite matrix containing a dispersion of hard second phases – martensite and/or bainite in addition to retained austenite in volume fractions greater than 5%. During deformation, the hard second phases create a high work hardening rate while the retained austenite transforms to martensite, increasing the work hardening rate further at higher strain levels. The complex phase (CP) steels consist of a very fine microstructure of ferrite and a higher volume fraction of hard phases that are further strengthened by fine precipitates. In the martensitic (MART) steels the austenite that exists during hot rolling or annealing is transformed almost entirely to martensite during quenching on the run-out table or in the cooling section of the annealing line. All AHSS are produced by controlling the cooling rate from the austenite or austenite plus ferrite phase, either on the run-out table of the hot mill (for hot rolled products) or in the cooling section of the continuous annealing furnace (continuously annealed or hot-dip coated products). AHSS cooling patterns and resultant microstructures are schematically illustrated on the continuous cooling-transformation diagram, available for examination in the AISI website. The cooling patterns are designed on the bases of mathematical models, which attempt to predict the structures and properties resulting from the processing technique.

Research is continuing in the development of twinning induced plasticity and lightweight steels with induced plasticity (Cornette et al., 2005; Gigacher et al., 2005)[1].

4.2.1 Very low carbon steels

The structure of these steels is fully ferritic. A micrograph, reproduced from the website http://www.mittalsteel.com is shown in Figure 4.1.

While the strength of these steels is very low, their very high formability makes them ideal candidates for parts that carry low loads but require high strain carrying ability during the forming process. Automotive components and motor lamination steels are potential uses.

4.2.2 Interstitial free (IF) steels

These steels contain less than 0.003% C. The nitrogen level is also reduced during their preparation and the remaining carbon and nitrogen are tied

[1] I am grateful to Dr G. Nadkarni, Mittal Steel, USA (Southfield, MI) and to Dr J. Tiley of Hatch & Associates, for bringing these grades of steels to my attention.

Figure 4.1 Microstructure of an extra deep drawing ferritic steel ×300 (reproduced from http://www.mittalsteel.com/).

up using small amounts of alloying elements, such as Ti or Nb. The steels are finish rolled above 950°C. Their strengths are low but they possess very high formability, especially after annealing (138–165 MPa yield strength, 41–45% elongation). Their structure is very similar to that shown in Figure 4.1.

4.2.3 Bake-hardening (BH) steels

The carbon content is even lower, 0.001% C. The steels harden during the paint-curing cycle, performed usually at 175°C, for 30 min. The hardening is caused by the precipitation of carbonitrides. The as-received yield strength of the steel is 210–310 MPa; after baking and a 2% pre-strain these rise to 280–365 MPa. There is little change of the tensile strength but the dent resistance is increased. These grades are extensively used in automotive outer body panels. A typical microstructure of a bake-hardening steel is given in Figure 4.2.

4.2.4 TRIP steel

These steels are highly alloyed and have been heat treated to produce metastable (that is, not fully stable with respect to transformation) austenite plus martensite. When subjected to permanent deformation, the austenite

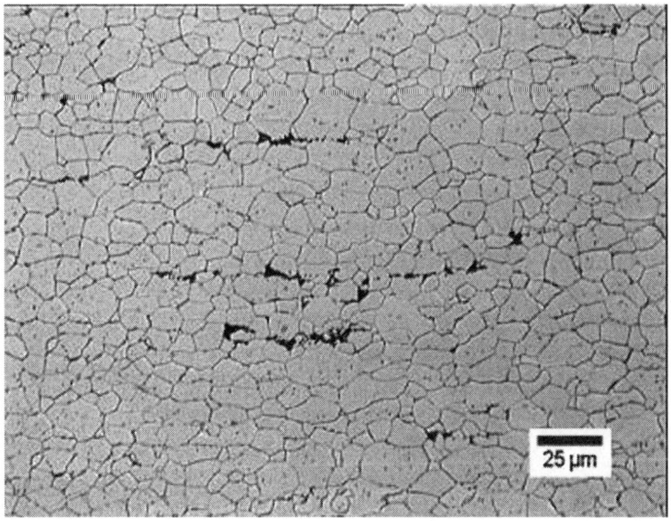

Figure 4.2 Bake-hardening steel (reproduced from http://www.mittalsteel.com/).

experiences strain-induced transformation to martensite. A tempering process may follow the transformation. The steels are highly ductile and strain rate sensitive. Their tensile strength can reach magnitudes as high as 800 MPa. They respond well to bake hardening and an extra 70 MPa strength is the result. These steels are one of the newest family of AHSS currently under development for the automotive industry[2]. The steels have a microstructure of soft ferrite grains with bainite and retained austenite. The hard martensite delays the onset of necking resulting in a product with high total elongation, excellent formability and high crash energy absorption. In addition, TRIP steels also exhibit extremely high fatigue endurance limit, thereby providing excellent durability performance. The micrograph of Mittal's TRIP steel is shown in Figure 4.3.

4.2.5 High strength low alloy (HSLA) steels

The HSLA steels, often referred to as microalloyed steels, are low carbon steels with the strength increased by small amounts of alloying elements such as niobium, vanadium, titanium, molybdenum or boron, singly or in combinations. Their tensile strength may reach 450 MPa and their ductility may be as high as 30%. Thermo-mechanical processing is used to affect their mechanical

[2] As indicated on the website of Mittal Steel.

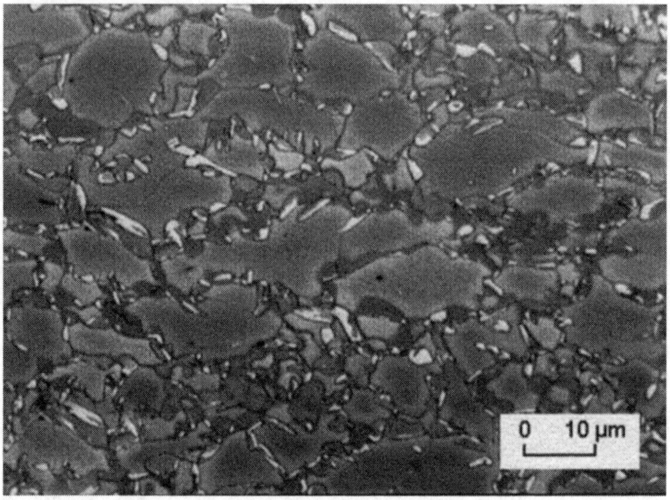

Figure 4.3 Transformation-induced plasticity steel (reproduced from http://www.mittalsteel.com/).

and metallurgical attributes. Arguably, one of the best collection of information concerning these alloys appears in the Proceedings of the International Conference on the Thermomechanical Processing of Microalloyed Austenite, held in Pittsburgh, in 1981. Micrographs of many of these steels under a large number of processing conditions have been published at that conference. In what follows, two examples are shown. Figure 4.4 shows two micrographs (Maki *et al.*, 1981) of a 0.2% C, 0.002% B steel, dynamically recrystallized under

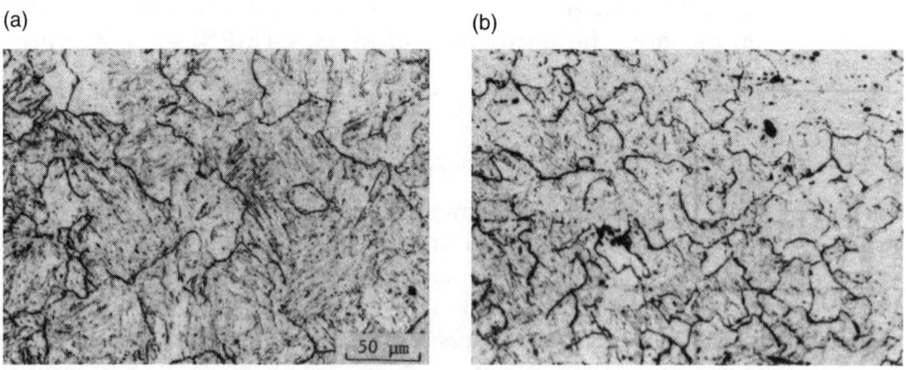

Figure 4.4 Micrographs of a boron steel, fully recrystallized. (a) $\dot{\varepsilon} = 1.7 \times 10^{-2} s^{-1}$ and $\varepsilon = 0.37$, (b) $\dot{\varepsilon} = 1.7 \times 10^{-1} s^{-1}$ and $\varepsilon = 0.45$ (Maki et al., 1981).

different deformation conditions. In both $T = 1273\,K$. In a) $\dot{\varepsilon} = 1.7 \times 10^{-2} s^{-1}$ and $\varepsilon = 0.37$ while in b) $\dot{\varepsilon} = 1.7 \times 10^{-1} s^{-1}$ and $\varepsilon = 0.45$.

4.2.6 Dual-phase (DP) steels

These are low alloy steels, similar to the HSLA steels. Their tensile strength are somewhat higher, 550 MPa. The structure, shown in Figure 4.5 contains approximately 20% martensite in a ductile ferrite matrix. As written on Mittal's website, DP steels are one of the important new AHSS products developed for the automotive industry. Their microstructure typically consists of a soft ferrite phase with dispersed islands of a hard martensite phase. The martensite phase is substantially stronger than the ferrite phase.

A compilation of the mechanical attributes of several materials is shown in Figure 4.6, reproduced, following Pleschiutschnigg et al. (2004). The figure gives the elongation and the yield strength measured at 0.2% offset, of deep drawing steels, austenitic stainless steels, BH steels, TRIP steels, duplex stainless steels, DP steels, HSLA steels in addition to aluminum and magnesium. It is noted that the deep drawing quality steels and the austenitic stainless steels offer the highest formability. BH and TRIP steels indicate similar elongation but the TRIP steels also provide much higher strength.

Aluminum is less strong and less formable but much lighter than the ferrous metals. Its competitiveness needs to be based on its superior strength to weight ratio.

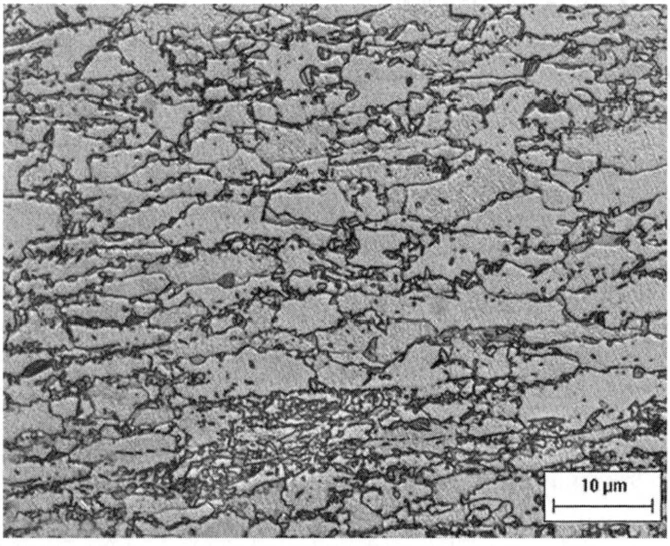

Figure 4.5 Dual-phase steel (reproduced from http://www.mittalsteel.com/).

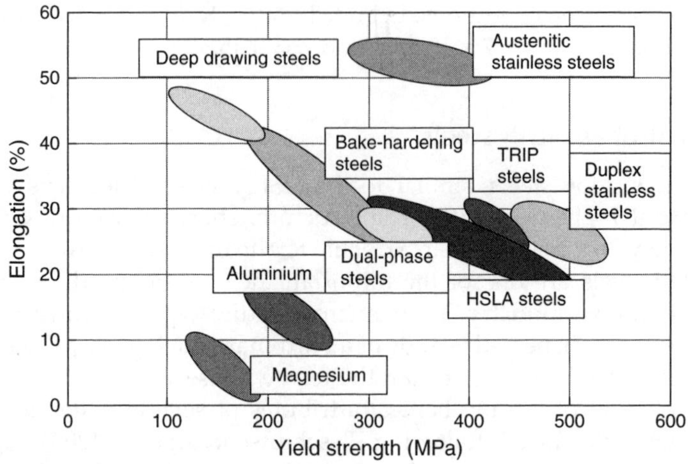

Figure 4.6 Compilation of materials (after Pleschiutschnigg et al., 2004).

The processing route that results in the TRIP steels and the DP steels is also discussed by Pleschiutschnigg et al. (2004). The rolling process is similar for both metals, the difference being the cooling rate; faster for the DP steels and slower for the TRIP steels. The authors indicate that the controlled rolling process, not the chemical composition, has a dominating influence on the results.

4.3 STEEL AND ALUMINUM

The competition between aluminum and steel alloys for use in the automotive industry is intense. The website of the U.S. Steel Company gives some of the data, indicating the advantages of the steel grades over aluminum. The information below is taken directly from the website http://xnet3.uss.com/auto/index.htm, in April, 2006.

A formability chart in the website compares the formability of several steels and aluminum. The winner as far as formability is concerned is the interstitial free steel, indicating up to 55% total elongation. Its tensile strength is low, however, at most 350 MPa. The strongest steel is the martensitic type, as expected, possessing a tensile strength of almost 1700 MPa, but at a total elongation of about 7% subsequent plastic forming processes need to be designed with extreme care. Aluminum appears to be somewhere in the middle, with elongation varying from a low of 7% up to 30–32% and a maximum tensile strength of 600 MPa. TRIP steels have a strength between 600 and 1250 MPa and elongations of 18–40%.

A very interesting compilation of stress–strain curves for several steel and aluminum alloys is also given in the USS website. Obtained at fairly low

(0.0005 s^{-1}) and somewhat higher rates of strain (9.8 s^{-1}), the figure indicates that the steel alloys' strengths increase with increasing rates of strain while those of the aluminum do not. Specifically, the maximum strength of the TRIP 590Y steel, at 9.8 s^{-1}, is near 750 MPa and at the lower rate of 0.0005 s^{-1} it is 620 MPa. In the same strain rate range, the strength of the DP steel increases by about 100 MPa as the rate of strain is increased and that of the deep drawing quality steel increases by about 150 MPa. The 5754-0 aluminum alloy indicates no rate sensitivity. Note, however, that there are several aluminum alloys whose strengths are in fact, strain rate sensitive; an example is the commercial purity aluminum alloy, 1100-H0.

Of major importance, also observable from the figure of USS, is the total strain sustained by each of the metals as this has a major impact on the design of subsequent forming processes and hence, it will affect productivity. The clear winner here is the deep drawing quality steel, deformed at the low strain rate – note that the 0.0005 s^{-1} is almost creep – deforming to a fracture strain of 45%. The fracture strain of the TRIP steel at high rates (~36%) is near that of the 5754-0 aluminum, strained at the lower rate. Most of the steels appear to be more formable than the single aluminum alloy.

Further data, also from the USS website, compares the strengths and the cost indices of the two metals using charts. A quotation, discussing the information is reproduced below:

> *The figure . . . shows common metallurgical grades undergoing a pre-strain of 2% and typical automotive paint bake cycle on the left compared to their prospective cost index shown on the right. Pricing for steel grades is based on seven combined typical market sources and the ULSAB-AVC cost model. Aluminum pricing was gathered based on 2002 publications from the MIT Material Systems Lab and typical market information, such as the American Metal Market (2002–2003).*

The cost index of the aluminum alloys is more than five times that of the steels. For example, the 6111-T4 alloys yield strength is given as approximately 240 MPa and its cost index is 5.9. This may be compared to that of the deep drawing quality steel, DP 600. The yield strength of this alloy is the highest among those shown, at 580 MPa, but its cost index is 1.1. The SAE grade 3 steel is demonstrating a yield strength of 250 MPa and a cost index of one.

Information regarding aluminum alloys is also easily available from the Internet. Alcan's website indicates the formation of a spin-off company, Novelis Inc., formed in 2005, and now dealing with rolled products and sheet metal operations. The website http://novelis.com lists the benefits of aluminum over that of other materials; the list is copied directly from the website:

The benefits of using lightweight aluminum sheet in transport applications are clear:

- *Aluminum offers high potential for weight savings, thus reducing emissions through the life of the vehicle, improving fuel efficiency and also handling;*
- *The metal is easily and widely recycled, saving energy and raw materials;*

- *It has very good deformation characteristics and manufacturing properties;*
- *Aluminum will absorb the same amount of crash energy as steel, at a little more than half the weight;*
- *It has good corrosion resistance.*

Novelis' product range for the transportation market includes: sheet for automotive vehicle structures and body panels; pre-painted and plain sheet for commercial vehicle applications such as dump bodies, cabs and trailer flooring; "shate" for ship hulls and decks, tippers, road tankers, etc; plain, heat-treated or painted slit strips and coils – customized to the needs of automotive part suppliers; high specification foil (industrial finstock) and brazing sheet for heat exchangers; and foil for insulation applications.

4.4 THE INDEPENDENT VARIABLES

The traditional approach in identifying the variables and parameters that affect the behaviour of metals is quite simple and works well in many cases. The usual formulations indicate that in the cold deformation regime the resistance to deformation is assumed to be an exclusive function of the strain, that is $\sigma_{cold} = f(\varepsilon)$, and in the high temperature region the only independent variable is the rate of strain, or $\sigma_{hot} = g(\dot{\varepsilon})$. While both relations are used regularly in the analyses of metal forming processes, they represent a much too simplified view of how the metals behave.

A list of independent variables that affect the material attributes is much longer. It may include the strain, strain rate, temperature and metallurgical parameters (for example, the grain size, Zener–Hollomon parameter, chemical composition, activation energy, precipitation potential, amount of recrystallization, volume fraction of various phases). Arguably, it may also include the dependence of the results on the testing technique as it is very rare to see a successful comparison of stress–strain curves of a metal obtained in tension, compression and torsion.

Of course, an equation that includes all of these variables would never be used by engineers; so, as always, a compromise is needed. As will be demonstrated below, adequate results are obtained when the resistance to deformation is given in terms of the strain, rate of strain, temperature and the activation energy:

$$\sigma = f(\varepsilon, \dot{\varepsilon}, T, Q) \tag{4.1}$$

where Q is the activation energy, to be discussed in more detail later in this chapter, and T is the temperature, usually expressed in Kelvin. There are two steps involved in determining an actual, useful and usable form of eq. 4.1. The first step requires a systematic testing program designed with the end-use of the resulting data in mind[3]. The multidimensional databank thus obtained is then employed to develop an appropriate mathematical model which describes the metal's resistance to deformation. In what follows, these two steps are discussed in some detail.

4.5 TRADITIONAL TESTING TECHNIQUES

The objectives of the tests are several and depend strongly on the objectives of the tester. The materials engineer wants to know how the sample of the metal will react to loads. The engineer dealing with metal forming wants to know how to use the results of the test in analysing a metal forming process in addition to deciding what test to use to enable that analysis. The objectives of the materials chemist and physicist also would have an effect on the choice of the test. While there are numerous experiments available for the determination of the metal's resistance to deformation for use in planning, designing and analysing metal forming processes, three of them are used most often. These are the

- Tension test
- Torsion test
- Compression test
  - Axially symmetrical sample
  - Plane sample (width > thickness)

In each of these, constant strain rates and constant temperature conditions need to be established so the only variables to be monitored remain the force and the deformation. As well, the strains are to be high enough to allow a direct comparison of the metals' behaviour in the tests with those required in the actual process.

4.5.1 Tension tests

These are the easiest and simplest to perform, using samples of cylindrical or rectangular cross-sections. The advantages are that

- there are no frictional problems to be considered and
- the tests are governed by ASTM codes (ASTM Standards, E 8 and E 8M, so interlaboratory variability is minimized.)

---

[3] These may include examining the behaviour of the material or the design and analysis of a metal forming process.

The disadvantages indicate that tension testing is not the most suitable when the information gathered is to be used to study metal forming processes. They are as follows:

- Only low strains are possible, at most 40–50%;
- The uniaxial nature of the stress distribution is lost when diffuse straining ends and localized straining begins, subjecting the necked down region to triaxial tension;
- In order to keep the rate of strain constant, increasing the cross-head velocity during the test is necessary and
- While it is possible, it is difficult to perform the test in isothermal conditions.

A schematic diagram of a tension testing setup is shown in Figure 4.7, reproduced from Schey (2000). A universal testing machine is shown, along with a flat sample, the actuator that moves the cross-head, a load cell, an extensometer and a recording device which plots, online, the force – deformation curve. In an updated, modern variant of this setup, the measurements would be collected using digital data acquisition, and a stress–strain curve would then be plotted in real time.

Attention needs to be paid to the manner in which the sample is attached to the cross-head. As shown in the figure, there are holes drilled in the sample and the attachment ensures the application of the force in the direction of its longitudinal centreline. At the same time, the effect of the stress concentration

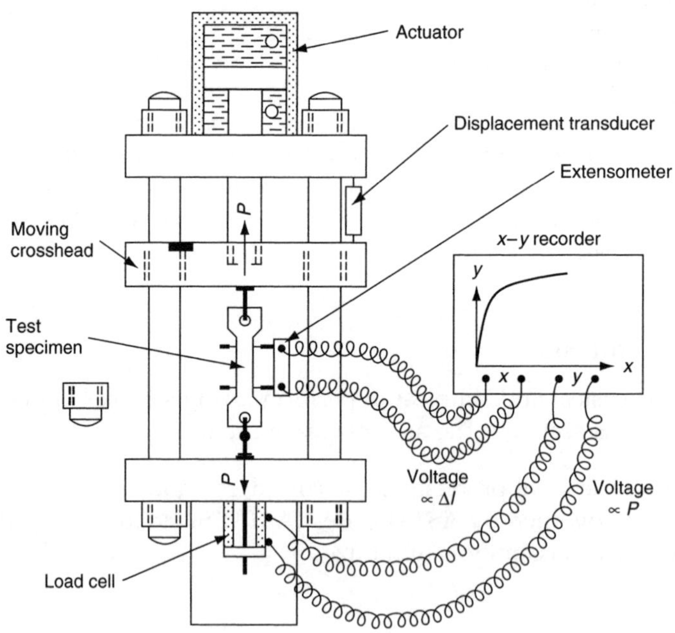

Figure 4.7 The setup of a tension test (Schey, 2000, reproduced with permission).

at the holes is to be considered carefully, and the sample should be designed such that fracture does not occur there. In many commercially available tensile testers, jaws attached to the machine through spherical seats are used to hold the sample and these also ensure appropriate alignment without the problems that may be caused by stress concentration.

The comment made above, concerning the difficulties associated with providing isothermal conditions, may be appreciated by examining the figure. There are several possibilities, none easy. An openable, split furnace may be used which would enclose most of the sample and the extensometer. This would necessitate the use of very expensive extensometers that are capable of performing within the high temperature furnace. A modern variation would make use of an optical device, focused on a deforming section of the sample, through a viewing hole of the furnace wall. In either case, the jaws would also need to withstand the high temperatures. The load cell and the rest of the testing machine would have to be protected from temperature damage, probably by installing inline heat exchangers. Another possibility to heat the sample is the use of induction heaters and a coil which would cover only the reduced, deforming portion of the sample; however, induction heaters are bulky and very expensive.

4.5.2 Compression testing

These may be performed on cylindrical or plane samples; for details, see ASTM Standards E9. Figure 4.8, again reproduced from Schey (2000), shows a schematic of the compression test, using a cylindrical sample (a) and the

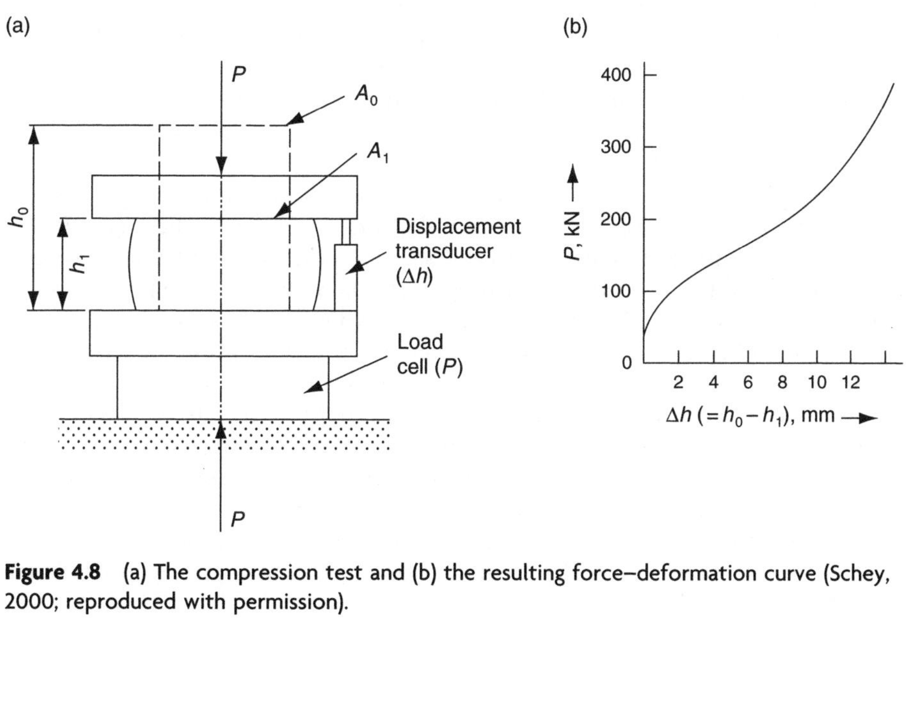

Figure 4.8 (a) The compression test and (b) the resulting force–deformation curve (Schey, 2000; reproduced with permission).

force–deformation curve that results (b). Note that the curve is increasing exponentially, reflecting the growing contact area and the attendant increase of frictional resistance. The advantages of the compression tests are

- larger strains are possible, typically 120–140% when cylinders are compressed and up to 200% when plane samples are tested and
- the state of stress is mostly compressive, as in bulk forming.

The disadvantages are that:

- Frictional forces at the ram–sample interface grow as the test progresses and their effects must be controlled and removed from the data;
- Tensile straining at the cylindrical surfaces or the edges of plane samples limits the level of straining (the circumferential strain may be calculated easily, making use of mass conservation);
- The achievement of constant true strain rates during the tests requires careful feedback control, making the use of a cam-plastometer or, in a modern setting, a computer-controlled servohydraulic testing system;
- The distribution of the strains in the normal direction is not uniform and
- When plane-strain compression is performed, isothermal conditions are difficult to achieve.

It is relatively simple to conduct a compression test on an axially symmetrical sample at high temperatures and to make sure that almost isothermal conditions exist within the furnace. An openable furnace is necessary with a fairly long heated length. The sample is to be compressed between flat platens, made of a material that retains its strength at the test temperature. For steels the platens are often made of silicon carbide. Various Inconel alloys may also be used. As well, it is important to place water-cooled heat exchangers between the compression platens and the rest of the testing system: the load cell and the actuator. The procedure followed is also of importance as is the location of the thermocouple or the temperature measuring system that controls the furnace temperature.

In the plane-strain compression test, shown schematically in Figure 4.9, a flat sample is compressed between two flat dies. As long as the shape factor in the plane-strain test is similar to the shape factor in a flat rolling process, the strain distributions in the deforming portions within the two processes are similar (Pietrzyk et al., 1993). This allows one to recommend that in order to develop a mathematical model of the resistance to deformation for use in a 1D model of flat rolling, the plane-strain compression test should be used.

An experimental difficulty in developing isothermal data in the plane-strain compression test is immediately evident. Enclosing the complete apparatus in a furnace is not practical. Heating the sample only is possible but not easy since the uniformity of the temperature distribution is difficult to maintain. Resistance heating as performed in the Gleeble machines or induction heating may be best; though in the latter, placement of the coils may cause further difficulties.

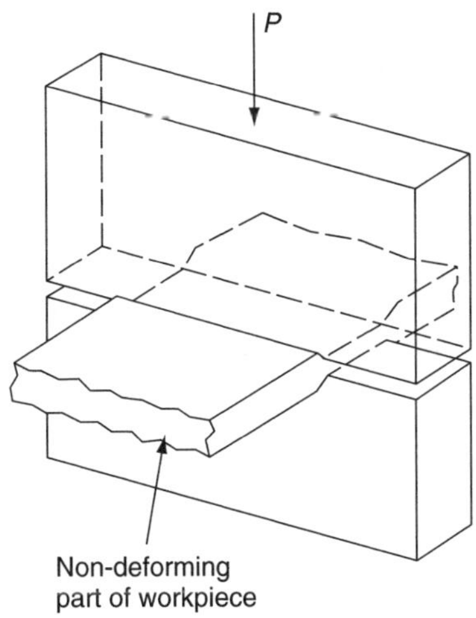

Figure 4.9 Schematic diagram of the plane-strain compression test (Schey, 2000; reproduced with permission).

4.5.3 Torsion testing

This type of testing materials is the most suitable when the data are to be used to analyse large-strain processes, such that a slab would experience during its journey through a hot strip mill, as it is being reduced from a thickness of about 300 mm to a final thickness of about 1–2 mm. Finite strains of 400–500% can be obtained easily, allowing the simulation of the complete history of hot rolling, including the phenomena at the roughing mill and the finishing train of hot strip mills. The advantages are:

- very large strains are possible;
- constant rate of strain is simple to achieve and
- no frictional problems are present.

The disadvantages are that

- the torsional stresses and strains vary over the cross-section and a considerable amount of analysis is necessary to extract the uniaxial normal stress–strain data and
- the variation in the time it takes for different locations of the cross-section to experience metallurgical phenomena, specifically dynamic recovery and recrystallization, may cause a non-homogeneous structure.

It is essential to allow the length of the sample to change without restrictions while the torsional testing proceeds as constraining the length would induce longitudinal stresses in addition to the shearing stresses.

4.6 POTENTIAL PROBLEMS ENCOUNTERED DURING THE TESTING PROCESS

The usual approach to determine the metal's resistance to deformation in order to simulate the hot or the cold rolling processes is to conduct compression tests, using plane or axially symmetrical samples[4]. While the test procedures are well understood and many of them are controlled by well-known standards, two areas of potential difficulties still exist: that of friction and temperature control. In what follows, these difficulties are discussed.

4.6.1 Friction control

This problem is encountered in the compression testing process, whether using axially symmetrical or plane samples. As the samples are being flattened, the contact area grows and continuously increasing effort must be devoted to overcome the frictional resistance at the compression platens. Baragar and Crawley (1984) show that frictional effects are not very pronounced when strains under approximately 0.7 are considered. Above that level of deformation, however, the increasing frictional effects must be removed from the force-deformation data in order to obtain uniaxial behaviour. When using cylindrical samples, this may be accomplished by adopting the relation:

$$p = \sigma_f \left(1 + \frac{m\, d}{3\sqrt{3}\, h}\right) \tag{4.2}$$

where the uniaxial flow strength is σ_f, p is the interfacial normal pressure, m is the friction factor and d and h are the current diameter and height of the sample, respectively. The friction factor is best determined in the ring compression test[5] (Male and DePierre, 1970). Avitzur (1968) quotes Kudo's (1960) formula, connecting the coefficient of Coulomb friction and the friction factor in the form $\mu \left(p_{ave}/\sigma_{fm}\right) = m_{ave}/\sqrt{3}$, presented earlier as eq. 3.60. In room temperature testing, it is possible to minimize frictional problems by using a double layer of teflon tape over the flat ends of the sample. In high temperature tests a glass powder–alcohol emulsion may be employed.

Removing the effects of friction while the data obtained from plane-strain compression testing are analysed is equally important.

In what follows, an example of the use of the above formula, eq. 4.2, is presented, considering the compression test, performed on a Nb-V microalloyed steel. Samples of the steel, measuring 10 mm in diameter and 15 mm

---

[4] This statement may cause an argument among material scientists, many of whom value the advantages provided by the torsion test more than the simplicity of the compression tests.

[5] The ring compression test will be discussed in detail in Chapter 5, Tribology.

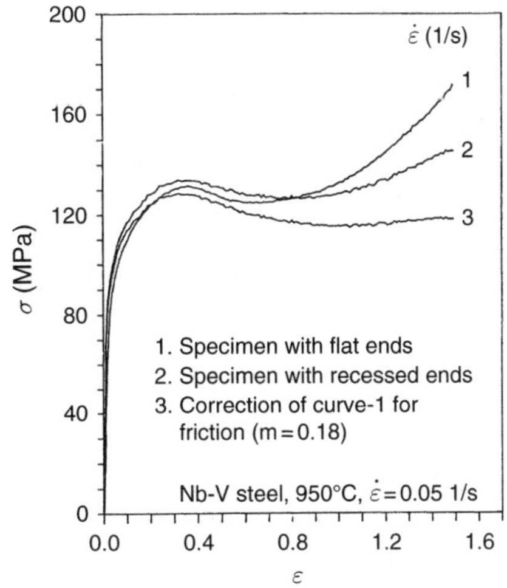

Figure 4.10 True stress–true strain curves of a Nb–V steel, at 950°C, under three different conditions (Wang, 1989).

long were compressed under nearly isothermal conditions, at a constant true strain rate of $0.05\,s^{-1}$. The temperature of the sample was 950°C. Three tests were conducted, the results of which are shown in Figure 4.10 (Wang, 1989). In all three tests, glass powder in an alcohol emulsion (Deltaglaze 19) was used as the lubricant. The first experiment used a sample prepared with its ends machined flat, and beyond a true strain of 0.8 the resulting stress–strain curve indicated a steep rise which, if no elevated temperatures were employed, may be confused with strain hardening. In the second test, the well-known Rastegaev technique (1940) was followed, indenting the ends of the sample to a depth of 0.1 mm and leaving a ridge of about the same dimension. The objective was to trap the lubricant at the ram/sample interface. The resulting curve still indicated some rise. (It is noted that researchers often employ very shallow, concentric or spiral grooves on the flat ends to achieve the same objectives. The present writers' experience indicates that multiple grooves are more difficult to machine without offering any significantly increased benefits over recessed ends in the reduction of friction.) In the third attempt, the value of the friction factor was determined, under the same conditions in a ring-test[6], to be 0.18. The uniaxial flow strength was calculated by eq. 4.2 and is shown in Fig. 4.10, identified as curve 3. The curve demonstrates the steady-state behaviour, expected of the steel, at the test temperature and strain rate.

[6] The ring compression test, for use in determining the friction factor, will be discussed in detail in Chapter 5, Tribology.

4.6.2 Temperature control

The need to control the temperature during the tests for strength is equally important at both low and high temperatures. It is equally difficult to measure the temperature of the sample accurately during the experiment. Overcoming the first difficulty is most important when the tests to determine the material's resistance to deformation are conducted. The second problem is of significance when the test results are reported and mathematical models for use in subsequent analyses are to be developed[7].

4.6.2.1 Isothermal conditions

The usual procedure in conducting a test is to pre-heat the furnace and the compression rams to the desired temperature. This is followed by opening the furnace door, placing the sample on the bottom ram for a sufficient length of time to reach a steady state, bringing the top ram in contact with the sample and starting the compression process. The rams are usually of a larger diameter than the compression sample and are of considerably larger thermal mass. They are connected to the load cell and the actuator through water-cooled heat exchangers, and their lengths are considerable, even if the heated length of the furnace is not very long. Because of the heat exchangers, the rams' temperatures are not uniform along their lengths and typically they are lower than that of the furnace. The furnace temperature is usually monitored by a thermocouple whose bead is a few millimetres away from the inner surface of the furnace's insulation. The control of the furnace temperature is achieved by monitoring the output of this thermocouple. The average temperature at the centre of the furnace is quite certainly lower than the indicated value. When the furnace is opened to allow the placing of the sample on the ram, considerable amount of cooling takes place.

While time consuming, expensive and labour intensive, thermocouples should always be embedded in the sample and in the loading rams. The thermocouple on the sample should be used to control the temperature of the furnace. The test is to commence when the sample and the ram temperatures are very close, within a pre-determined tolerance. (A modern alternative, of course, is the use of optical pyrometers through spyholes in the furnace walls, instead of thermocouples). The thermocouple in the sample will also indicate the temperature rise due to work done on it. In reporting the results this rise should be accounted for. Realizing that the work done per unit volume is

[7] When reporting the results of the tests, many writers are guilty of not describing the equipment and the procedure in minute detail. Both of these are necessary if the tests are to be duplicated.

almost exactly equal to the area under the true stress–true strain curve, the temperature rise may be estimated by

$$\Delta T = \frac{\int \sigma \, d\varepsilon}{c_p \rho} \tag{4.3}$$

where the specific heat is designated by c_p and the density by ρ, both of which are temperature dependent; see Touloukian and Buyco (1970). Corrections to develop the flow curve under isothermal conditions require the determination of the temperature as the sample is being compressed and interpolation and extrapolation to compute the appropriate values of the stresses. In these calculations it is assumed that all work done is converted into heat, an assumption which is close enough though not quite correct.

4.6.2.2 Monitoring the temperature

The potential accuracy of the temperature measurements should be considered, as well. Manufacturers' catalogues list the accuracy of a type K (chromel–alumel) thermocouple as ±0.5%, full scale[8]. If testing at 1000°C is considered, this indicates a potential error of 10°C. The effect of this error on the magnitude of the temperature needs to be understood in light of the temperature sensitivity of the resistance to deformation of steels. This varies over a large range, as shown in Figure 4.11, which indicates the dependence of the peak stresses of several steels on the temperature.

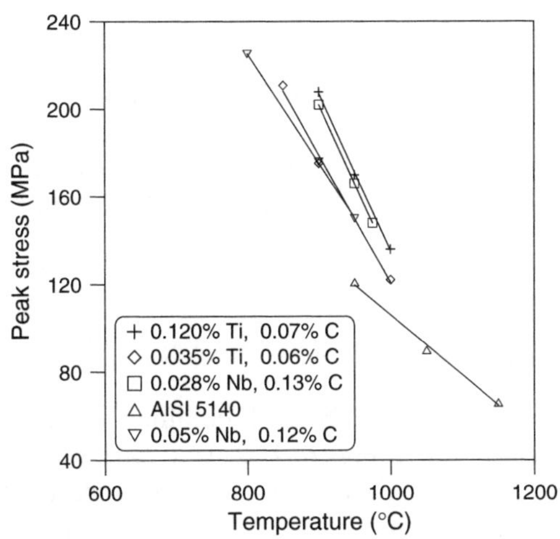

Figure 4.11 The temperature dependence of the peak stress of several steels (Lenard et al., 1999; reproduced with permission).

[8] Platinum–Rhodium thermocouples, Type R, could be used, of course, and these are considerably more accurate (~0.1%) than the Type K version. They are much more expensive, however.

In the worst-case scenario, consider the microalloyed steel, containing 0.028% Nb and 0.13% C. The graph shows a slope of 0.9 MPa/°C and the 10°C difference would then indicate an error in the strength of 9 MPa. As the steel's strength at that temperature is about 130 MPa, the very small error in temperature measurements creates a very much more significant error of about 7% in the strength data. The implications of this 7% are evident when considering the sensitivity of the predicted roll separating forces and roll torques to variations of the input.

Several high temperature furnaces are available with a spyhole, allowing the use of optical pyrometers which, when focused on the sample would monitor its temperature as the testing is proceeding. This approach is preferred over the use of thermocouples which require a hole to be drilled into the sample to house the thermocouple. While the stress concentration around the hole and the embedded thermocouple is not expected to affect the material's resistance to deformation in any significant manner, it is nevertheless an interruption and can be avoided by the use of optical devices. Further problems with the embedded thermocouples include the different strength of the bead and the sample in addition to the possible imperfect contact of the bead and the bottom of the hole. The latter may be eased somewhat, but not eliminated completely, by the use of high conductivity cement.

4.7 THE SHAPE OF STRESS–STRAIN CURVES

The stress–strain curves of metals differ greatly, depending on the temperature at which the test is performed. The two cases, low and high temperature behaviour, are treated below.

4.7.1 Low temperatures

Stress–strain curves of an AISI 1008 steel, obtained in uniaxial tension and at room temperature, are given in Figure 4.12. Two curves are shown. In the first, the steel was tested, as received. In the second, the results of cold working are evident, indicating that the steel's strength increased by approximately 40% as a result of one cold-rolling pass in a two-high rolling mill, causing 58.5% reduction. Both curves indicate strain hardening.

The steel is not highly ductile. Even in the annealed condition, the fracture occurred at a strain of 0.18. The stress–strain curve following the 58.5% cold reduction exhibits the yield point extension. Most of the ductility is lost as a result of the cold work done on the sample.

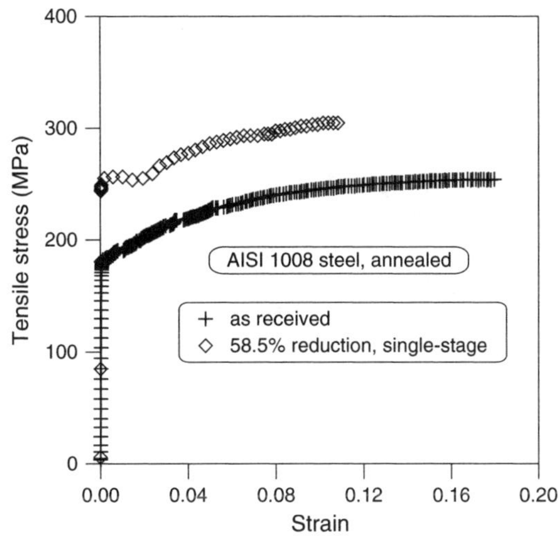

Figure 4.12 The stress–strain curves of an AISI 1008 steel.

4.7.2 High temperatures

The shape of a stress–strain curve, obtained in a test, conducted at high temperatures, differs significantly from that at low temperatures, showing the effects of metallurgical phenomena on the resistance of the material to deformation. The difference is shown in Figure 4.13, which shows the true stress–true strain curves, obtained in compression testing of a 0.1% C, 0.0877% Nb, 0.0795% V steel. During the compression test, a glass–alcohol emulsion was used to cover the ends of the samples to minimize interfacial friction. While the curves have not been corrected for temperature rise, this omission is not expected to cause significant errors, since the rates of strain were not excessive.

The curves demonstrate the traditionally expected behaviour of steels at high temperatures. Two metallurgical mechanisms affecting the shape of the curves are to be considered: these are the hardening and the restoration processes. The latter includes dynamic recovery and dynamic recrystallization. Since all samples were annealed prior to the tests, it may be safely assumed that the steels were initially fully recrystallized and that the austenite grains were uniform in size and were equiaxed[9]. As soon as the compression process begins, hardening due to the pancaking of the grains begins and at a fairly small strain, say 3–5%, dynamic recovery also starts. A micrograph, taken at that strain would indicate the flattened grains. The migration of the dislocations may also be observed but no changes to the grains, other than some

[9] A perfectly equiaxed figure is the circle with its diameter constant. An equiaxed grain is usually hexagonal; see, for example, Figure 4.1.

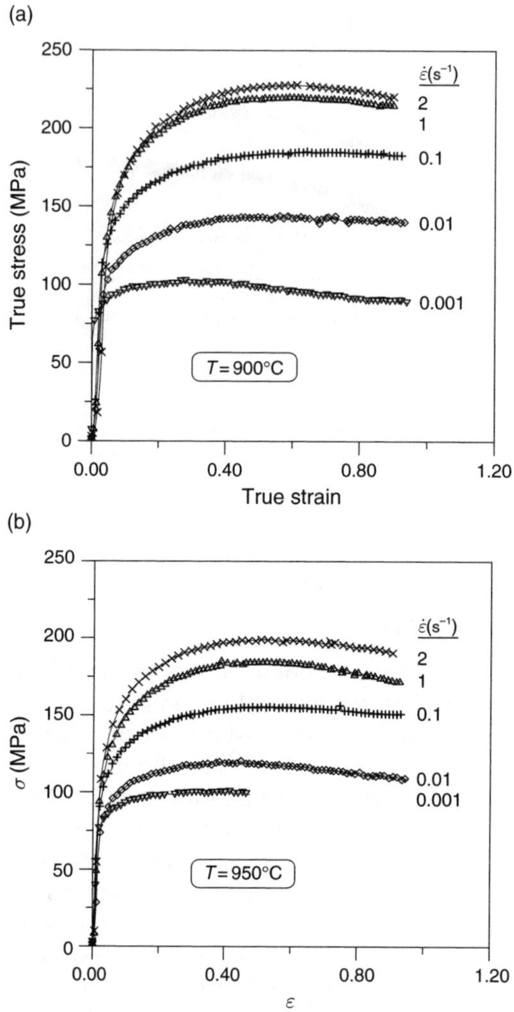

Figure 4.13 (a) The stress–strain curves at 900°C and (b) at 950°C.

flattening, are expected due to dynamic recovery. The loading is now continued, and the hardening and the softening processes occur simultaneously. When the rate of softening exceeds that of hardening, the slope of the stress–strain curve begins to decrease. At a particular strain, identified as the critical strain, usually denoted by ε_c, another restoration process, that of dynamic recrystallization, is initiated and the slope of the stress–strain curve drops even more. A micrograph, taken just beyond the critical strain, would show the new, strain-free grains nucleating, usually at the grain boundaries. The process

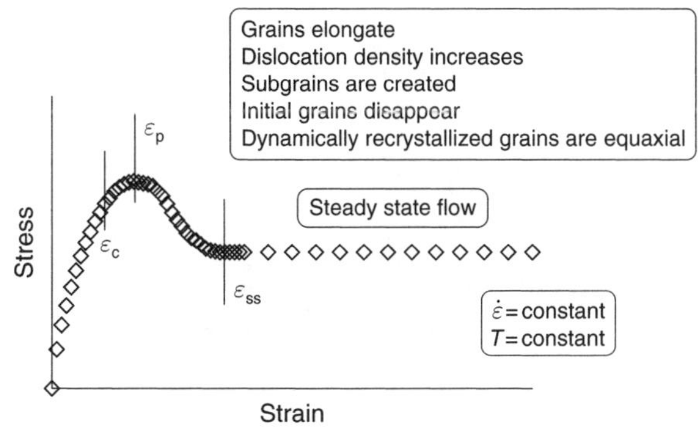

Figure 4.14 A schematic diagram of a stress–strain curve at high temperatures (reproduced from Lenard et al., 1999 with permission; some changes were introduced).

is still continuing and now all three metallurgical events are active at the same time. Further straining reaches the condition when the rate of hardening just equals the rate of softening and a plateau in the stress–strain curve is reached, identified as the strain corresponding to the peak, ε_p. The stress at that location is referred to as the peak stress and is usually designated by σ_p. Further loading causes the softening rate to exceed the hardening rate and the material's resistance to deformation falls until a steady state, at a strain identified as ε_{ss}, is reached. Beyond that strain the stress–strain curve becomes independent of the strain but is dependent on the strain rate and the temperature.

A schematic diagram of a true stress–true strain curve, obtained at high temperatures, is shown in Figure 4.14 (reproduced from Lenard et al., 1999, with some changes), with all three strains, ε_c, ε_p and ε_{ss} indicated.

Another set of micrographs, see Figures 4.15a–d, indicates the progress of dynamic recrystallization in a 0.1 C, 0.04 Nb steel (Cuddy, 1981). In Figure 4.15a the structure before testing is shown. The austenite grains are large, measuring 370 μm on average. The structure, after straining to 0.4 at a temperature of 900°C at a strain rate of $0.017\,s^{-1}$, is shown in Figure 4.15b. The recrystallization process has not started yet. Deformation to a strain of 0.43 at a higher temperature of 1000°C and a strain rate of $0.05\,s^{-1}$ caused partial dynamic recrystallization, see Figure 4.15c. The recrystallization process was completed when the sample was subjected to a strain of 0.55 at 1100°C and a strain rate of $0.17\,s^{-1}$, see Figure 4.15d.

4.8 MATHEMATICAL REPRESENTATION OF STRESS–STRAIN DATA

At this stage of the study, the necessary data on the metal's resistance to deformation – the stress, strain, rate of strain, temperature and hence, the

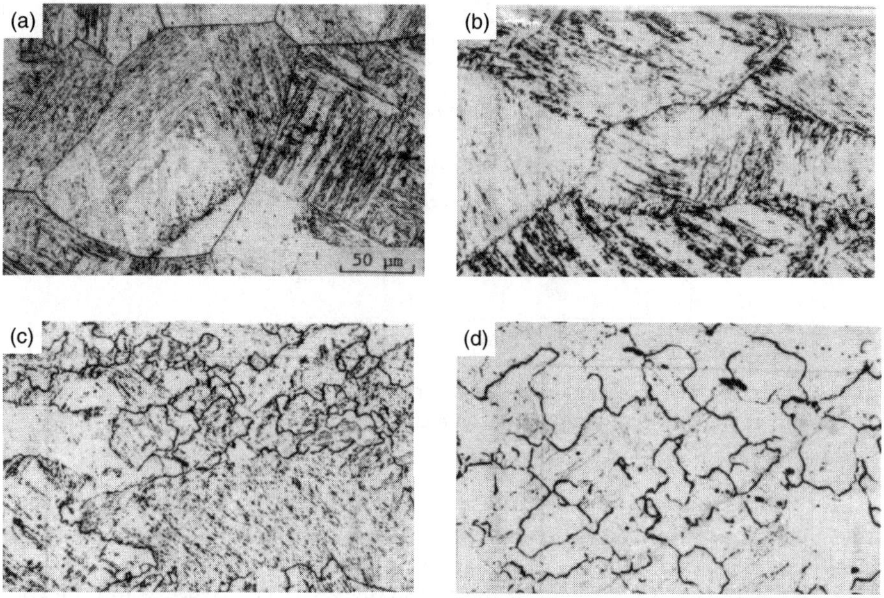

Figure 4.15 The progress of dynamic recrystallization in a 0.1 C, 0.04 Nb steel. (a) The structure before deformation, (b) no recrystallization, (c) partial recrystallization and (d) complete recrystallization (Cuddy, 1981).

Zener–Hollomon parameter, defined as $Z = \dot{\varepsilon}\exp(Q_d/RT)$, are in hand; in the expression Q_d represents the activation energy for plastic deformation, R is the universal gas constant, 8.314 J/m K and T is the temperature in Kelvin. The next step is to develop a mathematical model for further use in analysing a metal forming process. The traditional approach is to make use of non-linear regression analysis and to fit the experimental data, as best as possible, to a pre-determined relation. Another possibility is to place the experimental data in a multidimensional databank and when the stress values are needed in an application, interpolate or extrapolate for them, at the actual strain, rate of strain and temperature.

Two fairly recently developed possibilities to determine the material attributes have been presented, but so far they have not been employed extensively. One of them uses artificial intelligence, specifically neural networks, to estimate and predict the metals' behaviour. The other, parameter identification, is based on a combination of a finite-element simulation of a test – in the present instance, that would be a rolling pass – with the measurements of the overall parameters, such as the roll separating force or the roll torque (Gelin and Ghouati, 1994; Kusiak et al., 1995; Malinowski et al., 1995; Gavrus et al., 1995; Khoddam et al., 1996). The measurements of the process parameters are then compared with the predictions by the finite-element method. An error norm is defined as the vector of distances between these measured and calculated values. The minimization of the error norm is used to determine the unknown parameters in the constitutive law.

In the statistical method an equation is always obtained which can be used with more or less ease in the subsequent steps of the analysis. There are two difficulties. The first problem concerns the just-developed "best-fit curve" which may not fit all data points equally well, and therefore, some carry-on errors are unavoidable. The second problem is encountered when additional material data are developed. The non-linear regression analysis then needs to be repeated and a new relation that fits the new data as well as the original must be obtained.

The latter deficiency, that of repeating the statistical re-development of the empirical relation, is overcome by the ability of the neural networks to renew themselves. The disadvantage, often claimed by engineers, is that an equation is not obtained.

4.8.1 Material models: stress–strain relations

There is an infinity of possibilities in formulating the constitutive relations, both at high and low temperatures. These equations are just that; they are chosen in an arbitrary manner to describe the metals' observed behaviour. The choice of the form and the independent variables are up to the researcher. Some of the better-known and accepted forms are given below.

4.8.1.1 Relations for cold rolling

While the choice of the form for stress–strain relations is practically infinite, two equations have been used regularly by researchers. Both relate the metal's strength to strain only in addition to material constants which may depend on the rate of strain. The first is $\sigma = K\varepsilon^n$ where the constants, K and n can be determined for any particular stress–strain data, either by a least-squares minimization routine or by forcing the curve through two pairs of stress and strain values. Both approaches are acceptable. The other relation, more suitable for the analysis of metal forming and particularly for the rolling process, also relates the metal's strength to the strain in the form $\sigma = Y\,(1 + B\varepsilon)^n$ where the three material constants need to be determined by fitting to experimental data. This expression indicates that the metal possesses some strength at zero strain.

In addition to the strain, the strengths of some metals (titanium, for example) are also dependent on the rate of strain. A relationship that has been found useful in such cases is $\sigma = Y\,(1 + B\varepsilon)^n\,\dot{\varepsilon}^m$ where the exponent m is the strain rate hardening coefficient. Again, non-linear regression is needed to determine the coefficients and the exponents.

4.8.1.2 Relations for use in hot rolling

Statistical methods One of the simplest expressions, often used in the analysis of hot rolling problems, relates the metal's strength to the average rate of strain and two material constants, in the form $\sigma = C\dot{\varepsilon}^m$; values for the constants have been given by Altan and Boulger (1973) for a selection of ferrous and non-ferrous metals. An often quoted source for stress–strain–strain rate relations is Suzuki et al.'s (1968) compilation of experimental data. Stress–strain curves for a large number of ferrous and non-ferrous metals have been given, at various temperatures and rates of strain. The chemical compositions of the metals have also been provided.

Several, somewhat more complex equations were listed by Lenard et al. (1999), some of which are repeated below. One of these, based on the hyperbolic sine function, is due to Hatta et al. (1985). The hyperbolic sine law gives the strain rate in the form

$$\dot{\varepsilon} = c \sinh(\overline{\alpha}\sigma)^n \exp\left(-\frac{Q}{RT}\right) \tag{4.4}$$

Hatta et al. (1985) define the various terms in eq. 4.4, for a 0.16% C steel, as

$$c = \exp\{24.4 - 1.69 \ln[\mathrm{C}]\} \quad (\mathrm{s}^{-1}) \tag{4.5}$$

$$n = \exp\{1.63 - 0.0375 \ln[\mathrm{C}]\} \tag{4.6}$$

$$\overline{\alpha} = \exp\{-4.822 + 0.0616 \ln[\mathrm{C}]\} \quad (\text{in MPa}^{-1}) \tag{4.7}$$

and

$$Q = \exp\{5.566 - 0.0502 \ln[\mathrm{C}]\} \quad (\text{in kJ/mole}) \tag{4.8}$$

While Hatta et al. (1985) determined the activation energy by non-linear regression analysis, a somewhat more fundamental approach, making use of experimental data, is likely to lead to more physically realistic values. The recommendations are to follow these steps:

- re-write equation 4.4 in a different, simpler form: $\dot{\varepsilon} = A\sigma^n \exp\left(-\frac{Q}{RT}\right)$;
- perform a number of stress–strain tests at several temperatures and rates of strain;
- obtain the peak stresses and prepare a log–log plot of the peak stresses versus the temperatures;
- at an arbitrary stress level, obtain from the plot two temperatures and the corresponding rates of strain and
- determine the activation energy from the slope $Q \approx \frac{\Delta(\ln \dot{\varepsilon})}{\Delta(-1/RT)}$.

The activation energy, thus determined for a 0.1% C, 1.093% Mn, 0.088% Nb, 0.0795% V steel, was 483 kJ/mole (Lenard et al., 1999). In general, higher alloy content leads to larger values of the activation energy. It is noted that the strain does not appear in Hatta's relations, indicating that they are strictly applicable in the steady-state region. Wang and Lenard (1991) included the strain in the exponents of eq. 4.4 while developing a high temperature model for the deformation of a Nb–V steel.

Another set of empirical relations has been presented by Shida (1969), giving the metal's resistance to deformation as a function of the temperature, carbon content, strain and strain rate. These equations have been used successfully in a number of publications, concerned with hot rolling or hot forging of steels. The relations have been developed by Shida for carbon steels. It is expected that use of carbon equivalent instead of the carbon content may allow Shida's formulae to be used for alloy steels as well. The carbon equivalent may be calculated as a function of the alloy content of the steel from the relation[10]

$$C_{eq} = C + Mn/6 + (Cu + Ni)/15 + (Cr + Mo + V)/5 \tag{4.9}$$

The flow strength of the steel, in kg/mm^2, is given by Shida, in terms of the carbon content in %C, the rate of strain and the temperature, as given below:

$$\sigma = \sigma_f f \left(\frac{\dot{\varepsilon}}{10}\right)^m \tag{4.10}$$

The terms in eq. 4.10 are defined, depending on the temperature of deformation. For

$$T \geq 0.95 \frac{C + 0.41}{C + 0.32} \tag{4.11}$$

$$\sigma_f = 0.28 \exp\left(\frac{5}{T} - \frac{0.01}{C + 0.05}\right) \tag{4.12}$$

and

$$m = (-0.019\,C + 0.126)T + (0.075\,C - 0.05) \tag{4.13}$$

For temperatures below that defined by eq. 4.11

$$\sigma_f = 0.28\, q\,(C, T) \exp\left[\frac{C + 0.32}{0.19\,(C + 0.41)} - \frac{0.01}{C + 0.05}\right] \tag{4.14}$$

$$q\,(C, T) = 30(C + 0.9)\left(T - 0.95 \frac{C + 0.49}{C + 0.42}\right)^2 + \frac{C + 0.06}{C + 0.09} \tag{4.15}$$

---

[10] There are several formulae available for "carbon equivalent", mostly developed for the study and modelling of welding processes.

and

$$m = (0.081\,C - 0.154)\,T - 0.019\,C + 0.207 + \frac{0.027}{C + 0.32} \tag{4.16}$$

The remaining parameters are

$$f = 1.3\,(5\varepsilon)^n - 1.5\varepsilon;$$

$$n = 0.41 - 0.07\,C \quad \text{and}$$

$$T = (\overline{T} + 273)/1000.$$

In the above relations, $\overline{T}$ is the temperature in °C and C is the carbon content in weight %. The true strain is denoted by ε. Shida gives the limits of applicability of his empirical relations[11] as

- $C < 1.2\%$;
- $\overline{T}$ in between 700°C and 1200°C;
- $\dot{\varepsilon}$ in between 0.1 and 100 s^{-1} and $\varepsilon < 70\%$.

While these equations have been used successfully in many instances, some caution is needed in specific applications. The difficulties are shown in Figure 4.16 where the predictions of four previously published empirical

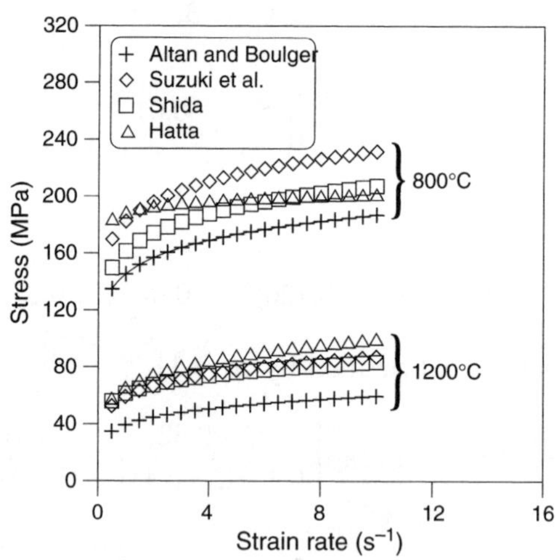

Figure 4.16 A comparison of the predictions of several emipirical relations, designed for high temperature behaviour.

[11] In several cases of empirical relations, developed to represent the metal's resistance to deformation, the limits of applicability are not given; a major omission.

equations are compared, by plotting the predicted strength as a function of the rate of strain, at a temperature of 800°C (close to the transformation temperature) and 1200°C, where the steel is fully in the austenitic state. A carbon steel is chosen for the comparison.

The curve denoted by the crosses is due to Altan and Boulger (1973). The equations for a steel, containing 0.15% C, are at a temperature of 800°C, $\sigma = 145.38\dot{\varepsilon}^{0.109}$ MPa and at a temperature of 1200°C, $\sigma = 39.27\dot{\varepsilon}^{0.181}$MPa. The steel closest to this and whose stress–strain curves are given by Suzuki et al. (1968) contains 0.147% C. Non-linear regression analysis gave the equations of the curves, plotted in Figure 4.16 (denoted by the diamonds) at 800°C, $\sigma = 182.34\varepsilon^{0.1039}$ MPa and at 1200°C, $\sigma = 59.06\varepsilon^{0.1698}$ MPa. The parameters of the equation due to Hatta et al. (1985) are given above, see eqs 4.4–4.8; this curve is given by the upward triangles. Finally, the curve obtained using Shida's relations for a steel containing 0.15% C is denoted by the squares.

It is observed that all four curves give the expected trends. The differences of the predictions are quite large, though, reaching up to 35%.

As mentioned above, Shida's relations have been used successfully in several instances and if no testing facilities are available, their use is recommended.

Developing a databank Use of a multidimensional databank was explored by Lenard et al. (1987). Stresses at particular values of strains, strain rates and temperatures were stored and retrieved as needed in the analysis of the flat rolling process. The results compared favourably with data, obtained by other approaches. While it is believed that using a large databank removes the need to develop arbitrary empirical relations and therefore it removes one error-prone step from the analysis of the rolling process, no extensive use of the approach is evident in the technical literature.

Artificial neural networks The predictive capabilities of the method are demonstrated by considering the hot compression testing of an aluminum alloy (Chun et al., 1999). Cylindrical samples of the Al 1100-H14 alloy of 20 mm diameter and 30 mm height have been used to determine the metal's resistance to deformation. The specimens have been machined from plates with the longitudinal direction parallel to the rolling direction. The flat ends of each specimen were machined to a depth of $0.1 \sim 0.2$ mm to retain the lubricant, boron nitride. A type K thermocouple in an INCONEL shield, with outside diameter of 1.54 mm and 0.26 mm thermocouple wires, was embedded centrally in each specimen. The compression tests were carried out on a servohydraulic testing system, at a true constant strain rate of 7.58 s^{-1} and at sample temperatures of 400, 450, and 500°C. The results are shown in Figure 4.17. The network was trained using the data at the temperatures of 400 and 500°C. Testing the network was performed by comparing the predictions to measurements obtained at 450°C. The good predictive ability of the network is evident.

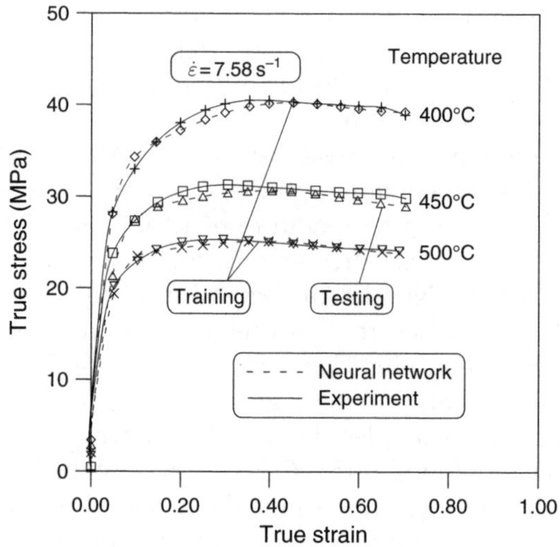

Figure 4.17 The predictions of the flow strength of a commercially pure aluminum alloy (Chun et al., 1999).

Parameter identification The parameter identification method has been developed in the last decade and applied to problems of metal forming (Gelin and Ghouati, 1994, 1999; Gavrus et al., 1995; Kusiak et al., 1995; Boyer and Massoni, 2001). The details of the technique have been reviewed by Lenard et al. (1999). Pietrzyk (2001) used the technique to determine and predict the high temperature behaviour of a low carbon steel and a 304 stainless steel; the predictions and the measurements compared very well. The ability of the method to predict the hot stress–strain curves of a harder aluminum alloy, when subjected to plane-strain compression is evident in Figure 4.18 (Lenard

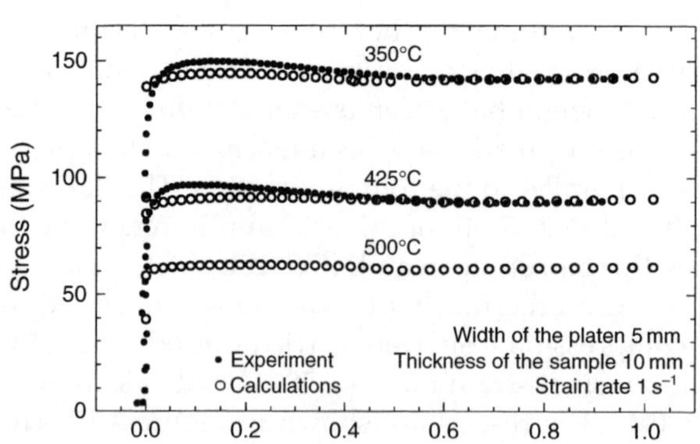

Figure 4.18 The stress–strain curves of an aluminum alloy, measured and compared to the calculations by the parameter identification technique (Lenard et al., 1999).

et al., 1999). A description of the experimental procedure and the results of the analysis are given by Pietrzyk and Tibbals (1995). The experiments have been carried out at temperatures of 350, 420 and 500°C and at a strain rate of $1\,s^{-1}$. The initial thickness of the aluminum alloy samples was 10 mm. The width of the platens and of the sample was 15 mm.

4.9 CHOOSING A STRESS–STRAIN RELATION FOR USE IN MODELLING THE ROLLING PROCESS

It is clear at this point that to satisfy the demands of its users the success of a mathematical model of the flat rolling process depends on how well the tribological and material attributes are treated. Tribology is to be discussed in the next chapter. The choice of a stress–strain equation will also contribute to success or failure, and extreme caution is advised when that choice is made. While the researchers have several possible avenues to follow, one of the approaches given below will likely be chosen:

- Conduct independent testing to determine the metal's mechanical attributes and use non-linear regression analysis to develop a model for later use.
- Search the existing technical literature for information on the attributes.

It is strongly advised, however, that if at all possible, the first approach in the list should be followed and the reasons are clearly demonstrated in Figure 4.16.

4.10 CONCLUSIONS

The metals' resistance to deformation was discussed in this chapter. First, the recently developed steels were presented and their micrographs were demonstrated. This was followed by a presentation of the perennial battle for supremacy between the steels and aluminum alloys. The most prevalent, traditional techniques available to test the metal's response to loading were given, including tension, compression and the torsion tests. Their advantages and disadvantages were listed. Approaches towards the treatment of constitutive data were presented. The mathematical models, arrived at by statistical techniques, parameter identification and artificial intelligence methods, designed to describe the behaviour of the materials, were also included. A simple approach to determine the activation energy was mentioned. Recommendations concerning the determination of the metal's resistance to deformation were given.

CHAPTER 5

Tribology

Abstract

The components of tribology – friction, lubrication and heat transfer – relevant to hot and cold flat rolling processes are discussed. Their combined effect, resulting in roll wear, is then considered. Costs associated with inappropriate application of tribological principles are mentioned. Independent variables affecting the quality of the surface of the rolled product, and the efficiency of the transfer of energy at the contact between the roll and the rolled strip are defined. The coefficient of friction and the friction factor are presented. Approaches available to obtain the coefficient of friction and the friction factor by experimental and analytical methods are considered. The concepts and ideas presented are tested in several case studies.

Lubrication by neat oils and emulsions is examined. The requirements for a well-lubricated contact are defined. The attributes of the lubricants, especially the sensitivity of the viscosity to the pressure and the temperature, are examined. Methods to measure and calculate the oil film thickness are described.

The dependence and interdependence of the coefficient of friction on independent variables are discussed.

The heat transfer coefficient is presented next. Experimental and analytical methods that lead to its determination are given. Its dependence on the process parameters is illustrated.

The combined effect of frictional and thermal effects at the interface – roll wear – is then examined, with attention to industrial and laboratory conditions.

Specific recommendations concerning the magnitude of the coefficient of friction and heat transfer under various conditions, close the chapter.

5.1 TRIBOLOGY – A GENERAL DISCUSSION

It is appropriate here to start with a quotation from Roberts (1997). He writes:

> *Of all the variables associated with rolling, none is more important than friction in the roll bite. Friction in rolling, as in many other mechanical processes can be a best friend or a mortal enemy, and its control within an optimum range for each process is essential.*

While Roberts wrote about friction, it is further appropriate and even necessary to replace that term with "tribology".

The study of surfaces in relative motion, in contact and under pressure – that of tribology – is a very broad subject. It has been studied by scientists and applied by engineers for thousands of years. The points of view of its practitioners are equally wide, encompassing the disciplines of tribochemistry, tribophysics, chemistry, chemical engineering, nanotribology, surface analysis, surface engineering, fluid mechanics, heat transfer, mathematics and mechanical engineering, and the list is far from complete. Attending a large, comprehensive conference entitled "Tribology" requires careful choice of the lectures to be attended and may easily lead to information overload.

In the present context, that is, flat rolling of metals, the focus is on three interconnected phenomena: friction, lubrication and heat transfer at the contact surfaces. These, in turn, create roll wear. In the metal rolling industry, the costs associated with wear problems account for about 10% of total production costs. The cost of inappropriate understanding or application of tribological principles has been estimated to be as high as 6% of GDP in the USA (Rabinowicz, 1982).

Interesting data have been given in a recently published book by Stachowiak and Batchelor (2005) concerning the same topic. They quote the Jost report (1966) which estimated that the correct application of basic principles of tribology would save the UK economy £515 million per annum. A report by Dake et al. (1986) indicated that in the USA about 11% of the total annual energy could be saved in the areas of transportation, turbo machinery, power generation and industrial processes.

A recent search on the Internet using the word "Tribology" yielded 237 000 results. It is, of course, not realistic to check and evaluate all of these. One that appeared of potential interest was the Virtual Tribology Institute, a group of European organizations that deal with all aspects of the subject. Members are located in several European countries while the manager of the Institute is located in Belgium. Also found by the search was the Center for Tribology, identifying itself as the largest tribology testing laboratory in the world, located in Campbell, California. Under "useful links", the site gives a list of universities where research on tribology is performed. Unfortunately, on checking the title "Northern American Universities", only schools in the USA could be located, Mexican and Canadian locations had been omitted.

Probably and arguably, the most complete website is the one provided by the University of Sheffield. The four sub-topics listed on the site are: Research, Teaching, Tools and Information and Consultancy. Clicking on Tools and Information, a plethora of useful items was found. The list of books dealing with the topic of tribology is likely the most complete available. The list of journals, periodicals and online resources are also most impressive. The list of "Tribologists Around the World" is very useful when one wants to know who is dealing with what in the field.

It is evident that the while the interactions of the components, parameters and variables of the field of tribology are beyond the capabilities of a single discipline, listing them is still valuable.

For the most complete listing of the attributes of an interface, in addition to their interactions, the reader is referred to the table shown in Fig. 3.2 of "Tribology in Metalworking" by Schey (1983). The table was first presented at a conference in 1980, but to the best of the present writer's knowledge, no better or more complete compilation has been given since. The three components of metal working system are identified as the die, the work piece and the interface between the two, which includes the lubricant. The table is reproduced below as Figure 5.1 and an examination of the interconnection of the parameters indicates the complexities of the process.

As mentioned above, in what follows, the phenomena of friction, lubrication and heat transfer will be discussed in turn, followed by a brief look at their combined effect: roll wear.

5.2 FRICTION

5.2.1 Real surfaces

An enlarged view of the cross-section of two surfaces is presented by Schey (1983) and the figure is reproduced here as Figure 5.2, clearly demonstrating the validity of the comment written by Batchelor and Stachowiak (1995), stating that surfaces are never clean. An adsorbed film and an oxide layer are always present in industry as well as a laboratory. It is of course possible, albeit difficult, to provide a controlled atmosphere while testing tribological attributes and this is often done. The results thus produced are of interest since the ability to control the independent variables is increased in a very significant manner; however, the applicability of such data to a real-life, industrial environment is highly questionable. The surfaces are also never perfectly smooth.[1]

5.2.2 The areas of contact

Two concepts need to be defined before any further discussion of the mechanisms of friction may be presented. The apparent area of contact, A is the first

[1] A strong criticism is offered here of the often repeated statement "the surface is perfectly smooth" to imply a lack of friction. This comment was written recently in a calculus text to be used in an introductory course. Exactly the opposite is correct as the smooth surfaces provide a large, real contact area and lead to high frictional resistance. Moving two perfectly smooth surfaces in contact, relative to one another, would be difficult as the resistance to overcome would be at the maximum. The proper terminology, if no friction is to be assumed, should be "the surfaces are perfectly lubricated", or better still, "the frictional resistance is taken to be zero".

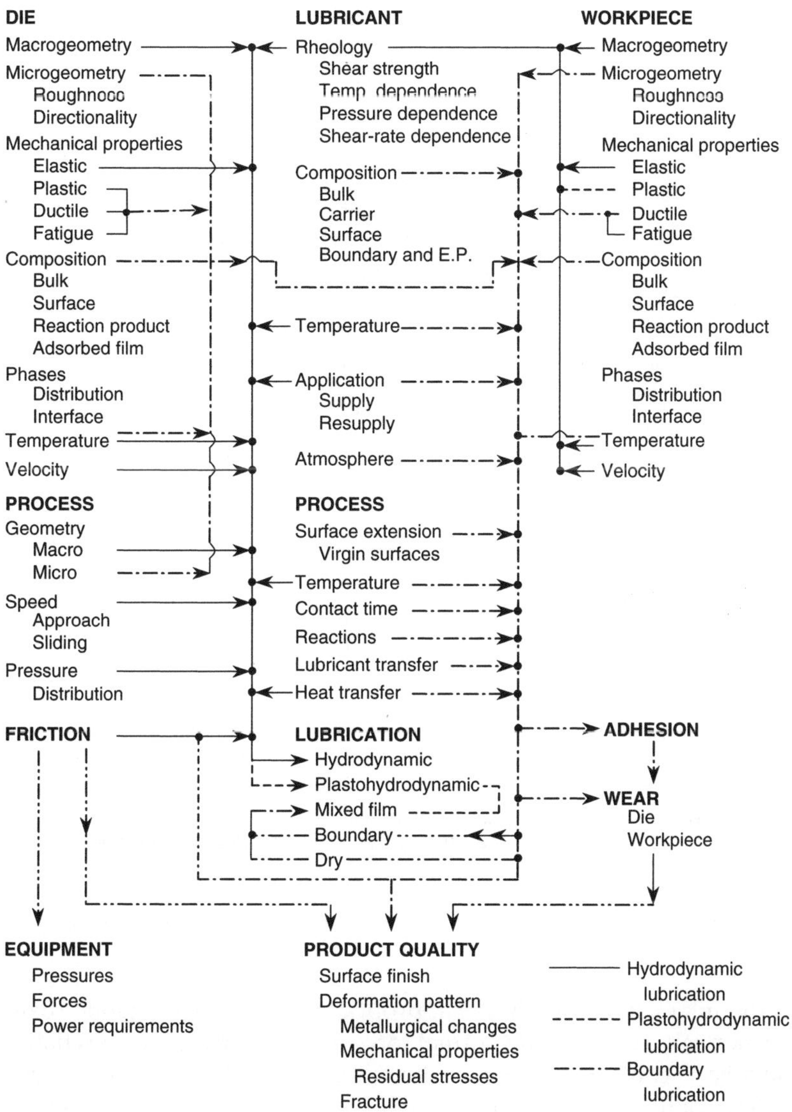

Figure 5.1 The tribological system (Schey, 1983, reproduced with permission).

and it is defined by the overall, outer dimensions of the contact surface. The real area of contact, usually denoted by A_r, affects the frictional phenomena in a much more fundamental manner and is defined as the totality of the areas in contact at the asperity tips. When the two surfaces approach one another, contact is first made at those tips, see Figure 5.3. It is to be realized that the shapes, dimensions, locations and contacts of the asperity tips are completely random;

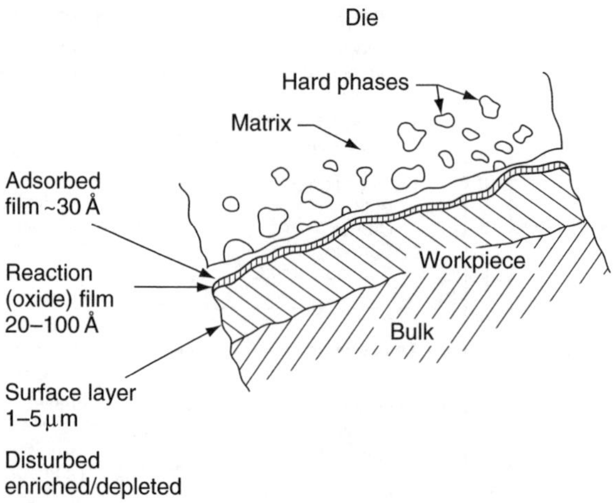

Figure 5.2 An enlarged view of the cross-section of two surfaces (Schey, 1983, reproduced with permission).

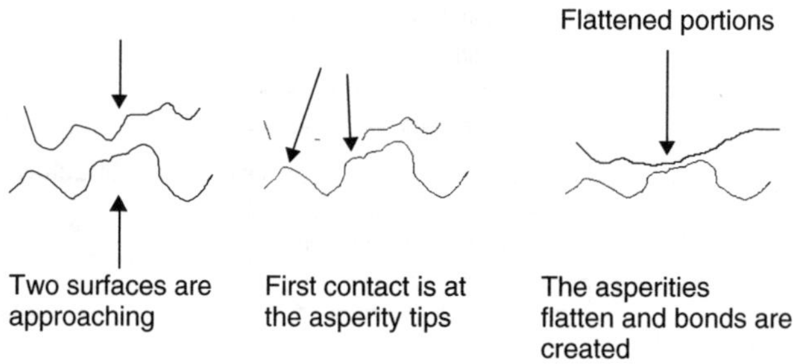

Figure 5.3 Asperities, valleys and the contact of two rough surfaces.

hence, the interactions may most appropriately be termed chaotic (Batchelor and Stachowiak, 1995). As the normal force increases, the asperities flatten and the real area of contact increases. If there is sufficient time – note that no more than a few milliseconds or even microseconds are needed – adhesive bonds are created, as described by the adhesion theory of Bowden and Tabor (1950) which gives the requirements for the establishment of adhesion: that the surfaces should be clean and close enough for interatomic contact. As the normal force increases and the asperity tips are flattened further, new, clean surfaces are created and the real area of contact approaches the apparent area. The rate of approach depends on the resistance to deformation and the formability of asperities. Relative movement of the contacting bodies is then possible only by applying a shear force, large enough to separate the contacting, flattened and bonded asperities.

5.2.2.1 The relationship of the apparent and the true areas of contact

The definitions of the two areas given above are quite clear. The interest at this point is how A and A_r are related and how the true area may approach the apparent area when circumstances change. The implication of this is also clear: as the asperities flatten while the loads are increased, the nature of the contact changes and this will affect all of the interfacial phenomena. Schey (1983) classifies the reaction of asperities to the normal and shear stresses at the contact surfaces in the following manner:

- The average normal stress is below the flow strength of the work piece and creates elastic stresses within the bulk, but the asperity tips experience permanent deformation and the true area approaches the apparent area;
- In addition to normal pressures, which are still below the flow strength, there is relative sliding between the die and the work piece;
- The normal pressures cause plastic flow of the bulk of the work piece.

Bowden and Tabor (1964) discuss the events that occur when the asperity tips come into contact and a certain amount of normal force is applied to the two bodies as they approach one another. While the average normal pressure may be well below the flow strength of either component, the stresses at the extremely small areas of the asperity tips will always be high enough to create permanent deformation. The tips will then deform and the load carried (W) will determine the true area of contact in terms of the flow strength of either body as:

$$A_{\mathrm{r}} = \frac{W}{\sigma_{\mathrm{fm}}} \tag{5.1}$$

where σ_{fm}, the flow strength may well increase as the deformation proceedes due to strain and strain rate hardening. The magnitude of the tangential force (F) required to move one of the bodies with respect to the other depends on the shear strength (τ) of the junction, which may be a combination of adhesive and ploughing forces depending on the nature of the contacting surfaces:

$$F = A_{\mathrm{r}}\tau = \frac{W}{\sigma_{\mathrm{fm}}}\tau \tag{5.2}$$

A more recent examination of the response of asperity tips to loads is reviewed by Stachowiak and Batchelor (2005). They quote the studies of Whitehouse and Archard (1970) and Onions and Archard (1973) which indicate that a large proportion of the contact between asperities is elastic under normal operating loads. They further mention that an exception – permanent asperity deformation – may occur at the contact surfaces in metal working processes. Since the normal pressures in bulk forming are high, plastic deformation of the asperities is an important contributor to surface phenomena.

The first step is to determine whether the contact is mostly elastic or a significant amount of plastic deformation of the tips of asperities exists. This is accomplished by calculating the plasticity index, ψ. Stachowiak and Batchelor (2005) present three definitions for the plasticity index. They identify the first as due to Greenwood and Williamson (1966), followed by that of Whitehouse and Archard (1970) and Bower and Johnson (1989). The three formulas give the index in terms of material and geometrical attributes. The easiest definition of plasticity index is the one given by Bower and Johnson (1989) in the form:

$$\psi_s = \left(\frac{E'}{p_s}\right)(\sigma\kappa)^{0.5} \tag{5.3}$$

where the plasticity index for repeated sliding is ψ_s, σ is the RMS surface roughness of the harder surface in m, κ is the curvature of the asperity tip in m^{-1}, and p_s is the shakedown pressure[2] of the softer surface. If the stresses are below the shakedown pressure, the deformation is elastic; otherwise plastic deformation is expected. The term E' is the composite Young's modulus for the two bodies in contact, a and b:

$$\frac{1}{E'} = \frac{1-\nu_a^2}{E_a} + \frac{1-\nu_b^2}{E_b} \tag{5.4}$$

and ν is Poisson's ratio.

The Whitehouse and Archard (1970) model defines the plasticity index as:

$$\psi^* = \left(\frac{E'}{H}\right)\left(\frac{\sigma}{\beta^*}\right) \tag{5.5}$$

where H is the hardness of the deforming surface, and β^* is the correlation distance. When $\psi^* < 0.6$, elastic deformations are expected. When $\psi^* > 1$, most of the contacts experience plastic deformation.

In the flat rolling process, especially where metals are concerned, significant plastic deformation of the asperities is certain to occur. To show that this is the case, consider the data of McConnell and Lenard (2000) where in each pass the average roll pressures are calculated to be in the range of 700–900 MPa, while the yield strength of the rolled metal is in the order of 250–300 MPa; permanent deformation will be present.

A relationship between the apparent area and the true area when plastic deformation takes place has been given by Majumdar and Bhushan (1991) in the form:

$$\frac{W}{AE'} = K\phi A_r^* \tag{5.6}$$

---

[2] The shakedown pressure is a limit; when the magnitudes of stresses are below it, elastic deformation is present, while above it, plastic deformation occurs (Stachowiak and Batchelor, 2005).

where $K = H/\sigma_y$, $\phi = \sigma_y/E'$ and $A_r^* = A_r/A$, and σ_y is the yield strength. Re-substitution of these terms into eq. 5.6 gives the true area of contact as a function of the load, and the hardness of the deforming body in a form similar to the Bowden and Tabor (1964) model, with the hardness replacing the flow strength:

$$A_r = \frac{W}{H} \tag{5.7}$$

Korzekwa et al. (1992), in stating the need for quantitative understanding of friction, present a model for the evolution of the contact area in a sheet undergoing a plastic-forming process. A rate dependent material subjected to large range of strains is considered along with the effect of bulk deformation on asperity flattening, which is modelled as the indentation of a flat surface by a rigid punch. A viscoplastic finite-element model is used in the calculations of the changing true contact area as the deformation continues. While the results concentrate on low contact pressures which are appropriate in sheet metal-forming operations, it is believed that with increasing loads the trends may not change markedly. The data are presented in the form of graphs, reproduced here as Figure 5.4, showing the changing contact area fraction, defined as the ratio of the half width of the rigid indenter to half the distance between the centres of the indenters, as a function of the bulk effective strain. The deformation of stainless steel 304L was considered.

The results indicate that the true area of contact increases as the normal pressures and asperity slopes increase. The straining directions also have a significant effect on the growth of the true contact area.

Sutcliffe and co-workers have considered the problems associated with asperity deformation and the true and apparent contact areas in the flat rolling of metals. Their work is innovative and during the writing of the present manuscript, the most up-to-date.

Sutcliffe (2000) lists two factors that affect frictional conditions: the manner in which the contacting surface asperities conform to each other while in contact and the frictional mechanisms at those contacting areas and the valleys in between. He writes that in considering the deformation of the asperities, the effects of bulk deformation and wavelength need to be taken into account as well. He writes that when sub-surface deformation is accounted for – the realistic approach, especially in bulk metal forming and flat rolling – the asperities are shown to flatten more.

Sutcliffe (2002) presents the rate of change of the ratio of contact areas as a function of the bulk strain. When the roll roughness is in the direction of rolling:

$$\frac{\mathrm{d}A}{\mathrm{d}\varepsilon} = \frac{\dot{W}}{\tan\theta} \tag{5.8}$$

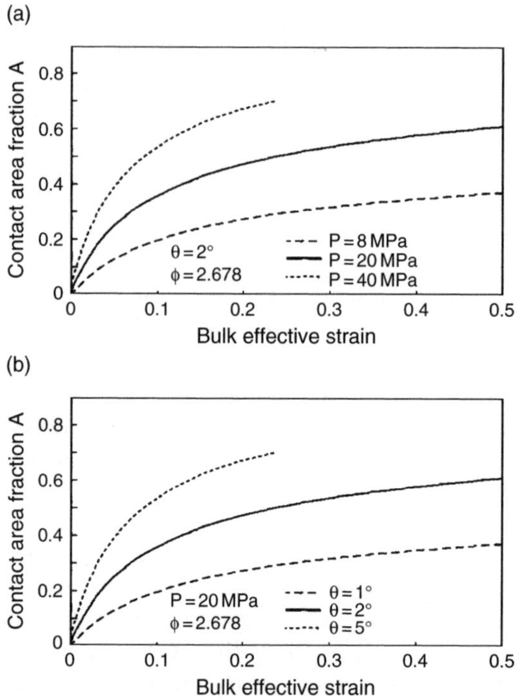

Figure 5.4 Contact area fraction as a function of the bulk strain for a range of normal pressures (a); asperity slopes (b); (Korzekwa et al., 1992).

and when it is in the transverse direction:

$$\frac{dA}{d\varepsilon} = \frac{1}{1+\varepsilon}\left(\frac{\dot{W}}{\tan\theta} - A\right) \tag{5.9}$$

where the bulk strain is designated by ε, the flattening rate is $\dot{W}$, and the slope of the asperities is θ. Integration of these relations yields the area of contact ratio as a function of the bulk strain and the normal pressure, shown in Figure 5.5 below.

Reasonable agreement of the predictions and the measurements is observed in the figure.

Stancu-Niederkorn et al. (1993) list some of the experimental techniques available to determine the real contact area. They classify them in two categories: off-process and in-process inspections. Off-process approaches, which cannot measure the elastic deformation, include measuring the profile after deformation by inspection or interferometry. The authors describe an experimental technique to measure the real contact area while the bulk of the work piece undergoes plastic deformation using ultrasound waves, the in-process

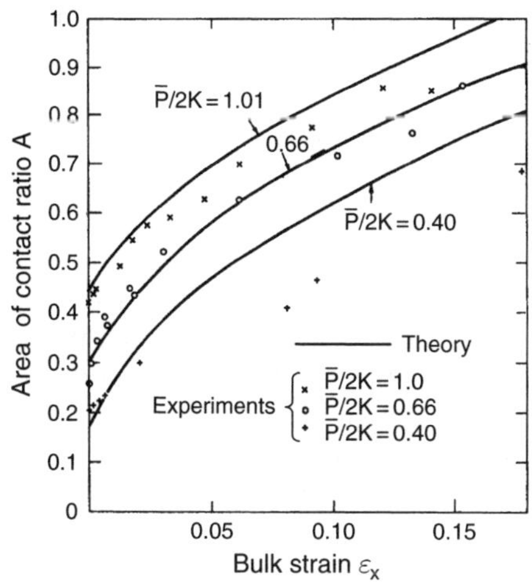

Figure 5.5 The change of the area of contact ratio with bulk strain (Sutcliffe, 2002, reproduced with permission).

inspection. Measurements were taken in free upsetting of steel samples and closed-die upsetting using copper specimens. Dry and lubricated conditions were examined. In the free upsetting tests, the real area of contact increased fast with increasing loads. In the closed-die upsetting, the real contact area reached about 95% of the apparent area at a normal load of 1100 MPa. Azushima (2000) uses the finite-element method to analyse the deformation of the hills and valleys as a result of the pressure of the entrapped oil. He plots the dependence of the contact area ratio on the reduction of the height of the asperities and finds that without the oil the flattening is much more pronounced. The area ratio is found to remain relatively constant when the lubricant is entrapped in the valleys.

Siegert et al. (1999) describe the development of optical measurement techniques and computer workstation technology, using which they characterize the topography of sheet surfaces in 3D. The instrument is expected to be usable directly at the press shop.

5.2.3 Definitions of frictional resistance

There are two traditional approaches to express the frictional phenomena between two surfaces in contact, in relative movement and under pressure. In one of these, the coefficient of friction, as defined by Amonton and Coulomb and applied in most analyses of problems of metal forming, is given as the ratio of the interfacial shear stress to the normal pressure:

$$\mu = \tau/p \tag{5.10}$$

The friction factor, on the other hand, is given as the ratio of the interfacial shear stress and the yield strength in shear of the softer material in the contact:

$$m = \tau / k \quad \text{where} \quad 0 \leq m \leq 1 \tag{5.11}$$

The existence of perfect lubrication is indicated when $m = 0$ while $m = 1$ points to sticking conditions. In developing mathematical models of bulk metal forming processes, either coefficient may be used; however, both describe the interactions at the interface in a highly simplified manner and both involve some conceptual difficulties. Schey (1983) points out that the Amonton–Coulomb definition becomes meaningless when the normal pressure is several times the flow strength of the metal. This is because the interfacial shear stress cannot rise beyond the yield strength in pure shear of the materials in contact and its ratio to the increasing normal pressure would continue to decrease and hence, the coefficient would also decrease. Mróz and Stupkiewicz (1998) agree with Schey (1983), writing:

> ... *the classical Amonton-Coulomb model is not suitable for most metal forming processes* ...

The difficulty with the application of the friction factor is the lack of precise knowledge of the meaning of k, originally defined as the yield strength in pure shear of the softer material of the pair in contact, while ideally it should represent the strength of the interface. Again as pointed out by Schey (1983), the properties of the interface are not necessarily identical to the properties of the materials in contact. By examining Figure 5.2, which shows a realistic view of an interface, involving surface layers, oxides and adsorbed films, equating k of the interface to that of one of the materials may indeed be troublesome.

Wanheim (1973) was among the first researchers to write that the usual Coulomb–Amonton model doesn't apply at high normal pressures which exist in bulk forming processes. In those cases, he suggests that the frictional stress should be taken as a function of the normal pressure, surface topography, length of sliding, viscosity and the compressibility of the lubricant. Wanheim et al. (1974) and Wanheim and Bay (1978) propose a general friction model using the above mentioned ideas. In their model, Coulomb friction is taken to be valid at low normal pressures whereas the friction stress approaches a constant value at high normal pressures. The approach was applied successfully to model the pressure distribution in plate rolling and the cross shear plate rolling process (Zhang et al., 1995). A mixed friction model was also used by Tamano and Yanagimoto (1978) with Coulomb friction at low pressures and sticking friction at high pressures. Another approach, mostly used in finite-element modelling, is to introduce a "friction layer" in between the contacting surfaces. Montmitonnet et al. (2000) discuss wear mechanisms and the differential hardness of the tool and the work piece that create a third body in between the die and the worked metal, identified as the transfer layer.

5.2.4 The mechanisms of friction

Mechanisms of the interface contact have been discussed with the aid of a very well-prepared figure by Batchelor and Stachowiak (1995), reproduced here as Figure 5.6.

They identify adhesion, ploughing and viscous shear as the main contributors to frictional resistance. In bulk forming processes, the former two are the most likely events to occur, as complete separation of the surfaces and thus full hydrodynamic conditions are rarely realized in practice. Montmittonet et al. (2000) discuss surface interactions further and mention the possibility that particles will be detached from one of the contacting bodies, possibly resulting in micro-cutting. Further, a wave may be pushed along the surface, creating a bulge; or repeated contact may cause fatigue failure. The relative magnitudes of adhesion and ploughing have been examined by Mróz and Stupkiewicz (1998). While developing a constitutive model for friction in metal forming processes, the authors indicate that friction forces include both adhesion and ploughing. They present a combined friction model, which simulates the interaction of the tool's asperities with that of the work piece. In the mathematical model, the effect of bulk plastic deformation is, however, neglected and as they propose, experimental verification of the predictions is still required.

Much depends on the angle of attack between the contacting surface asperities and on how the harder surface of the tool is prepared. Grinding, the traditional approach in preparing the rolls in the metal rolling industry, would produce relatively shallow angles while sand blasting would result in sharp asperities. As indicated while examining the effect of progressively rougher, sand blasted rolls on the coefficient of friction and the resulting rolled surfaces,

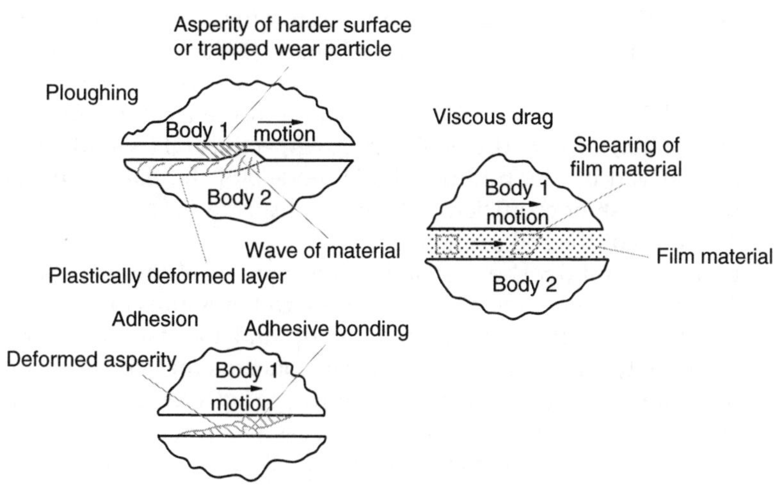

Figure 5.6 The major mechanisms of friction (Batchelor and Stachowiak, 1995).

ploughing appeared to be the major component, contributing to frictional forces and overwhelming the effects of adhesion (Lenard, 2004; Dick and Lenard, 2005).

5.3 DETERMINING THE COEFFICIENT OF FRICTION OR THE FRICTION FACTOR

Since the Coulomb coefficient of friction is defined as a ratio of forces and the friction factor is defined as a ratio of stresses, neither can be measured directly. Several experimental approaches are available, however, to determine various experimental parameters and thereby deduce the magnitude of the coefficient or the factor.

5.3.1 Experimental methods

Several methods for measuring interfacial frictional forces during plastic deformation have been developed, some of which have been listed by Wang and Lenard (1992). A more comprehensive list, applicable to other metal forming processes, including bulk and sheet metal forming, has been presented by Schey (1983). Some of the more useful approaches are described below.

5.3.1.1 The embedded pin – transducer technique

Originally suggested by Siebel and Lueg (1933) and adapted by van Rooyen and Backofen (1960) and Al-Salehi et al. (1973), the method has been applied to measure interfacial conditions in cold flat rolling (Karagiozis and Lenard, 1985; Lim and Lenard, 1984), warm rolling (Lenard and Malinowski, 1993) and hot rolling of steels (Lu et al., 2002) and aluminum (Hum et al. 1996). Variations of this procedure have been presented by Lenard (1990, 1991) and Yoneyama and Hatamura (1987). Typical results obtained by this approach are shown in Figures 5.7 and 5.8 for warm rolling of aluminum (Lenard and Malinowski, 1993) and hot rolling of steel (Lu et al., 2002), respectively. It is evident that the friction hill, derived by the traditional, 1D models of the flat rolling process and employing constant coefficients of friction, leads to unrealistic predictions of the distribution of the roll pressures.

As Figures 5.7 and 5.8 show, detailed information of the distributions of interfacial frictional shear stresses and the work roll pressures may be obtained by these methods, but the experimental set-up and the data acquisition are elaborate and costly. Since the major criticism concerns the possibility of some metal particle or oxide intruding into the clearance between the pins and their housing and invalidating the data, it is necessary to substantiate the resulting coefficients of friction by independent means. This substantiation has been performed successfully in several instances (see, for example, Hum et al., 1996). In that study, the coefficients of friction, determined by the pin-transducer

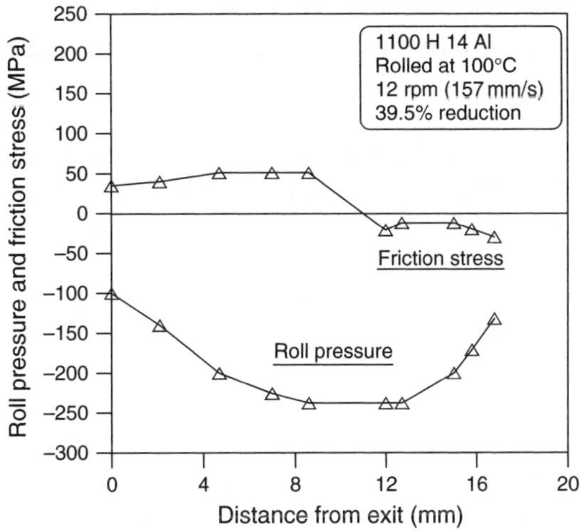

Figure 5.7 Roll pressure and friction stress during warm rolling of an aluminum strip, obtained with the use of pins and transducers embedded in the work roll (Lenard and Malinowski, 1993).

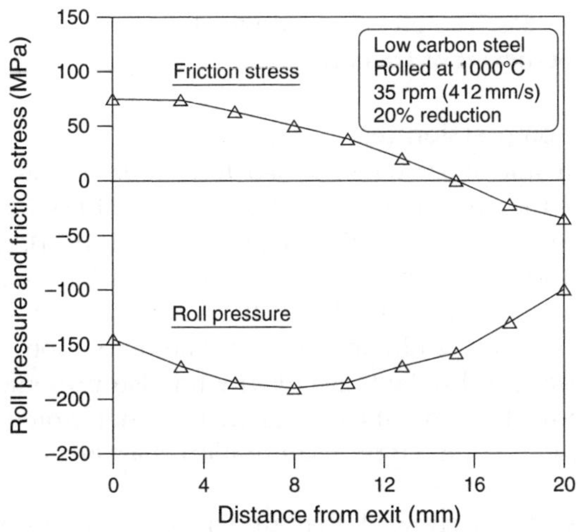

Figure 5.8 Roll pressure and friction stress during hot rolling of steel, obtained with the use of pins and transducers embedded in the work roll (Lu et al., 2002).

technique, were used in a model of the rolling process. The model calculated roll forces and roll torques which compared very well to the measured values, demonstrating that the technique leads to reliable data.

Another difficulty encountered when the embedded pins are used is the interruption of the surface of the roll at the pins. At present, the magnitude of the effect of this interruption is unknown, but considering the above-described substantiation of the measurements, it is not expected to be significant. The use of pins and transducers has been reviewed quite some time ago by Cole and Sansome (1968). The authors concluded that the approach can provide reliable data as long as care is taken in the design, manufacture and calibration of the apparatus.

A cantilever, machined out of the roll such that its tip is in the contact zone and fitted with strain gauges, and its various refinements were presented by Banerji and Rice (1972) and Jeswiet (1991).

5.3.1.2 The refusal technique

Januszkiewicz and Sulek (1988) used the "refusal technique" to monitor the coefficient of friction necessary to initiate the entry of the strip into the roll gap in a study of the effects of contaminants on the lubricating properties of lubricants. This approach makes use of the minimum coefficient of friction required to initiate the rolling process. Recalling eq. 2.1, the coefficient needed to allow the entry is dependent only on the bite angle. At small reductions, the bite angle is small and the required coefficient is also small. This fact is employed in the rolling process in which progressively larger reductions are attempted in each pass. The bite angle at which entry is first successful is then reported as the coefficient of friction.[3]

5.3.1.3 The ring compression test

The most popular and most widely used technique to establish the friction factor, however, is the ring compression test (Kunogi, 1954; Male and Cockroft, 1964; Male and DePierre, 1970; DePierre and Gurney, 1974). In the test, a ring of specific dimensions is compressed in between flat dies and the changes of its dimensions are related directly to the friction factor. Using calibration curves, the friction factor is obtained easily. The derivation of the calibration curves is well described by Avitzur (1968) who also presented a detailed set of calculations, indicating how the curves are to be determined. The schematic diagram of the ring test and a typical calibration chart are shown in Figures 5.9 and 5.10.

Bhattacharyya (1981) showed that under some circumstances the compressed rings develop tapering, with the top and bottom surfaces deforming in a different manner, most likely due to different tribological conditions on the two surfaces. The tapering disappeared when the samples were

[3]Caution is advised here. The dependence of the coefficient of friction on the reduction will be examined in Section 5.5.1. It will be shown that, depending on several circumstances, increasing reductions may cause lower or higher coefficient.

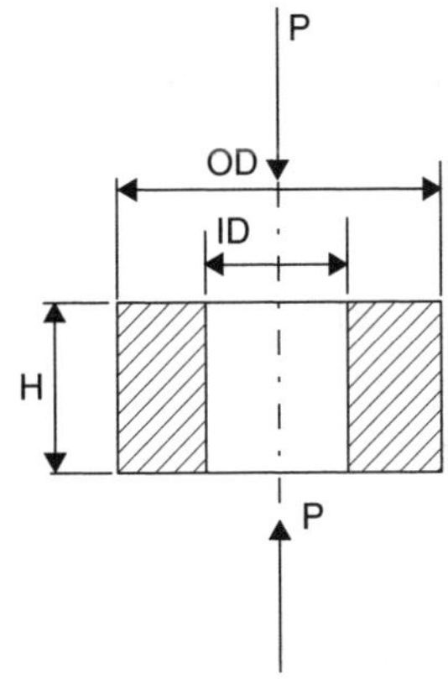

Figure 5.9 The ring compression test.

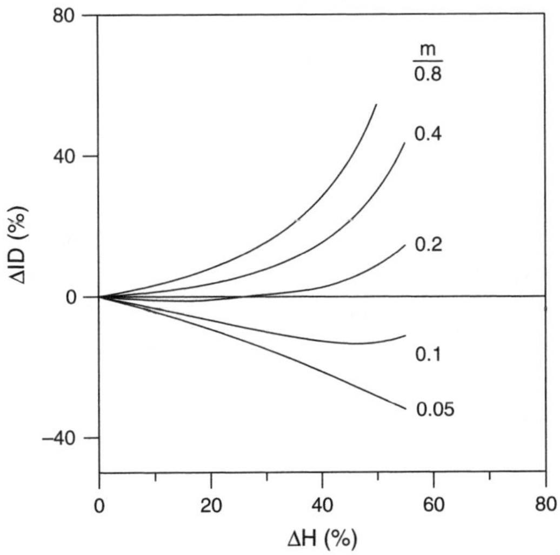

Figure 5.10 A typical calibration chart for the ring test.

pre-compressed, and the true areas of the contact at the top and the bottom surfaces converged. Tan et al. (1998) used different ring geometries to study the ring compression process. Concave, rectangular and convex shaped cross-sections were employed. The results indicated that the influence of strain hardening on friction is complicated. Friction was affected by the normal pressure in a significant manner.

Szyndler et al. (2000) compressed austenitic stainless steel rings at high temperatures without lubrication, and used the parameter identification method to determine the friction factor. The factor's dependence on the temperature was found to be well described by the relation $m = 3.511 \times 10^{-4}T - 0.01846$, where the temperature is in °C,[4] indicating that the friction factor increases with increasing temperatures.

5.3.2 Semi-analytical methods

Numerous attempts to relate the coefficient of friction or the friction factor to various parameters have been presented in the literature, too many to be reviewed here. Only some, considered to be among the more useful, are presented below.

5.3.2.1 Forward slip – coefficient of friction relations

Several formulae, connecting the forward slip to the average coefficient of friction have been published in the technical literature. The predictive abilities of these relations have been studied (Lenard, 1992), and the results have been compared to data produced by the embedded-pin technique. While the conclusions indicated that the reviewed equations don't work very well, they are presented below for completeness. It can be recalled that the forward slip is given by:

$$S_f = \frac{v_{exit} - v_r}{v_r} \tag{5.12}$$

where v_{exit} stands for the exit velocity of the rolled strip and v_r designates the surface velocity of the work roll.

Sims' formula (1952) connects the forward slip to the reduction, r, the flattened roll radius, R', the exit thickness, h_{exit}, and the coefficient of friction, μ:

$$\tan^{-1}\sqrt{S_f} = \frac{1}{2}\left[\tan^{-1}\sqrt{\frac{r}{1-r}} - \frac{1}{2a}\ln\left(\frac{1}{1-r}\right)\right] \tag{5.13}$$

where $a = \dfrac{\mu}{1-r}\sqrt{\dfrac{R'}{h_{entry}}}$.

Ekelund's (1933) formula gives the coefficient of friction in terms of the bite angle, ϕ_1, the roll radius and the forward slip as:

$$\mu = \frac{(\phi_1/2)^2}{\phi_1/2^2 - \sqrt{\frac{2S_f}{\frac{2R}{h_{exit}} - 1}}} \tag{5.14}$$

[4] The effect of the temperature on the coefficient of friction will be discussed in more detail in Section 5.4.2.4 below.

Bland and Ford (1948) use similar variables to give the coefficient of friction as:

$$\mu = \frac{h_{\text{entry}} - h_{\text{exit}}}{2\sqrt{R'\left(h_{\text{entry}} - h_{\text{exit}}\right)} - 4\sqrt{S_f R' h_{\text{exit}}}} \tag{5.15}$$

Roberts (1978) includes the roll force (P_r) and the torque for one roll ($M/2$) in addition to the reduction and the roll radius to define the coefficient of friction:

$$\mu = \frac{M/2}{P_r R'\left[1 - 2\sqrt{\frac{S_f(1-r)}{r}}\right]} \tag{5.16}$$

Another relationship, due to Inhaber (1966), includes the roll force and the roll torque, in addition to the neutral angle (ϕ_n), the arc of contact, the maximum pressure, P_2, and the pressures at entry and the exit, P_1 and P_0, respectively. Note that in the absence of external tensions, P_1 equals the yield strength of the strip at the entry and the pressure at exit, and P_0 equals the yield strength there. The equation is:

$$P_r + \frac{M/2}{R\mu} = 2R'(\phi_1 - \phi_n)\frac{P_2 - P_1}{P_2/P_1} \tag{5.17}$$

where the neutral angle is given in terms of the forward slip

$$\phi_n = \sqrt{S_f h_{\text{exit}}/R'} \tag{5.18}$$

and the maximum pressure is calculated from

$$P_2 = \exp\left(\frac{c_0 \ln P_0 - c_1 \ln P_1 + P_1 - P_0}{c_0 - c_1}\right) \tag{5.19}$$

where

$$c_0 = \frac{1}{2R'}\left(\frac{P_r}{\phi_n} - \frac{M/2}{R\phi_n\mu}\right) \tag{5.20}$$

and

$$c_1 = \frac{P_r + M/2R'\mu}{2R'(\phi_1 - \phi_n)} \tag{5.21}$$

In order to test the predictive abilities of the above listed formulae, the forward slip needs to be measured. This may be accomplished in several ways. As mentioned above, one of the often used methods is to mark the roll surface with fine lines, parallel to the roll axis and placed uniformly around the roll. As the strip is rolled, these marks create impressions on the rolled surface. The forward slip may then be calculated by:

$$S_f = \frac{l_1 - l}{l} \tag{5.22}$$

where l_1 is the average of the distances between the marks on the strip's surface, and l is the distance between the lines on the roll's surface. It is expected that as long as the lines are created carefully, using sharp, hard tools, the resulting data on the forward slip is reasonably accurate.

The predictions of equs 5.13–5.17 were compared to measurements of the coefficient friction and forward slip while rolling strips of commercially pure aluminum (Lenard, 1992). The coefficients of friction were determined by force transducers embedded in the work rolls and the forward slip was obtained by the use of lines on the roll surface and by eq. 5.22. The results are shown in Figure 5.11, plotting the coefficient of friction versus the forward slip.

As the figure shows, all relations predict realistic values for the coefficient of friction. However, the trends are not predicted well. Ekelund (1933), Sims (1952), Roberts (1978) and Bland and Ford (1948) all predict an increasing trend, reaching a plateau and then dropping; the measurements indicate an upward exponential. It is concluded that most of the formulae are not successful in providing reliable and consistent predictions. Arguably the best, though not perfect, approach is due to Inhaber (1966).

Marking the roll's surface, however carefully, may affect the interfacial conditions and hence, the frictional forces, albeit, as concluded when the embedded pins were used, these effects may not be very large. If lubricants are also present, the need to distribute them over the contact may also be compromised somewhat since the marks will act to retain some of the oils. There are alternatives for researchers who don't want to mark the roll's surface and these make use of optical devices.

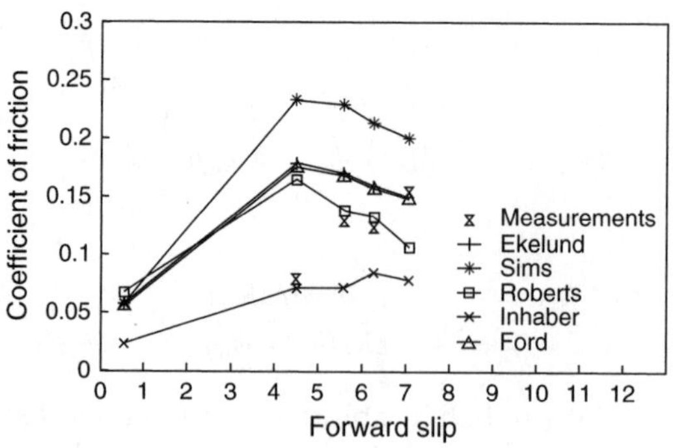

Figure 5.11 Comparison of the calculated and the measured forward slip (Lenard, 1992).

McConnell and Lenard (2000) used two photodiodes, located a known distance apart, to monitor the exit velocity of the rolled strip. The time interval between the signals of the diodes allowed the determination of the exit velocity, and this in turn allowed the use of the original definition of the forward slip in terms of the roll's and the strip's speeds. Li et al. (2003) used laser Doppler velocimetry to measure the relative velocities of the roll and the strip.

It must be noted that measurements of the forward slip are often error prone. A simple examination of either eq. 5.12 or 5.22 illustrates the difficulties. For example, if the roll's surface velocity is 900 mm/s and the strip's exit velocity is 1000 mm/s, the forward slip is determined to be 0.11. If, however, the roll velocity is mistakenly measured to be 909 mm/s – an 1% error – the corresponding error of the forward slip is almost 10%, almost an order of magnitude higher.

An interesting approach to determine the coefficient of friction using the forward and the backward slip was given by Silk and Li (1999). They used the original definition of the forward slip (see eq. 5.12), and the backward slip as the relative difference between the entry speed of the strip and speed of the roll:

$$S_{\mathrm{b}} = \frac{v_{\mathrm{r}} \cos \phi_1 - v_{\mathrm{entry}}}{v_{\mathrm{r}} \cos \phi_1} \tag{5.23}$$

The authors make several assumptions which lead to simple expressions for the coefficient of friction in the forward slip zone, defined as the region between the neutral point and the exit, μ_{f}, and in the backward slip zone which is the region between the entry and the neutral point, μ_{b}:

$$\mu_{\mathrm{f}} = \frac{\Delta h/R'}{2\sqrt{\frac{\Delta h}{R'}} - 4\sqrt{\frac{2S_{\mathrm{f}} h_{\mathrm{exit}}}{2R' - h_{\mathrm{exit}}}}} \tag{5.24}$$

and

$$\mu_{\mathrm{b}} = \frac{\Delta h/R'}{2\sqrt{\frac{\Delta h}{R'}} - 4\sqrt{(1 - S_{\mathrm{b}})\frac{h_{\mathrm{entry}} \cos \phi_1}{2R'} - \frac{h_{\mathrm{exit}}}{R'}}} \tag{5.25}$$

These relations are subject to the following assumptions:

- Coulomb friction exists in between the roll and the rolled metal;
- The roll pressure is constant over the contact area; and
- The angles are small compared to unity.

In order to determine the coefficients, numerical values for the forward and backward slips are necessary. Silk and Li (1999) use information obtained from the instrumented loopers of the hot strip mill of Hoogovens; their results refer stands F2 and F3 of the Hoogovens hot strip mill. The magnitudes of the coefficients of friction vary from 0.15 to 0.23 in both stands, realistic values, considering the efficient lubrication applied in the mill.

5.3.2.2 Empirical equations – cold rolling

The three well-known formulas, connecting the coefficient of friction to the roll separating force rely on matching the measured and calculated forces and choosing the coefficient of friction to allow that match. One of the often-used formulae, given by Hill is quoted by Hoffman and Sachs (1953) in the form:

$$\mu = \frac{\frac{P_r}{\sigma_{fm}\sqrt{R'\Delta h}} - 1.08 + 1.02\left(1 - \frac{h_{exit}}{h_{entry}}\right)}{1.79\left(1 - \frac{h_{exit}}{h_{entry}}\right)\sqrt{\frac{R'}{h_{entry}}}} \tag{5.26}$$

where P_r is the roll separating force per unit width, σ_{fm} is the average plane-strain flow strength in the pass, and R' is the radius of the flattened roll.

Roberts (1967) derived a relationship for the coefficient of friction in terms of the roll separating force P_r, the radius of the flattened roll R', the reduction r, the average of the tensile stresses at the entry and exit σ_1, the average flow strength of the metal in the pass, σ_{fm}, and the entry thickness of the strip, h_{entry}:

$$\mu = 2\sqrt{\frac{h_{entry}}{R'r}}\left[\frac{P_r(1-r)}{\sigma_{fm} - \sigma_1}\sqrt{\frac{1}{R'h_{entry}r}} - 1 + \frac{5r}{4}\right] \tag{5.27}$$

Ekelund's equation, given by Rowe (1977) in the form of the roll separating force in terms of material and geometrical parameters and the coefficient of friction may be inverted to yield the coefficient of friction:

$$\mu = \frac{\left[\left(\frac{P_r}{\sigma_{fm}\sqrt{R'\Delta h}} - 1\right)\left(h_{entry} + h_{exit}\right) + 1.2\Delta h\right]}{1.6\sqrt{R'\Delta h}} \tag{5.28}$$

A comparison of the predicted magnitudes of the coefficient by these formulae is shown in Figure 5.12, using data obtained while cold rolling steel strips lubricated with a light mineral seal oil (McConnell and Lenard, 2000). Two nominal reductions are considered. The first is for 15% and the second for 50% of originally 0.96 mm thick, 25 mm wide, AISI 1005 carbon steel strips. The metal's uniaxial flow strength, in MPa, is $\sigma = 150\,(1 + 234\,\varepsilon)^{0.251}$.

The tests were repeated at progressively increasing roll surface velocities, from a low of 0.2 m/s to a high of 2.4 m/s. Care was taken to apply the same amount of lubricant in each test; ten drops of the oil on each side of the strip, spread evenly.

In the figure, the coefficient of friction is plotted versus the roll surface velocity, which does not appear in any of the above formulae in an explicit manner. However, the effect of increasing speed is felt by the roll force, which, as expected, is reduced as the relative velocity at the contact surfaces increases. Increasing velocity is expected to bring more lubricant to the entry to the contact zone. The dropping frictional resistance indicates the efficient entrainment

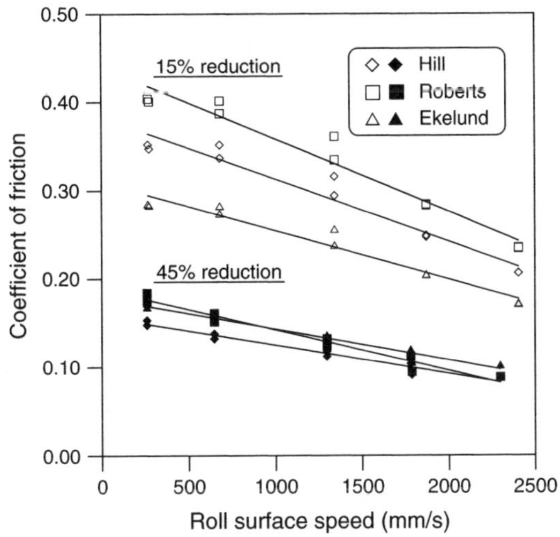

Figure 5.12 The coefficient of friction, as predicted by Hill's, Roberts' and Ekelund's formulae, for cold rolling of a low carbon steel (Lenard et al., 1999).

of the lubricant and its distribution in between the roll and the strip surfaces. No starvation of the contacting surfaces is observed.

All three formulas give realistic, albeit somewhat high numbers for the coefficient of friction and all predict the expected trend of lower frictional resistance with increasing velocity. As well, the coefficient of friction is indicated to decrease as the reductions increase, demonstrating the combined effects of the increasing number of contact points, the increasing temperature and the increasing normal pressures.

The first two phenomena result in increasing frictional resistance with reduction. The third causes increasing viscosity and hence, decreasing friction and, as shown by the data, it has the dominant effect on the coefficient of friction. The magnitudes vary over a wide range, however, indicating that the mathematical models influence the results in a significant manner.

An analytical approach to determine the coefficient of friction has been presented by Li (1999). Two approaches were given, both subjected to five, *a priori* assumptions: homogeneous plane-strain compression is present; the coefficient of friction is constant in the arc of contact; the strip is rigid-plastic; the neutral plane is within the arc of contact; and the rate of strain is low. In the first approach, the roll pressure distribution, as predicted by Bland and Ford (1948), is used, the rolling strain is determined and the minimum coefficient of friction for steady-state rolling without skidding is obtained as:

$$\mu_c = \frac{\varepsilon - a}{h_{entry}} \tag{5.29}$$

where a is defined as a tension parameter in terms of the front and back tensions (σ_{entry} and σ_{exit}, respectively) and the rolled strip's strength at the entry and exit ($\overline{\sigma}_{\text{entry}}$ and $\overline{\sigma}_{\text{exit}}$, respectively):

$$a = \ln\left(1 - \frac{\sigma_{\text{entry}}}{\overline{\sigma}_{\text{entry}}}\right) - \ln\left(1 - \frac{\sigma_{\text{exit}}}{\overline{\sigma}_{\text{exit}}}\right) \tag{5.30}$$

The coefficient of friction is then obtained in an iterative manner, until the measured and calculated rolling strains agree.

In the second approach, the forward slip is used. A relationship is then derived, in terms of the coefficient of friction, μ, the thickness at the entry and the exit, h_{entry} and h_{exit}, the bite angle, ϕ_1, and the forward slip, S_f:

$$\left[\frac{h_{\text{exit}}^{\mu\cot\phi_1} + \sqrt{h_{\text{exit}}^{\mu\cot\phi_1} - (1-\mu^2\cot^2\phi_1)\, h_{\text{entry}}^{\mu\cot\phi_1} h_{\text{exit}}^{\mu\cot\phi_1}}}{1+\mu\cot\phi_1}\right]^{\frac{1}{\mu\cot\phi_1}} = (S_f+1)\, h_{\text{exit}} \tag{5.31}$$

Using the measured forward slip and an iterative approach, the coefficient of friction is calculated. The magnitudes of the calculated coefficients of friction are as measured or predicted elsewhere. The results indicate that as the rolling strain increases, the coefficient also increases – a finding valid for cold rolling of aluminum but contradicting most experimental data obtained while rolling steel, which indicate that increasing reductions result in dropping coefficients of friction.

Beynon et al. (2000) studied friction and the formation of scales on the surfaces of hot rolled steels. They determined the coefficient of friction by using the forward and backward slip data, measuring the speed of the roll and of the strip simultaneously and using the model of Li (1999), described just above. Their results indicate that mixed sliding/sticking conditions exist at the contact zone. Further, a neutral zone, rather than a neutral plane, is present there.

Martin et al. (1999) analysed friction during finish rolling of steel strips. A combination of experimental data and a coupled thermo-mechanical model of the rolling process were used. Their conclusions are interesting in that they contradict existing experience: the frictional conditions exhibited weak correlation with rolling speed, temperature and reduction.

5.3.2.3 The study of Tabary et al. (1994)

A somewhat different approach is followed by Tabary et al. (1994) in determining the coefficient of friction during cold rolling of fairly soft, 1200 aluminum alloy strips. The authors mention the difficulties associated with the determination of the coefficient of friction in the roll bite. One of the difficulties is the changing direction of the friction force in the roll gap, aiding the movement of the strip until the no-slip region is reached, and retarding its movement beyond until the exit is reached. The location of the no-slip region may be manipulated by applying external tensions and that is the approach followed in this study. Using external tensions the neutral point is forced to be at the exit, causing the friction forces in the deformation zone to act only in one direction. The von Karman differential equation of equilibrium is then integrated with assumed values for the coefficient of friction and the inlet yield strength. Both of these are adjusted until the calculated and measured roll forces match and the boundary condition at the exit is satisfied, so the method is essentially one of the inverse analyses.

A rare and most welcome section of Tabary et al.'s paper is the analysis of the errors in the reported values of the coefficient of friction. They also account for the contribution of the hydrodynamic action to the coefficient of friction, μ_h, according to the relation:

$$\mu_h = \frac{\tau_0}{p_{ave}} \sinh^{-1}\left(\frac{\eta \Delta v}{\tau_0 h_s}\right) \tag{5.32}$$

where p_{ave} is the average roll pressure and η is the viscosity at p_{ave}. The average relative speed is Δv, the smooth oil film thickness is h_s and τ_0 is the Eyring shear stress, estimated to be 2 MPa.

The results indicate that the coefficient of friction is a strong function of the reduction and the ratio of the smooth lubricant film thickness to the combined rms roughness (Λ). The coefficients increase with increasing reduction, and drop with increasing Λ, varying from a high of approximately 0.08 to a low of 0.02.

5.3.2.4 Empirical equations and experimental data – hot rolling

Formulas, specifically intended for use in the analyses of flat, hot rolling of steel have also been published. Those given by Roberts (1983) and by Geleji, as quoted by Wusatowski (1969), are presented below. Roberts' formula indicates that the coefficient of friction increases with temperature. Geleji's relations indicate the opposite trend. Roberts (1983) combined the data obtained from an experimental two-high mill, an 84 inch hot strip mill and a 132 inch hot strip mill, all rolling well-descaled strips, and used a simple mathematical model to calculate the frictional coefficient. Linear regression analysis then led to the relation:

$$\mu = 2.7 \times 10^{-4} T - 0.08 \tag{5.33}$$

where T is the temperature of the workpiece in °F. Geleji's formulae, given below, have also been obtained by the inverse method, matching the measured and calculated roll forces. For steel rolls, the coefficient of friction is given by:

$$\mu = 1.05 - 0.0005T - 0.056 v_r \tag{5.34}$$

where the temperature is T, given here in °C and v_r is the rolling velocity in m/s. For double poured and cast rolls, the relevant formula is:

$$\mu = 0.94 - 0.0005T - 0.056v_r \quad (5.35)$$

and for ground steel rolls:

$$\mu = 0.82 - 0.0005T - 0.056v_r \quad (5.36)$$

It is observed that Geleji's relations, indicating decreasing frictional resistance with increasing temperature and rolling speed, confirm experimental trends. Also note that the predictions of Roberts' – see eq. 5.33 – indicate the opposite trend, agreeing with the results of Szyndler et al. (2000), who obtained the friction factor as a function of the temperature during unlubricated ring compression of stainless steel samples.

Rowe (1977) also gives Ekelund's formula for the coefficient of friction in hot rolling of steel:

$$\mu = 0.84 - 0.0004T \quad (5.37)$$

where the temperature is to be in excess of 700°C, again indicating that increasing temperatures lead to lower values of the coefficient of friction.

Underwood (1950) attributes another equation to Ekelund, similar to those above, giving the coefficient of friction as:

$$\mu = 1.05 - 0.0005T \quad (5.38)$$

Roberts (1977) presents a relationship for the coefficient of friction for a well-descaled strip of steel, in terms of the strip's temperature T in °F, obtained by fitting an empirical relationship to data, obtained by inverse calculations:

$$\mu = 2.77 \times 10^4 \exp\left[\frac{-2.61 \times 10^4}{459 + T}\right] + 0.21 \quad (5.39)$$

A comparison of the predictions indicates that the relations may not be completely reliable in all instances. For example, using a steel work roll and a rolling a steel strip at a temperature of 1000°C and at a velocity of 3 m/s, Roberts' equation, eq. 5.33, predicts a coefficient of friction of 0.415 while Geleji's relation gives 0.382, indicating that while the numbers are close, the difference, almost 8%, is not insignificant. When 900°C is considered, Roberts' coefficient becomes 0.366 and Geleji's increases to 0.432, creating a large difference. Ekelund's predictions are 0.44 and 0.48, at 1000°C and 900°C, respectively. It is difficult to recommend any of these relations for use.

Lenard and Barbulovic-Nad (2002) hot rolled low carbon steel strips at entry temperatures varying from a low of 800°C to 1100°C, using an emulsion of Imperial Oil 8581 and distilled water, at a ratio of 1:1000. During heating, the strips were held in a furnace which was purged using oxygen-free nitrogen, allowing close control of the scale thickness. The roll separating forces, the roll torques, the roll speed and the entry and exit strip surface temperatures were measured in each pass. The coefficient of friction was obtained by inverse calculations, using the refined 1D model, presented in Chapter 3, Section 3.4. Non-linear regression analysis led to the relationship:

$$\mu = -0.183 - 0.636\left(\frac{\dot{\phi}_{\mathrm{roll}}}{\dot{\varepsilon}}\right)\exp\left(-0.279\frac{\Delta T}{T}\right) + 0.248\left(\frac{p_{\mathrm{ave}}}{\sigma_{\mathrm{fm}}}\right)^{1.2} \tag{5.40}$$

where the parameters are the ratio of the roll speed ($\dot{\phi}_{\mathrm{roll}}$ in revolutions/s), to the strain rate of the rolled strip ($\dot{\varepsilon}$ in s^{-1}), the ratio of the surface temperature drop (ΔT in K) to the average temperature in the deformation zone (T in K), and the ratio of the average interfacial pressure (p_{ave} in MPa) to the metal's resistance to deformation (σ_{fm} in MPa).

Some of the data of Lenard and Barbulovic-Nad (2002) are shown in Figure 5.13, giving the coefficient of friction, calculated by Hill's formula, as a function of the average temperature of the strip in the deformation zone. The two graphs demonstrate the difficulties in attempting to arrive at some definite conclusions. At low speeds and low reductions, the coefficient appears to drop with increasing temperatures while at higher reductions and speeds, the opposite is noted.

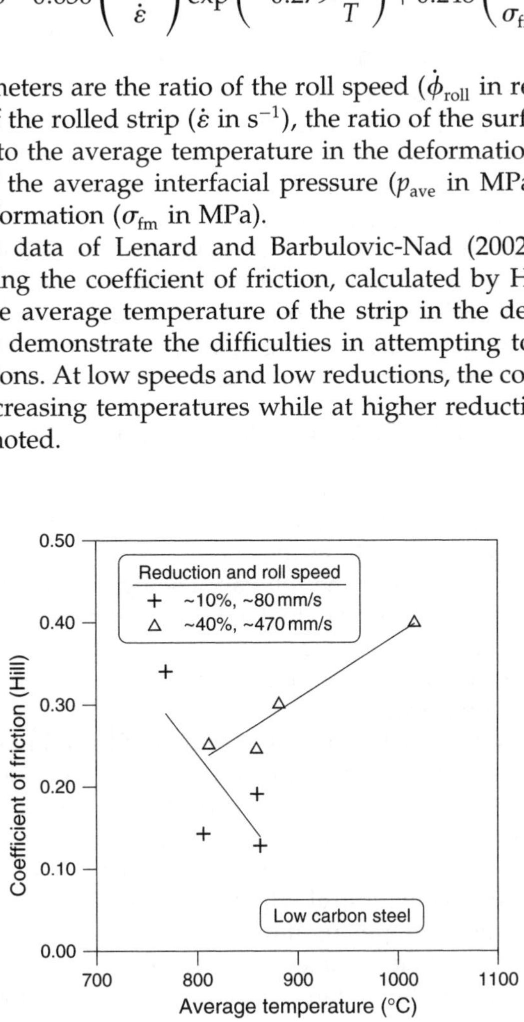

Figure 5.13 The dependence of Hill's coefficient of friction on the temperature at low speeds and reductions and at high speeds and reductions (Lenard and Barbulovic-Nad, 2002).

The behaviour of the coefficient of friction is most likely affected by the thickness of the layer of scale. Since the pre-test heating process was the same, the scale thickness at entry was also the same in both sets of experiments. At lower speeds, however, there is sufficient time for the thickness of the layer of scale to grow during the pass, and as has been pointed out often, thicker scales lead to lower frictional resistance.

Wusatowski's data (1969) may be used to clarify some of the apparent contradictions. He determined the coefficients of friction applicable during industrial hot rolling of carbon steels, using the inverse method. While he showed that the coefficient is strongly dependent on the temperature, the dependence was not linear. The coefficient increased with the temperature from about 750 to 900°C and after reaching a plateau there, it dropped when the temperature increased further. The temperatures where the change of slopes occurred also depended on the effective carbon content of the steels. It may be concluded that the strength of the layer of scale and the adhesion between it and the parent metal also affect the coefficient of friction. The effect of the strength of the scale layer on surface interactions is clearly indicated by Li and Sellars (1999) who showed how the scale may break-up and the hot steel may extrude through the cracks and contact the roll surface.

Jin et al. (2002) developed a relation for the coefficient of friction (as calculated by Hill's formula, eq. 5.26), using data obtained while hot rolling ferritic stainless steel strips with careful control of the development of surface scales:

$$\mu = 1.0728\,(p_{\text{ave}}/\sigma_{\text{fm}})^{0.4242} - 0.9014\,(T/1000)^{0.3197} + 0.0016\,t_{\text{scale}} - 0.0014\,\dot{\varepsilon} \quad (5.41)$$

where T is the entry temperature in K, t_{scale} is the thickness of the layer of scale at the entry in μm, and $\dot{\varepsilon}$ is the strain rate. The results of Jin et al. (2002) are shown in Figure 5.14, plotting the coefficient of friction as a function of the temperature at the entry to the roll gap. The figure includes data obtained at temperatures varying from 900 to 1100°C, scale thickness before the pass from 1.5 to 11 μm and roll surface speeds of 750–980 mm/s. In spite of the broad scatter, the downward trend of the coefficient of friction with increasing temperature is clearly present.

While these results may be compared to that of Szyndler et al. (2000) who found the opposite: the friction factor in their ring tests increased with the temperature, the comparison may be one of those of comparing apples and oranges. Szyndler et al. used ring compression tests, an austenitic stainless steel, no lubrication and no special attention to scale formation; Jin et al. (2002) used rolling, a ferritic stainless steel, a light mineral seal oil as the lubricant and careful control of the development of the layer of scale. Considering the often-mentioned interaction of all parameters in creating a particular magnitude of the coefficient of friction, the different trends, while not easily predictable, are not really surprising.

The discussions presented above indicate that while the coefficient of friction is dependent on the temperature in a significant manner, relations that

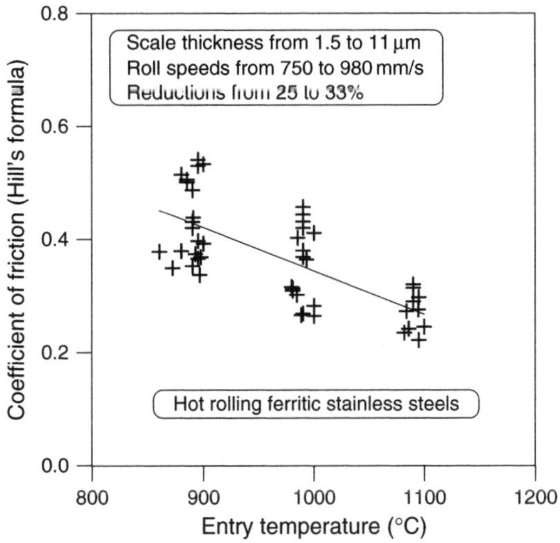

Figure 5.14 The coefficient of friction as a function of the temperature during hot rolling of ferritic stainless steel strips (Jin et al., 2002).

attempt to use only some of the independent variables inevitably lead to errors. The interactions of the variables and the parameters need to be understood before reliable functional connections are established.

5.3.2.5 Inverse calculations

A method often followed to determine the coefficient of friction relies on conducting experiments during which a parameter that depends on the coefficient in a well-known manner is measured. In the mathematical model of the process, the coefficient of friction then becomes the only unknown and is determined such that the measured and the calculated parameters match.

This approach was followed by McConnell and Lenard (2000). The coefficient of friction was calculated by Hill's equation (eq. 5.26) in addition to the use of the 1D model of Roychoudhury and Lenard (1984), referred to in Chapter 3 as the refined version of Orowan's model.[5] The results are shown in Figure 5.15, plotting the 1D coefficient of friction on the ordinate and Hill's coefficient on the abscissa. It appears that the two coefficients are linearly related with the 1D model's results approximately 40% below those of Hill. The relation of the two coefficients, obtained by non-linear regression analysis, is given by eq. 5.42:

$$\mu_{1D} = 0.594\,\mu_{Hill} + 0.0165 \tag{5.42}$$

[5] See Section 3.4, Refinements of Orowan's model.

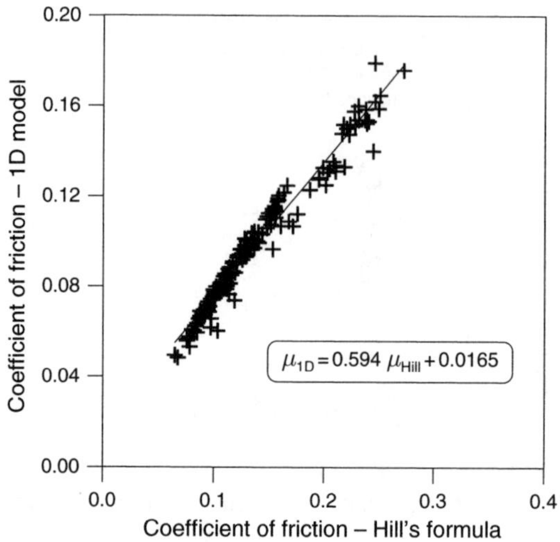

Figure 5.15 The coefficient of friction, obtained by Hill's formula and by inverse analysis, using a 1D model of the rolling process.

5.3.2.6 Negative forward slip

It has been observed in several experiments (Shirizly and Lenard, 2000; Shirizly et al., 2002) that at higher rolling speeds and larger reductions the forward slip becomes negative, indicating that the surface velocity of the exiting strip is less than that of the work roll. In these cases there is neither a neutral point nor a neutral region. Realizing that the mathematical models, presented above, include the idea that the rolled strip exits at a velocity higher than the surface velocity of the roll and the no-slip location is between the entry and the exit, a different approach is needed to analyse the instances when the forward slip is negative.

Avitzur's upper bound formulation (Avitzur, 1968) is adopted in the present work. The power to reduce the strip can be obtained from the measured roll torque and the roll speed:

$$\text{Power} = \frac{\text{Torque} \times \text{roll surface velocity}}{\text{Roll radius}} \tag{5.43}$$

and this is equated to the power obtained using the kinematically admissible velocity field of Avitzur (see eq. 3.61, Chapter 3). The friction factor, which is the only unknown in eq. 5.43, can then be determined. The coefficient of friction is then evaluated using the relationship (Kudo, 1960):

$$\mu = \frac{m/\sqrt{3}}{p_{\text{ave}}/\sigma_{\text{fm}}} \tag{5.44}$$

where the average pressure is obtained from the roll separating force divided by the projected contact area. Using the experimentally obtained power and the exit velocity of the strip, the average coefficient of friction in the roll gap is then obtained directly.

5.3.2.7 The correlation of the coefficient of friction, determined in the laboratory and in industry

Munther (1997) conducted hot rolling experiments on a small laboratory mill, using low carbon and high strength low alloy steel strips. In each test, the roll separating force, the roll torque, the reduction, the thickness of the layer of scale, the speed and the temperatures at the entry and exit were measured. The coefficient of friction values were determined by inverse calculations, using the 2D finite-element code, Elroll.

In addition, mill logs were obtained from Dofasco Inc., giving all necessary data to allow the determination of the coefficient of friction, again by inverse calculations.

The results are illustrated in Figure 5.16, plotting the coefficient of friction values against the dimensionless group $\Delta h\,\sigma_{fm}/P_r$; in the plot the coefficients of friction from the laboratory mill were corrected to allow for the effect of

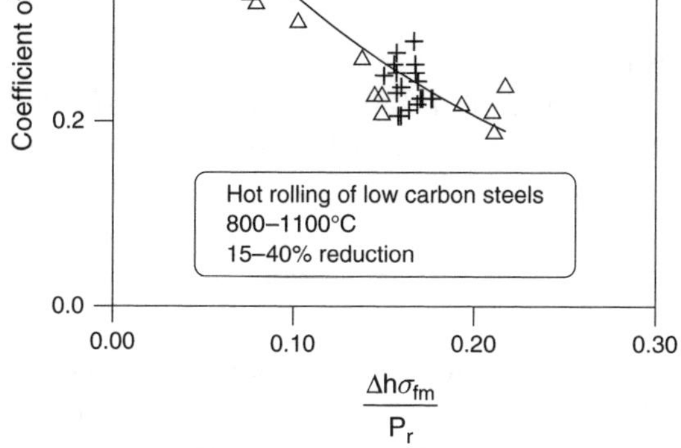

Figure 5.16 A comparison of the corrected values of the coefficient of friction from a laboratory mill and those from industry (Munther, 1997).

geometry according to the square root of the ratio of the respective work roll radii:

$$\mu_{\text{lab,corr.}} = \mu \sqrt{\frac{R_{\text{lab.}}}{R_{\text{ind.}}}} \tag{5.45}$$

It is noted that while there is some scatter, both the laboratory and the industrial data fall on the same trend line.

5.4 LUBRICATION

The objectives of using lubricants and emulsions in the flat rolling process include energy conservation, protection of the work roll surfaces, control of the coefficient of friction, control of the resulting surface parameters as well as cooling. Each of these depends on several interacting variables, and arguably one of the most important among these is the coefficient of friction. In what follows, the basic concepts of friction in lubricated flat rolling are discussed.

5.4.1 The lubricant

Heshmat et al. (1995) reviewed modelling of friction, interface tribology and wear for powder-lubricated systems and solid contacts, and stated that anything in between the contacting surfaces is a lubricant, be it a powder, a contaminant, layer of scale or in fact, oil. In the present context, the lubricant considered is oil, of course, either in the neat form or as an emulsion, usually in water.

5.4.1.1 The viscosity

The behaviour of the lubricant in the contact zone is affected by its viscosity, defined as the factor of proportionality between the shear stress within the oil, τ, and the shear strain rate, $\dot{\gamma}$:

$$\tau = \eta\dot{\gamma} \tag{5.46}$$

The factor of proportionality thus defined is referred to as the dynamic viscosity and its units are, in the SI system, Pa s. The kinematic viscosity, μ is obtained by dividing the dynamic viscosity by the density, ρ:

$$\mu = \eta/\rho \tag{5.47}$$

and if the density is in kg/m^3, the units of the kinematic viscosity are m^2/s[6]. The viscosity of Newtonian fluids is taken to be a constant. Non-constant viscosity indicates a non-Newtonian fluid. In what follows, Newtonian behaviour only will be considered.

5.4.1.2 The viscosity–pressure relationship

The effect of pressure on viscosity, and in turn on the coefficient of friction or the friction factor, is significant and cannot be ignored. The Barus equation, used frequently, gives the viscosity at an elevated pressure, η, in terms of the viscosity at atmospheric pressure, η_0, the pressure–viscosity coefficient, γ, and the pressure, p:

$$\eta = \eta_0 \exp(\gamma p) \tag{5.48}$$

The equation is simple to use, but the user should be cautious: Stachowiak and Batchelor (2005) quote Sargent (1983) who wrote that the Barus equation leads to errors when applied at pressures in excess of 500 MPa. These errors have also been mentioned by Cameron (1966), showing that at high temperatures and pressures the exponential law can overestimate the viscosity by a factor of 500. The warning is repeated by Szeri (1998) who, however, limits the applicability of the Barus equation to pressures of only 0.5 MPa; nevertheless, the low number is most likely in error. The fact that during flat rolling of steels the normal pressures will easily exceed 500 MPa has also been mentioned above and for realistic estimates of the viscosity other relations need to be employed. Cameron (1966) presents an equation for the viscosity under higher pressures in the form:

$$\eta = \eta_0 (1 + Cp)^n \tag{5.49}$$

where C and n are constants. The exponent n is taken to equal 16 and C is expressed in the form:

$$C = 10^a (1 - b) \eta_0^{(b-1)} \tag{5.50}$$

where $b = 0.938$ and $a = -(0.4 + °F/400)$, where the temperature of the oil is to be expressed in Fahrenheit, the viscosity is to be in centipoises and the pressure in psi. The relation is expected to be valid under 300°F. Stachowiak and Batchelor (2005) give the value of C in Pa^{-1}, in the form of a plot of

[6] Unfortunately, much of the information on viscosity is not given in these units. Instead, often poise (dyne s/cm^2) is used for dynamic viscosity and centi Stoke (cS) for kinematic viscosity. To convert from poise to Pa s, multiply it by 0.1. When dealing with kinematic viscosity, use the following: cS equals mm^2/s.

C versus η_0. The viscosity in the figure is in cP and the temperature of the lubricant is in °C .

For mineral oils, the viscosity–pressure coefficient is given by Wooster and quoted by Stachowiak and Batchelor (2005) as:

$$\gamma = (0.6 + 0.965 \log_{10} \eta_0) \times 10^{-8} \tag{5.51}$$

where the viscosity at zero pressure, η_0 is in centipoise and the pressure–viscosity coefficient is in Pa^{-1}.

The pressure–viscosity coefficients can also be calculated following the study of Wu et al. (1989) who quote the formula of So and Klaus (1980) in a slightly revised form, giving the coefficient as:

$$\gamma = 1.030 + 3.509 (\log \mu_0)^{3.0627} + 2.412 \times 10^{-4} m_0^{5.1903} (\log \mu_0)^{1.5976} - 3.387 (\log \mu_0)^{3.0975} \rho^{0.1162} \tag{5.52}$$

in units of $\mathrm{kPa}^{-1} \times 10^5$; the predictions are shown by the authors to be very accurate. While in eq. 5.51 the coefficient is given as a function of the viscosity only, in eq. 5.52 the density as well as the temperature also affect it in addition to the viscosity.

The constant m_0 is defined as the viscosity–temperature property, given by the ASTM slope divided by 0.2. The slope is given by Briant et al. (1989) and the coefficient m_0 is then obtained from:

$$m_0 = \left({}^{1}\!/_{0.2}\right) \frac{[\log \log(\mu_0 + 0.7) - \log \log(\mu + 0.7)]}{\log T_0 - \log T} \tag{5.53}$$

The lubricant density is also dependent on the pressure. Szeri (1998) quotes the relationship of Dowson and Higginson (1977):

$$\rho = \rho_0 \left(1 + \frac{0.6 \times 10^{-9} p}{1 + 1.7 \times 10^{-9} p}\right) \tag{5.54}$$

where ρ_0 is the density at atmospheric pressure and the pressure is in Pa.

The pressure–viscosity coefficients, as predicted by eqs 5.51 and 5.52, may be tested by employing the data presented by Reid and Schey (1984). They used SAE 30 oil and gave its properties: the dynamic viscosity at 38°C was 0.11 Pa s, and at 110°C, 0.01 Pa s. The density is given by Booser (1984) as 875 kg/m^3. Reid and Schey (1978) give the pressure–viscosity coefficient as 0.02 MPa^{-1}; eq. 5.51 yields 0.0249 MPa^{-1} while eq. 5.52 gives 0.0257 MPa^{-1}, all relatively close. In spite of some complexities in computation, the use of eq. 5.52 is, however, recommended as the authors show its ability to determine the pressure–viscosity coefficient accurately in a large number of instances.

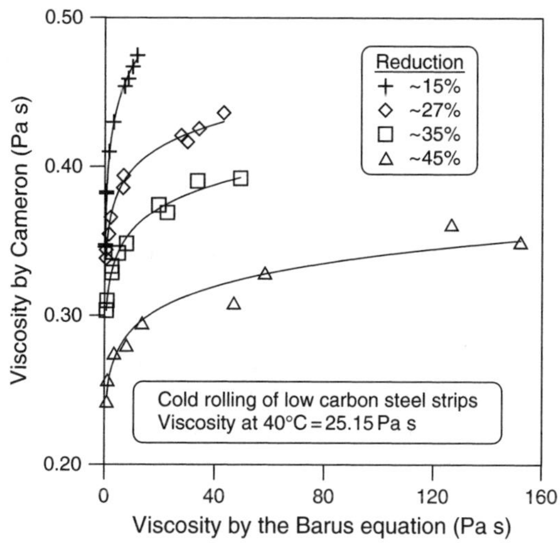

Figure 5.17 The viscosity as predicted by the Barus and Cameron equations, eqs. 5.48 and 5.49, respectively.

A comparison of the dependence of viscosity on pressure, as predicted by the Barus and Cameron's equations is given in Figure 5.17, using an oil *Exxcut 225*, prepared with petroleum base oils, sulfurized hydrocarbons, fats and esters as the additives, with a viscosity of 25.15 mm^2/s at 40°C and a density of 869.3 kg/m^3. It appears that the predictions of the two relations diverge wildly as the reduction increases. The predictions of the Barus equation become highly unrealistic at higher pressures.

5.4.1.3 The viscosity–temperature relationship

Stachowiak and Batchelor (2005) show four different relationships for the temperature and the viscosity. These are due to Reynolds in the form:

$$\eta = b\exp(-aT) \tag{5.55}$$

expected to be accurate for a limited temperature range; one due to Slotte:

$$\eta = a/(b+T)^c \tag{5.56}$$

which may be more useful than Reynolds' formula. The Walther equation, which forms the basis of the ASTM viscosity–temperature chart,[7] is given next as

$$\mu + a = bd^{\frac{1}{T^c}} \tag{5.57}$$

[7] The chart is part of ASTM method D 341–77.

where μ is the kinematic viscosity in m^2/s. When the ASTM chart is used, the constant d is taken to be 10 and the constant a is 0.6. The last relation is the Vogel equation, identified as the most accurate:

$$\eta = a \exp\left(\frac{b}{T-c}\right) \tag{5.58}$$

In these equations a, b, c and d are constants. In each of eqs 5.55–5.58, η is in Pa s, μ is in m^2/s and T is in K.

5.4.1.4 The combined effect of temperature and pressure on viscosity

Most researchers use the form $\eta = \eta_0 \exp(\gamma p - \beta T)$ to estimate the combined effects of pressure and temperature on the viscosity; the pressure–viscosity coefficient is γ in units that match the pressure, p, and β is the temperature–viscosity coefficient, also in units which match that of the temperature. When the magnitudes of the viscosity at two temperatures are available, determination of the viscosity–temperature coefficient is simple, as long as one assumes a linear variation in between the temperatures.

Sa and Wilson (1994) use a somewhat more complex relation for the viscosity in the form $\eta = \eta_0 \exp(\gamma p - \beta T - \delta T p)$ where δ is identified as a cross-coefficient. The establishment of accurate values of that coefficient may not be easy or straightforward.

5.4.2 The lubrication regimes

Arguably the best and most widely known approach characterizing the lubrication regimes is with the aid of the Stribeck diagram (see Figure 5.18), first developed by Stribeck, a German railway engineer who studied friction in the journal bearings of railcar wheels. The diagram plots the coefficient of friction against a dimensionless group of parameters, identified as the Sommerfeld number, used in the design of journal bearings; it is the product of the dynamic viscosity and the relative velocity, divided by the normal pressure (Faires, 1955). Schey (1983) writes that the use of the term "Sommerfeld number" is, however, somewhat incorrect.

Knowing the magnitude of the coefficient of friction, the curve allows one to determine the extent of various lubricating regimes in a metal forming process. In the first portion, where the viscosity of the lubricant and the relative velocity of the contacting surfaces are low and the interfacial pressure is high, boundary lubrication is observed, in which metal-to-metal contact is predominant in addition to some lubricant-to-metal contact. The roughness of the resulting surfaces will approach that of the forming die, that is, the work roll. As oils of higher viscosity are introduced in the contact zone at higher relative speeds, the boundary regime changes as more lubricant is drawn and more lubricating pockets are created in the valleys in between the asperities,

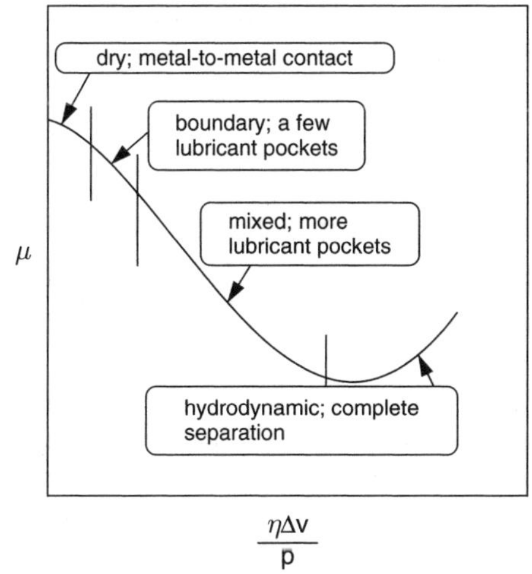

Figure 5.18 The Stribeck curve.

and the "mixed mode" of lubrication, involving less metal-to-metal contact, is found. Moving further toward the right along the axis of the Sommerfeld number, the hydrodynamic regime is located, characterized by complete separation of the contacting surfaces. In this regime, the increase of the coefficient of friction is a result of increasing frictional resistance in the oil film, usually characterized as a Newtonian fluid, separating the surfaces. In this region, the product surface roughens after rolling because of the free plastic deformation of the grains near and at the surface.

The nature of the lubrication regimes is often defined in terms of the thickness of the lubricant film and the combined roughness of the work roll and the rolled strip. The ratio:

$$\lambda = \frac{h_{\text{oil film}}}{\Lambda} \tag{5.59}$$

where $h_{\text{oil film}}$ is the oil film thickness[8] and Λ is the r.m.s. roughness of the two surfaces, given by:

$$\Lambda = \sqrt{R_{q1}^2 + R_{q2}^2} \tag{5.60}$$

and R_{q1} and R_{q2} are the r.m.s. surface roughness values of the two surfaces. When the oil film thickness to surface roughness ratio is less than unity, boundary lubrication is present. When $1 \leq \lambda \leq 3$ mixed lubrication prevails

[8] The oil film thickness will be discussed in Section 5.4.7.

while for a ratio over three, hydrodynamic conditions and full separation of the contacting surfaces exist.

In the flat rolling process, mixed or boundary lubrication regimes are usually prevalent.

5.4.3 A well-lubricated contact in flat rolling

Lubricants and emulsions are used to optimize the frictional events in the rolling process in addition to controlling the quality and the temperature of the resulting surfaces. Effective lubrication is essential to control the tribological interactions between the work rolls and the work piece in the flat rolling process. The interactions include four phenomena, the requirements of all of which need to be satisfied to create well-lubricated contacts. First, sufficient amounts of the lubricant or emulsion must be made available at the entry to the roll bite; the emphasis being on the word "sufficient amounts", implying that too little or too much can be equally counterproductive. The lubricant must then be entrained, that is, the oil or its droplets must be captured and drawn into the contact zone between the work roll and the rolled strip. After successful entry, the lubricant must be spread through the contact uniformly well so that all of the surfaces are covered evenly. The lubricant must then travel through the deformation zone to the exit and should not be squeezed out at the sides.

The current industrial practice during the rolling of steel or aluminum strips is to use oil-in-water emulsions, which, if the above requirements are satisfied, have been shown to create good lubricating conditions and acceptable surfaces in addition to efficient cooling. In some isolated cases neat oils are used.

Delivering sufficient amount of the lubricant or emulsion is dependent on the hardware and on the practice followed by the operators of a particular mill. In laboratory experiments, three possible approaches are employed. The emulsion may be sprayed either on the work rolls, or on the top and bottom surfaces of the entering strip or directly at the point of entry. Schmid and Wilson (1995) found that the film thickness was greatest when the emulsion nozzles were directed into the gap. The volume flow is usually carefully controlled and kept constant. If neat oil is used, it is usually applied by a pipette and then spread over the surfaces by a clean roller. In either case it is necessary to measure the weight of the strip to be rolled before and after the application of the lubricant, so the actual pre-rolling oil film thickness should be well known. In industry, the lubricant delivery systems vary from mill to mill. Roberts (1978, 1983) describes many of these in detail, for both hot and cold rolling.

The other three events – entrainment, uniform cover of the surface and travel through the contact zone – depend on the interaction of process and material parameters, including the rolling speed, the reduction and the resistance of the rolled metal to deformation. As well, and arguably more important than the attributes already mentioned, the surface roughness parameters of the roll and the strip to be rolled affect the lubrication process in a most significant manner. Since the roll is very much harder than the strip, the latter's asperities are expected to flatten shortly after entry, implying that the real contact area reaches its maximum fast. While under industrial conditions the surface of the work roll changes due to wear and roll pick-up, in a laboratory, where the volume of the rolled metal is much less, the roll's surface roughness is not expected to change in any significant measure. Hence, the nature of the surface roughness of the work roll is considered to be among the most important contributors to the success of the last three requirements of the lubricating system.

These comments are equally valid whether neat oils or emulsions are used. Since the use of emulsions is increasing in the industry, the discussions in what follows will concern mostly them. An additional phenomenon needs to be considered when oil-in-water emulsions are used: that of the behaviour of the droplets when they encounter the entry region of the roll-strip conjunction.

5.4.4 Neat oils or emulsions?

It is well known that the use of neat oils results in significant reduction of the loads on the mill. In order to make an intelligent choice between using neat oils or oil-in-water emulsions, an answer to the question is needed: How do emulsions affect the roll separating force and the roll torque?

Cold rolling experiments on low carbon steel strips were conducted to provide an answer (Shirizly and Lenard, 2000). The actions of four lubricants were compared, in addition to dry rolling and using water only, for their abilities to affect the roll forces and the torques on the mill and the frictional conditions. The lubricants included the SAE 10 and SAE 60 automotive oils, the SAE 10 base oil with 5% oleic acid added as a boundary additive and the SAE 10 base oil, emulsified, using water and polyoxyethylene lauryl alcohol as the emulsifier, 4% by volume. Oleic acid was chosen as the boundary additive since it was shown to react to pressure and temperature less than several other fatty oils (Schey, 2000). While the automotive lubricants were not formulated for use in the flat rolling process, their properties are well known, and that is the reason for their choice in this comparative study. Reid and Schey (2000) also used automotive oils in their study of full film lubrication during rolling of aluminum alloys.

5.4.4.1 Roll force and roll torque

Typical roll separating force and roll torque data, as a function of the reduction and at various roll surface speeds (20 and 160 rpm, leading to surface velocities of 262 and 2094 mm/s) are shown in Figures 5.19–5.21, respectively. The lubricants and the emulsions used are also indicated in the figures. As expected, the roll forces and the torques increase as the reduction is increased,

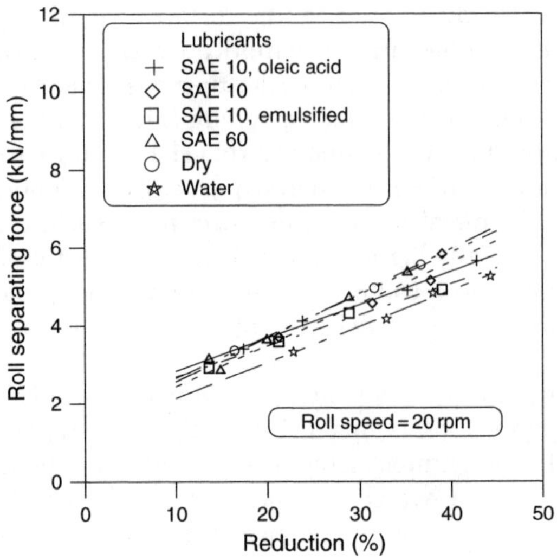

Figure 5.19 The roll separating force at 20 rpm.

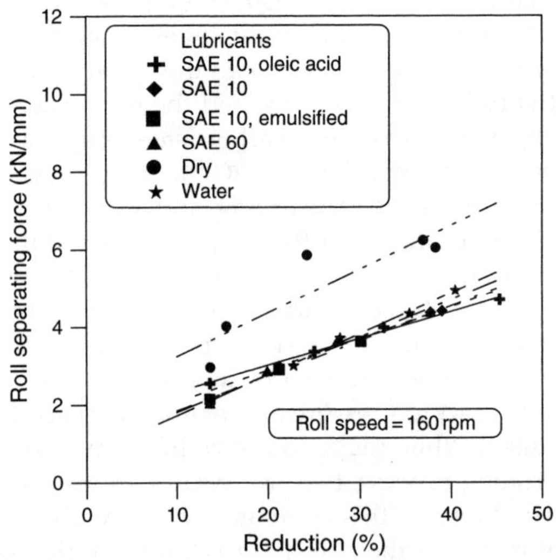

Figure 5.20 The roll separating force at a speed of 160 rpm.

in a fairly linear fashion. Increasing the speed of rolling is expected to create more favourable lubricating conditions in the roll gap as more oil is drawn into the contact zone, at least in the tests where neat oils have been used. While the data for 20 rpm and 160 rpm were plotted separately, the lower forces at the higher speeds are clearly observable. Also, it was expected that dry conditions

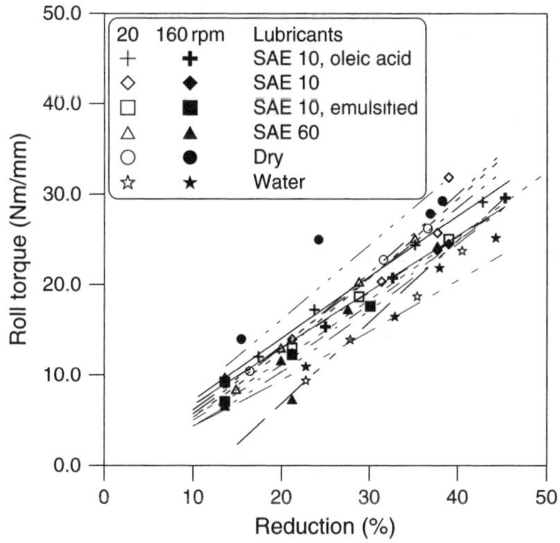

Figure 5.21 The roll torque at 20 and 160 rpm.

will require the largest forces and torques to reduce the strip and as noted, this expectation was realized at both rolling speeds. The lubrication effect is much more pronounced at high speeds. At 20 rpm, while dry conditions produced larger forces and torques, these are of the same order of magnitude as those caused by some of the other lubricants. Also, the effect of water only on the forces and torques was surprising. At both speeds, use of water created favourable conditions as far as the loads on the mill were concerned. While the roll torques were not the lowest with water only, they were among the lowest. It appears that the lubricating effects of water are comparable to that of some of the lubricants.

The roll force appears to be much more sensitive to the variation of the rolling velocity than the roll torque. There is a clear drop of the force as the speed is increased, of about 25% magnitude. The torques, however, are not affected by the change of speed in any significant measure. There is one observable, interesting trend in the roll force data: as the reduction is increased, the forces required to roll the steel become more affected by the lubricant type at the lower speed of 20 rpm. At the higher speed, the opposite trend is present. At low reductions, the oil type has a noticeable effect on the forces. This effect is less evident as the reduction is increased.

In general, however, no significant effect of the type of lubricant or emulsion on the forces and the torques is observed in the data. In prior studies concerning cold rolling of commercially pure aluminum using neat oils (Lenard and Zhang, 1997), there was a clear drop of the forces and torques when the SAE 5 oil was replaced by the much more viscous SAE 30. The expectations in the present work, that of significantly lower loads on the mill as the viscosity is increased, albeit smaller than with the aluminum, were not realized. The data of Lin et al. (1991) appears to support this observation. The authors used four lubricants with viscosity indexes varying from a low of 97 to a high of 115 and found that the loads on the mill were not affected in any significant measure.

5.4.4.2 The coefficient of friction

Hill's formula is used to determine the magnitudes of the coefficient of friction in the cold rolling process. The results are shown in Figures 5.22 and 5.23, for 20 and 160 rpm, respectively, where the coefficient of friction is plotted against the reduction, for all lubricants, dry conditions and water only.

In general and as expected, the highest frictional resistance is observed at low speeds and dry conditions. At 20 rpm, the lowest magnitudes of the coefficient of friction are produced by water only as the lubricant, confirming the trend noted above with the loads on the mill. No significant differences in frictional resistance are noted when any of the oils, neat or emulsified, are used. In all cases the coefficient of friction is reduced as the reduction is increased. Rolling in the dry condition resulted in frictional values that are among the highest.

There is an unmistakable, albeit not very pronounced, dependence of the frictional resistance on the lubricant viscosity at the 160 rpm rolling speed. Of the four lubricants, the most viscous, SAE 60 appears to yield the lowest coefficient of friction and the highest values are obtained under dry

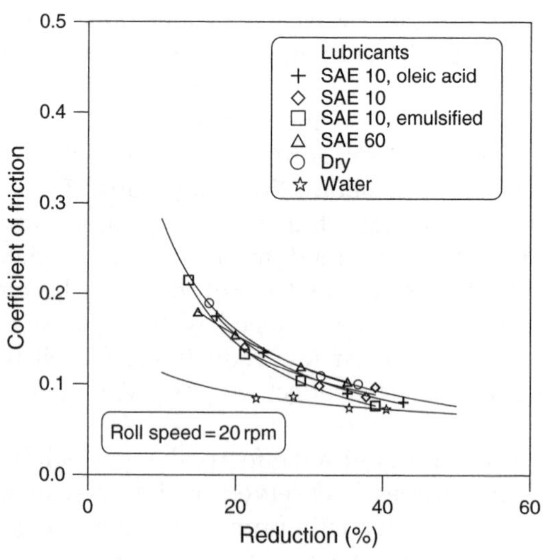

Figure 5.22 The coefficient of friction as a function of the reduction at 20 rpm.

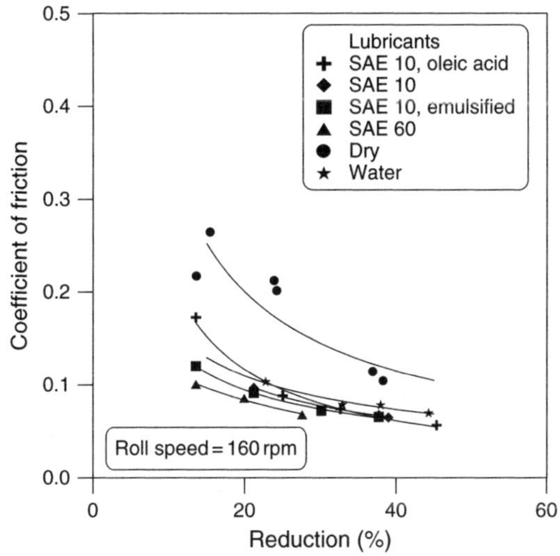

Figure 5.23 The coefficient of friction as a function of the reduction at 160 rpm.

conditions. The magnitude of frictional resistance with the SAE 10, containing the oleic acid additive, is significantly lower than those rolled dry, as expected. The SAE 10, neat or emulsified, leads to friction values that are practically identical and not very much different from SAE 10 and the boundary additive. When using the four lubricants, the coefficient of friction reduces with increasing reduction. As well, the coefficient of friction is observed to decrease as the speed of rolling increases, under both dry and lubricated conditions.

Based on these results, the recommendation is: use emulsions as often as possible.

5.4.5 Oil-in-water emulsions

Most metalworking emulsions are oil-in-water (O/W) systems where oil is the dispersed phase and water the continuous phase. Emulsions are composed of three primary ingredients: the oily phase, the emulsifier and water. The emulsions used in rolling are composed of a water phase in which spherical micelles of oil, with diameters ranging from 1 to 10 μm, are dispersed.

To keep these micelles from coalescing, an emulsifier, sometimes referred to as a surfactant, is used. Emulsifiers are composed of a molecular structure having two distinct ends. The hydrophilic (water loving) end is made of polar covalent bonds and is therefore soluble in water. The lipophilic (oil loving) end is soluble in natural and synthetic oils. When the emulsion is formed, the hydrophilic groups will orient towards the water phase and the lipophilic hydrocarbon chain will orient toward the oil phase.

5.4.5.1 Behaviour of the droplets

Kumar et al. (1997) wrote that the fundamental problem in the use of emulsions is the behaviour of the oil particles, the capture of which by the entering strip or the roll surfaces is not yet fully understood.

A study, sometime before these comments, clarified the mechanisms of droplet capture. A transparent, translating plate, against which a stainless steel roll of 0.8 μm surface roughness was pressed, and a high-speed camera were used to study the droplets in O/W emulsions (Nakahara et al., 1988). Three types of droplets were identified. The first types penetrate the contact zone, called penetration droplets. Some droplets enter but don't travel through to the exit and these are identified as the stay droplets. The remaining are the droplets that are rejected completely, called the reverse droplets. A lubricant feed rate of 1.2 cc/s and a normal load of 49 N were used in the tests. The relative velocity was varied, from 5 to 20 mm/s. Both the oil film thickness and the oil concentration at the entry, where an oil rich pool was observed, were dependent on the emulsion concentration, also found by Zhu et al. (1994) and Kimura and Okada (1989). The lubrication mechanisms were considered to be velocity dependent. In the low speed range, the oily pool at the entry provided the lubrication while at higher speeds a "fine oil-in-water emulsion" produced the oil films. As the relative velocity increased, the number of penetration droplets decreased. The minimum size of the particles observed was 50 μm, however, significantly larger than those in most practical rolling processes. As well, the load and the relative speeds in practice are much larger than in the tests of Nakahara et al. (1988). The observations of Nakahara et al. (1988) imply that under some circumstances no oil particles will travel through the deformation zone and starvation at the exit may result.

In order to get a better explanation of the lubrication conditions, Kumar et al. (1997) offered a potential explanation for the experimental observations (Nakahara et al., 1988) that particles very close to the tooling are rejected and play no part in the lubrication process. Kumar et al. tried to explain that kind of behaviour through a computational fluid dynamics model of a rigid particle in the inlet zone. Their theory applied to emulsion lubrication, slurry lubrication and wear involving solid particles suspended in a liquid. The dimensionless results for different particle sizes were obtained for symmetrical and unsymmetrical inlet zones. The results indicate that the segregation location of a particle is closer to the centre of the gap as the particle size increases. That suggests that larger particles are pushed into the back-flowing centre region and are rejected. Larger particles segregate closer to the roll surface than smaller ones. Small particles that are in the back-flowing regions will be rejected from the inlet zone and will have their clearance with the tooling increase, thus they will not be entrained later.

5.4.5.2 Entrainment of the emulsion

The background here is the division of the entry zone to the roll gap into three regions: the supply region, the concentration region and the pressurization region. In the supply region, the oil droplets are isolated from the surfaces and each other. In the concentration region, ambient pressure exists and the local concentration of the oil times the film thickness remains constant. In the pressurization region, the water is trapped within the oil film and no further concentration is possible.

Plate-out, dynamic concentration and the mixture theory have been used to explain the supply and subsequent entrainment of the oil particles into the contact zone. In the plate-out process, the particles adhere to and coat the surfaces (Schey, 1983). The droplets adsorb onto the surface and spread to the wetting angle. Several droplets eventually cover the complete surface and are available to enter the conjunction. The usual criticism of the plate-out theory concerns its potential inapplicability under industrial conditions. With an increase in speed, which may reach 20–30 m/s, there may not be sufficient time for the plate-out process to occur and the film thickness at the entry will decrease or disappear entirely as a result of oil starvation.

As the pressure in between the roll and the strip increases, the oil droplets are flattened and, because of their higher viscosity, are drawn into the inlet zone (Wilson et al., 1994). The concentration of the oil is therefore increasing at the inlet, leading to the dynamic concentration theory, which hypothesizes that the oil-in-water emulsion inverts to become a water-in-oil emulsion as the pressure in the contact zone increases. Larger droplets are more likely to be entrained (Schmid, 1997), an observation that is opposite to that of Nakahara et al. (1988). Once they have penetrated the contact zone to the point where the gap and the droplet size are of similar magnitude, they are irreversibly captured. While the mixture theory is the last observable lubrication regime at very high speeds (Schmid and Wilson, 1996), it appears unsuitable under realistic rolling situations when the film thickness is much less than the droplet diameter (Schmid and Wilson, 1995).

Yan and Kuroda's (1997a) model shows that there is a velocity difference between two phases of the emulsion and this causes a variable concentration of the oil phase in the lubricating film. At low entraining speeds the oil pool is formed in the inlet zone, so the film thickness is obtained primarily by the oil phase. At high entraining speeds, the increment of oil concentration becomes slow and both the oil and water phases are entrained into the contact zone.

These results agree qualitatively with the experimental observations of Zhu et al. (1994), who showed a set of experimental results of the elastohydrodynamic lubrication film thickness with oil-in-water emulsion in a wide range of rolling speeds for different oil concentration and pH values. Experimental observations indicated that the phase inversion/oil pool formation mechanism around the inlet zone takes place only at very low speeds, which are most likely far below particle speed ranges for major industrial applications.

In a follow-up paper, Yan and Kuroda (1997b) extended their previous work (1997a) and discussed the variation of the oil concentration in the lubricant film. They concluded that the reason that elastohydrodynamic lubrication film thickness of an emulsion is of the same order of magnitude as that of the neat oil is because the general elastohydrodynamic film thickness of an emulsion is smaller than the droplet size, and the increase in the oil concentration makes it the same order as that of the neat oils.

5.4.5.3 The emulsion in the contact zone

Schmid and Wilson (1995) indicated that the mechanism of lubrication with O/W emulsion is highly dependent on speed effects which influence the frictional conditions (boundary lubrication, mixed, hydrodynamic, etc.). Wilson and Chang (1994) introduced a simple model of mixed lubrication of bulk metal forming processes under low speed conditions, where the inlet zone doesn't contribute significantly to hydrodynamic pressure generation. The model showed that relatively high hydrodynamic pressures can be generated in the work zone under conditions where it was previously considered that hydrodynamic effects were unimportant. The outlet film thickness predicted by the model was found to be much larger than those predicted using the full-film or the high speed mixed lubrication theories.

Dubey et al. (2005) cold rolled a low carbon steel, using oil-in-water emulsions. They concluded that larger oil droplets created thicker oil films. They confirmed earlier data, indicating that increasing speeds reduce the coefficient of friction.

It appears that the behaviour of the droplets in the emulsion, their contribution to the oil film thickness, their entrainment or rejection, as well as the speed dependence of their effects on friction are still unclear. At this point, considering the studies of Shirizly and Lenard (2000), the dynamic concentration theory is considered to be the most applicable to the flat rolling process.

5.4.6 A physical model of the contact of the roll and the strip

Sutcliffe (2002) considers the lubrication mechanisms in between the work roll and the rolled strip. His figure is reproduced here as Figure 5.24. He shows the two significant contributors to frictional resistance and relative motion of the roll and the strip: contact at the asperities and contact at the lubricant-filled valleys. His figure emphasizes the importance of the changes to the asperities as the rolling process is continuing, as both contributions change when high normal and shear stresses act on the bodies in contact.

The mechanism at the contact is usually referred to as micro-plasto-hydrodynamic lubrication (MPHL). One important aspect of this mechanism is the oil drawn out of the valleys due to the sliding action of the roll and the strip. An effective coefficient of friction may then be defined:

$$\mu = (A_r/A)\mu_c + [1 - (A_r/A)]\mu_v \tag{5.61}$$

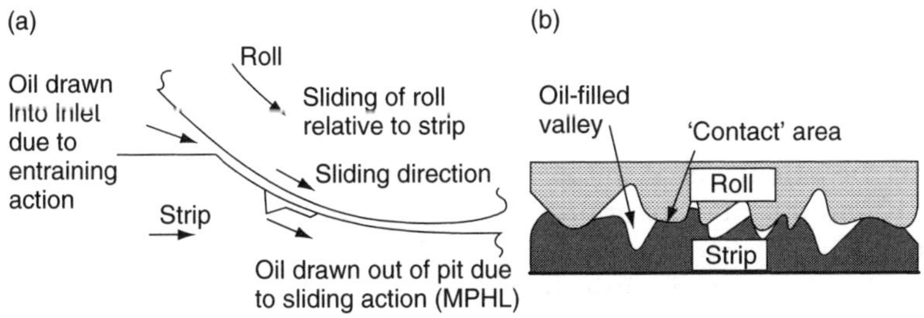

Figure 5.24 Schematic diagram of the lubrication mechanisms in flat rolling (a) and details of the contact, showing asperities and the lubricant-filled valleys (b); (Sutcliffe, 2002, reproduced with permission).

where μ_c is the coefficient of friction at the contacts and μ_ν is the coefficient of friction at the valleys. When the behaviour of the lubricant is assumed to be Newtonian, the frictional stress and hence, μ_ν may be estimated in terms of the dynamic viscosity of the lubricant and its shear strain rate, as given by eq. 5.46, above.

For non-Newtonian lubricant behaviour, the Eyring model is more appropriate:

$$\tau = \tau_0 \sinh^{-1}\left(\frac{\eta\dot{\gamma}}{\tau_0}\right) \tag{5.62}$$

where τ_0 is the stress at which non-linearity starts. Under the mixed or boundary lubrication regimes, the contribution to friction is mostly from the contact areas. Zhang (2005) points out, however, that dry contact may occur even under elastohydrodynamic conditions.

5.4.7 The thickness of the oil film

One of the parameters, considered to have a very significant effect on the forces of friction, is the thickness of the oil film in between the contacting surfaces, which in the present context, refers to the roll/strip contact. The results of an experiment, conducted using a flat-die apparatus[9], indicate the dependence of the coefficient of friction on the volume of the oil. It is realized that the geometry of the test is significantly different from that of flat rolling

---

[9] This technique involves compressing a lubricated sheet of material between two flat dies and drawing the sheet through, while monitoring the normal and the horizontal forces. Constant velocity in the longitudinal drawing direction and a constant normal load are maintained throughout the test. The coefficient of friction is then taken as the average of the draw force divided by the normal force.

and therefore the numbers quoted here are not relevant to the rolling process. The trends, however, are expected to be similar.

Since industrial practice in sheet metal forming indicates that 2–2.5 g/m^2 of lubricant on the sheet surface will ensure good tribological phenomena, that was the amount used initially in a set of experiments. This amount of oil creates a film thickness of approximately 2–3 μm in the area of contact in the test, and a film thickness ratio, λ = film thickness/effective surface roughness, of about 2–3, so a mixed lubrication regime would be expected. Loud metallic noises heard during the preliminary tests and some die damage, however, indicated the presence of mostly boundary lubrication. Weighing the sheets after the tests and comparing the weights to the pre-test values suggested that squeezing-out of the lubricant at the sides was most likely not the cause of the noise. Experiments were then conducted in which the amount of the oil was systematically increased. The results are shown in Figure 5.25, where the amount of the oil is given on the abscissa and the resulting coefficient of friction is shown on the ordinate. The graphs appear similar to the Stribeck curve, showing a possible approach to a hydrodynamic regime. Mixed tribological conditions appear to be present up to approximately 8 g/m^2.

5.4.7.1 Measurement of the thickness of the oil film

One of the earlier attempts to measure the thickness of the oil film was by Whetzel and Rodman (1959). They dissolved the lubricant that stayed on the rolled strip after the pass in a solvent, evaporated the solvent and thus determined the volume of the remaining oil. They assumed that the surfaces of

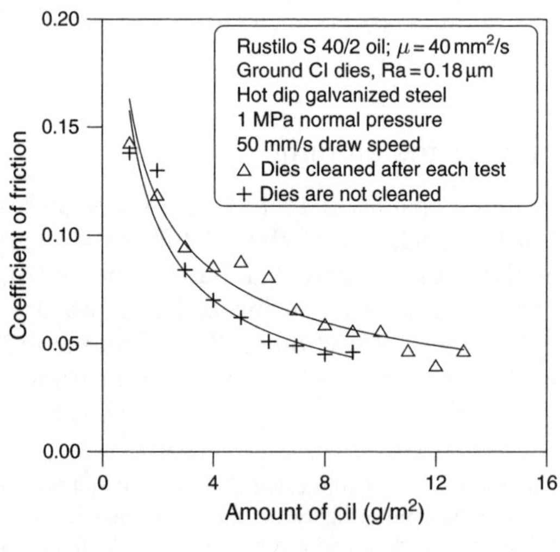

Figure 5.25 The dependence of the coefficient of friction on the film thickness in the flat-die test (Kosanov et al., 2006).

the strip were covered in a uniform manner. Another technique to determine the thickness of the oil film, that of the "oil drop" approach, was probably first introduced by Saeki and Hashimoto (1967). They measured the weight of the strip before rolling, added a drop of the lubricant, measured the weight after the pass and the area covered by the oil drop, leading to the thickness of the film. Azushima (1978) also used the "oil drop" approach to determine the thickness of the oil film. He rolled 1 mm thick stainless steel strips at speeds which varied from a low of 4 m/min to a high of 850 m/min and used three lubricants of 61, 30 and 4 cSt kinematic viscosities (at 38°C), respectively. The thickness of the oil film decreased with increasing reduction and increased with increasing rolling speeds, indicating again that the speed effect on droplet behaviour is an important consideration.

Sutcliffe (1990) measured both the average film thickness and the area of contact ratio, using two aluminum alloys and lead, lubricated with a mineral oil of 1.951 Pa s dynamic viscosity. Two rolls with a rough and a smooth finish were employed. He followed two methods. The first was that of Azushima (1978) and the oil-drop technique. In the other, he used the roughness data of the rolled strips, obtained while rolled with the smooth roll. Using the theory, developed by Sutcliffe and Johnson (1990), he presented the results in terms of the ratio of the mean film thickness to the combined r.m.s roughness of the *rough* rolls and the strip, plotted against mean film thickness to the combined r.m.s roughness of the *smooth* rolls and the strip. The predictions and the measurements were remarkably close.

Zhu et al. (1994) used either a steel ball or a steel roller, rotating against a glass disk. The surfaces of the disk and both the ball and the roller were prepared to be extremely smooth. The flow rate and the lubricant temperature were closely controlled during the experiments. Optical interferometry was used to establish the thickness of the lubricating films. Neat oil, pure water and six oil-in-water emulsions were tested, with the viscosity, at 40°C, changing from a low of 0.66 cSt (for water) to a high of 296.15 cSt for the emulsion, containing 40% oil. The droplet dimensions in the emulsions didn't differ by much, varying from 0.44 to 0.55 μm. The range of speeds was remarkably broad, from a low of 0.001 m/s to a high of 20 m/s. Using the neat oil, the film thickness increased linearly with the speed. The film thickness increased, dropped and then increased again with all six emulsions when the speed was increased.

Lo and Yang (2001) presented an analytical method to determine the thickness of the oil film in cold rolling. They assumed that the work roll surface is smooth and a mixed lubrication regime governs the contact between the roll and the rolled metal. Under these conditions the average oil film thickness is approximately the same as the average depth of the valleys on the rolled strip; hence, measurements of the surface roughness of the rolled strips leads to the thickness of the oil film.

Trijssenaar (2002) stated clearly that "the *a priori* assumption that the cold rolling lubrication film exists of pure oil, is not correct". She used the oil drop method to estimate the film thickness while lubricating the strip with an oil-in-water emulsion. She concluded that the Wilson and Walowit (1972) equation – see below, Section 5.4.7.2, eq. 5.63 – needs to be corrected for the conditions of her experiments, since it doesn't account for the surface roughness. When the corrections were introduced, the data and the predictions agreed very well.

A portable infrared analyser to measure the oil film thickness was discussed in Nordic Steel and Mining Review (1998), developed by Spectra-Physics Vision Tech.

5.4.7.2 Calculation of the oil film thickness

Neat oils: Wilson and Walowit (1972) developed a mathematical model to study the lubrication conditions in strip rolling under hydrodynamic conditions. They used several simplifying assumptions to allow the integration of Reynold's equation and obtained the often-used equation for the thickness of the oil film at the entry to the roll gap:

$$h_{\text{oil film,entry}} = \frac{3\eta_0\gamma\left(v_{\text{entry}} + v_{\text{r}}\right)R}{L\left\{1 - \exp\left[-\gamma\left(\sigma_{\text{fm}} - \sigma_{\text{entry}}\right)\right]\right\}} \tag{5.63}$$

where η_0 is the dynamic viscosity at 38°C, in Pa s, and γ is the pressure–viscosity coefficient in Pa^{-1}. The radius of the roll is designated by R in m, the roll surface velocity is v_r, the entry velocity of the strip is v_{entry}, both in m/s and L stands for the projected contact length, also in m. The average flow strength in the pass is given by σ_{fm}, and the tensile stress at the entry is σ_{entry}, in units that match those of γ. The relationship predicts that the film thickness at the entry will increase as the viscosity, the viscosity–pressure coefficient, the velocity and the roll radius increase. Assuming that the rolls and the strip are rigid at the inlet, the oil film thickness in the contact zone can also be determined as

$$h = h_{\text{oil film,entry}} + \frac{x^2 - x_1^2}{2R} \tag{5.64}$$

where x is the distance from the line connecting the roll centres and x_1 is the location of the entry to the deformation zone.

Emulsions: The model that comes closest to reality is due to Schmid and Wilson (1995). The authors derive a simple equation for the inlet oil film thickness and claim that the predictions of the model are supported by experiments. They also make the statement that the experiments seem to suggest that the efficiency of oil droplet capture increases with increasing rolling speed. However, use of the model depends on certain assumptions whose validity is proven by comparing the predictions to measurements. Thus, the model is semi-empirical.

The two assumptions that must be made involve a capture coefficient (C) and the oil concentration at which inversion of the emulsion occurs. The assumption is that inversion will occur at a concentration of 0.907.

The model next calculates the oil film thickness in the case of an infinite pressurization region, using the expression developed by Wilson and Walowit (1972). The expression is:

$$h_{0w} = \frac{6\eta_0 \overline{U}\gamma}{\theta[1-\exp(-\gamma\sigma_{fm})]} \tag{5.65}$$

where η_0 is the oil viscosity, Pa s, $\overline{U}$ is the average of the strip inlet and the roll surface velocities, γ is the pressure–viscosity coefficient, θ is the inlet angle and σ_{fm} is the plane-strain flow strength of the rolled metal.

The emulsion availability $(\overline{A})$ is calculated next, using the assumed value of the capture coefficient:

$$\overline{A} = \frac{C\phi_s d_s}{\phi_i h_{0w}} \tag{5.66}$$

In the expression d_s is the oil droplet size, ϕ_s is the oil concentration in the original emulsion and ϕ_i is the oil concentration at the inversion. The non-dimensional oil film thickness $H = h_0/h_{0w}$ is obtained from

$$H^2 - \left(\overline{A}^2 + 2\overline{A}\right)H + \overline{A}^2 = 0 \tag{5.67}$$

The thickness of the oil film at the entry to the pressurization region, h_0, can now be determined.

5.5 DEPENDENCE OF THE COEFFICIENT OF FRICTION OR THE ROLL SEPARATING FORCE ON THE INDEPENDENT VARIABLES

During the last several decades thousands of rolling experiments were conducted in the writer's laboratory. Steel and aluminum alloys were rolled with the independent variables being the rolling speed, the reduction, the roll diameter and its surface roughness, the lubricant and the temperature. The roll separating forces, roll torques, the temperatures, the roll speed and the resulting reductions were measured. The magnitudes of the coefficient of friction were calculated using Hill's formula, eq. 5.26, which is based on equating the measured and the calculated roll separating forces. In what follows, these data are presented to illustrate the dependence of the coefficient of friction or the roll separating force on some of the independent variables. It is noted, of course, that the coefficient of friction thus determined is not the actual value since, as mentioned above, a large number of variables, parameters and their interaction affect its magnitude. Hill's numbers, however, serve well in a relative sense and are good for the comparison of the trends.

5.5.1 The dependence of coefficient on reduction

There appears to be a general agreement that the coefficient of friction decreases as the reduction increases. This agreement is confirmed by the results shown in Figure 5.26, obtained while rolling low carbon steel strips, where the reduction is plotted on the abscissa. Two neat oils are included in the figure as well as two rolling speeds. Under all conditions, the coefficient of friction demonstrates a downward trend. The lubricant viscosity appears not to have a major effect on the coefficient but the rolling speed does; see the Section 5.5.2.

The behaviour of the coefficient of friction as dependent on the reduction, and hence, the normal pressure, changes when strips made of softer materials are rolled. This is demonstrated in Figure 5.27, which shows various aluminum alloys, rolled at room temperatures (Karagiozis and Lenard, 1985). The coefficients of friction in these experiments were determined by the pins and transducers embedded in the work roll (Lim and Lenard, 1984). As shown, the coefficient of friction increases with the reduction, indicating the faster rate of asperity flattening of the softer metals, leading to a faster rate of increase of the real area of contact, agreeing with the results of Tabary et al. (1994), reviewed above, see Section 5.3.2.3.

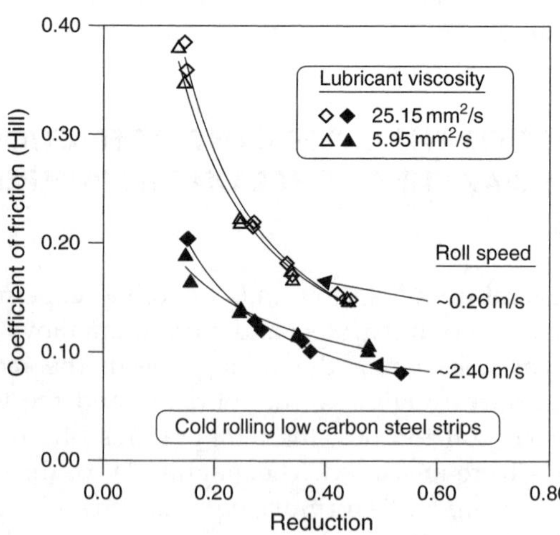

Figure 5.26 The dependence of the coefficient of friction on the reduction – low carbon steel strips.

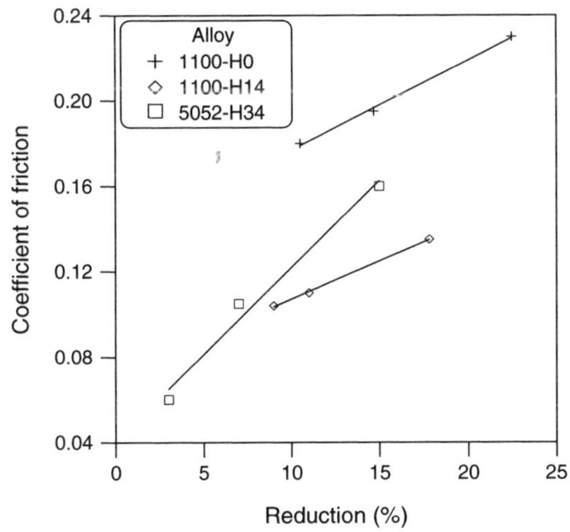

Figure 5.27 The dependence of the coefficient of friction on the reduction – aluminum alloys.

5.5.2 The dependence of coefficient on speed

Under most circumstances and provided that the four conditions for a well-lubricated contact are met (see Section 5.4.3), increasing velocity results in dropping coefficients of friction. This is aided by two events: the increased amount of lubricant being drawn into the contact zone and the time dependent nature of the formation of the adhesive bonds.[10] While Figure 5.26 already indicated the dependence of the coefficient of friction on the rolling speed, a more explicit figure underlines the issue, see Figure 5.28. Here, the speed is plotted on the abscissa. Again the same two lubricants are used as in Figure 5.26. Two reductions are indicated and in both cases the decreasing coefficient is evident.

The situation may change when the conditions for efficient lubrication are not met. Increasing the roughness of the work roll and changing the roll material results in increasing coefficients of friction with increasing rolling speeds, see Figure 5.29 (Lenard, 2004). The figure gives the coefficient of friction, obtained while rolling 6061-T6 aluminum alloy strips, reduced by approximately 55%. The roll speed is given on the abscissa and the results are given for three values of the roll roughness. At the highest roughness of $R_a = 2.4\,\mu m$, the coefficient of friction increases with the speed. Reich et al. (2001) examine the slopes of the forward slip-speed plots obtained while cold rolling 3004

[10] The adhesion hypothesis (Bowden and Tabor, 1950) states that the resistance to relative motion is caused by adhesive bonds formed between the contacting asperity tips, which are an interatomic distance apart.

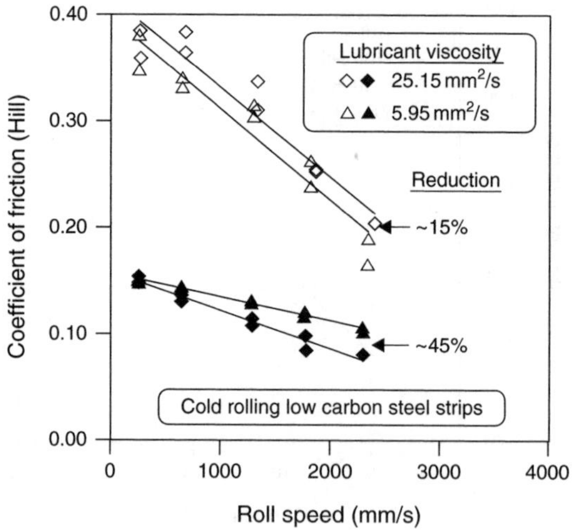

Figure 5.28 The dependence of the coefficient of friction on the rolling speed – low carbon steel strips.

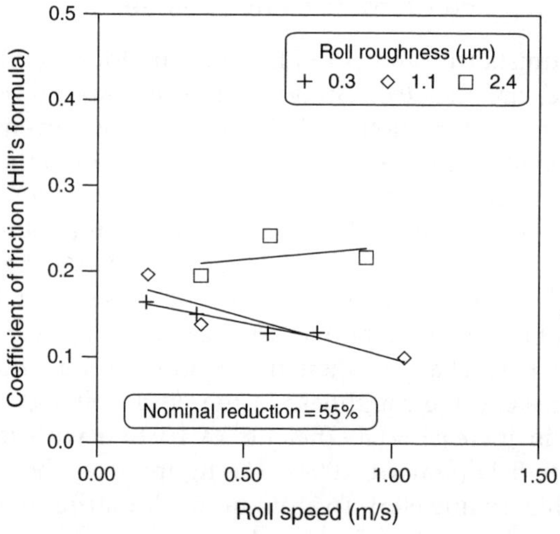

Figure 5.29 The coefficient of friction as a function of the roll speed and the surface roughness of the work roll; 6061-T6 aluminum alloy strips are rolled (Lenard, 2004).

aluminum alloy strips using an oil-in-water emulsion. They conclude that increasing values of the forward slip with increasing rolling speeds may indicate that the contact zone is starved of adequate amounts of the emulsion. As demonstrated above (see eqs 5.13–5.17), increasing forward slip indicates an increasing coefficient of friction and that is given in Figure 5.29, leading to the possibility of starvation in the contact zone.

5.5.3 The dependence of coefficient on the surface roughness of the roll

Dick and Lenard (2005) conducted cold rolling experiments on low carbon steel strips, using progressively rougher rolls in a STANAT two-high variable speed mill. The strips were lubricated by oil-in-water emulsions, delivered at a rate of 3 L/m.

Four kinds of roll surfaces were prepared. In the first instance, the rolls were ground in the traditional manner to a surface roughness of approximately $R_a = 0.3\,\mu m$ in the direction around and along the roll. The next three surfaces were prepared by sand-blasting, expected to create a random roughness direction. Using Blasto-Lite glass beads BT-11 resulted in a surface roughness nearly identical to that of the ground rolls, $R_a = 0.35\,\mu m$. The next surface was prepared using larger glass beads of grit #24, creating a randomly oriented surface, approximately $R_a = 0.91\,\mu m$. Following this, using #60 Lionblast oxide grit resulted in surface roughness of $1.31\,\mu m$ and using BEI Pecal EG – 12, another oxide grit, created surface roughness of approximately $1.76\,\mu m$.

Three lubricants, supplied by Imperial Oil, were used in an oil-in-water emulsion. Walzoel M3 is a low viscosity, high VI oil with synthetic ester lubricity agents and phosphorus containing antiwear agents. Its kinematic viscosity is $8.65\,mm^2/s$ at 40°C and $2.34\,mm^2/s$ at 100°C. Kutwell 40 is a medium viscosity and medium VI paraffinic oil with sodium sulfonate surfactant and antirust additives, and no lubricity ester or antiwear agents with a viscosity of $37\,mm^2/s$ at 40°C. Oil FSG is a high viscosity, high VI oil with natural ester lubricity agents and zinc and phosphorus containing antiwear agent. Its viscosity is $185\,mm^2/s$ at 40°C and $16.75\,mm^2/s$ 100°C. The supplier estimates the droplets to be in between 5 and $10\,\mu m$ in size.

The results indicate that the roll separating forces depend on the roughness of the work roll in a very significant manner, as shown in Figure 5.30. Two sets of data are given in the figure, both for high reduction. The empty symbols indicate the forces at low rolling speeds, while the full symbols indicate the same at higher rolling velocities. The forces increase almost in a linear fashion as the roll roughness is increasing. The speed effect is also observable in the figure and as above, under most conditions the forces drop as the speed increases.

While cold rolling of 6061-T6 aluminum alloy strips, using a low viscosity mineral seal oil and progressively rougher work rolls, the slopes of the roll force–roll roughness plots indicated sudden increases of the slope at approximately $1\,\mu m$ R_a (Lenard, 2004). These changes revealed the relative contributions of the adhesive and ploughing forces to friction and indicated that the

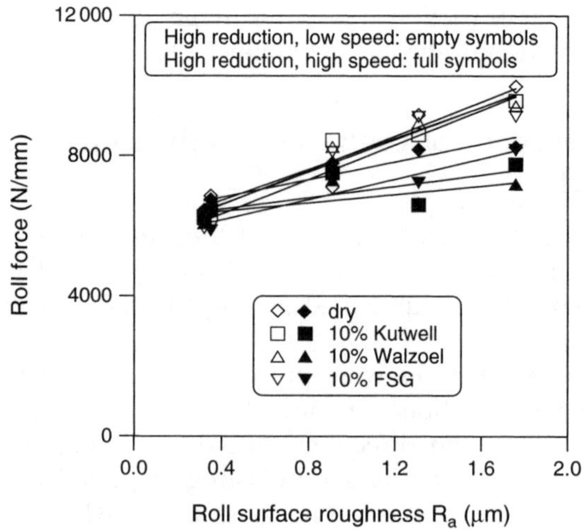

Figure 5.30 The roll separating force as a function of the roll surface roughness; low carbon steel strips are rolled (Dick and Lenard, 2005).

effect of ploughing overwhelms that of adhesion. It is possible that increasing the roll roughness beyond 1.76 μm R_a would lead to similar behaviour while rolling the steel strips, as no further sudden changes of the slopes are demonstrated. The different observations result from the significant differences of the viscosities of the lubricants and the different metals used. In the study of Lenard (2004), the mineral seal oil's viscosity was 4.4 mm^2/s while with the steel the lightest oil's viscosity was twice that. The low viscosity created a very low film thickness and the sharp asperities of the sand-blasted work roll must have pierced through the film as soon as contact was established at the entry. The sharp asperities must also have pierced the oil particles while the steel strips were rolled but because of the higher viscosities, these likely have occurred later and to a lesser extent.

5.5.4 The dependence of the roll separating force on the lubricant's viscosity

In the study mentioned above (Dick and Lenard, 2005), the effect of the viscosity of the emulsion was also examined.[11] The details of the emulsions were given above. The dependence of the roll force per unit width – and by its

[11] Considering here that the viscosity of the oil is the same as in the emulsion is legitimate since the dynamic concentration theory is expected to hold, implying that the strip/roll contact is lubricated by almost-neat oils.

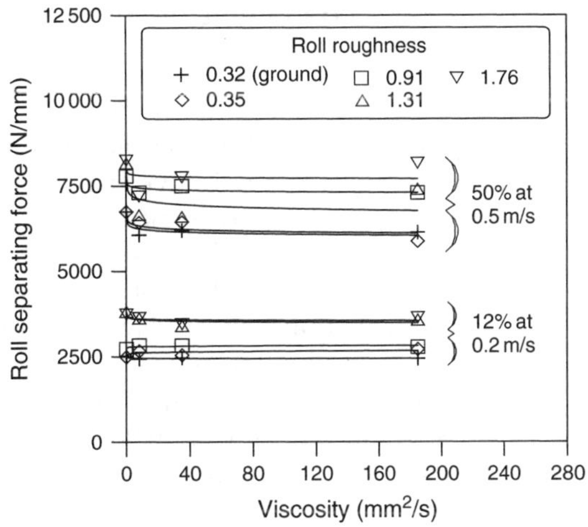

Figure 5.31 The roll separating force as a function of the lubricant viscosity; low carbon steel strips are rolled (Dick and Lenard, 2005).

association, the coefficient of friction, as well – on the viscosity of the oils in the emulsions is shown in Figure 5.31. The origin at a viscosity of zero indicates dry conditions. The roll forces under two process conditions are shown: high reduction at high speeds (corresponding to a nominal reduction of 50% and roll surface velocity of 0.5 m/s) and low reduction at low speeds (corresponding to a nominal reduction of 12% and a roll surface velocity of 0.2 m/s). As expected, and as predicted by the Stribeck curve, increasing viscosity should lead to lower loads, at least in the boundary and in the mixed lubrication regimes. The figure leads to a surprising observation: the viscosity appears not to affect the loads on the mill at the low speed and at the lower reduction. There is a minor drop of the forces as any emulsion is introduced but no meaningful change is observed. At the higher loads and speeds the effect of viscosity is clear. The behaviour of the forces appears to depend on both the viscosity and the roll roughness. The drop on the forces from dry conditions to any lubricant is evident once again. Further, as long as the roughness is under 1 μm, increasing viscosity leads to lower forces. When the roughness of the work roll was increased to 1.31 and 1.76 μm, increasing viscosity created increasing forces. This condition implies the presence of a hydrodynamic lubrication regime, but measurements of the roughness of the rolled strips contradict this possibility as the roughness of the rolled strips was always lower than they were before the pass. The implication is that the lubrication regime was close to hydrodynamic but was still in the mixed region. There must have been a large number of lubricating pockets and only some metal-to-metal contact. It is recalled that similar results were obtained by Shirizly and Lenard (2000) while rolling low carbon steel strips.

5.5.5 The dependence of the coefficient of friction on temperature

While hot rolling low carbon steels, the coefficient of friction has been shown to decrease with increasing temperature (Munther and Lenard, 1999), caused partly by the decreasing strength of the adhesive bonds between the scale-covered strip and the rolls which are easier to break at the higher temperatures, and the thickness of the scale. The coefficient decreased with increasing velocity and the attendant lower time available for the formation of bonds between the strip and the roll surfaces. Friction increased with increasing reduction, where the increasing size of the deformation zone and the longer contact time caused lower surface temperatures, higher strength and more adhesive bonds. Jin et al. (2002) hot rolled 430 ferritic stainless steel strips – their results are shown in Figure 5.14. While there is a large scatter in the figure, one may conclude that the coefficient drops with increasing speeds, temperatures and reduction.

5.5.5.1 The layer of scale

Munther (1997) studied the frictional conditions during hot rolling of steels. The roll/metal interface in hot rolling of steels always includes a layer of scale. The secondary scale formed during and after roughing is removed before the bar enters the finishing train. The 5–15 seconds that separate the scale breakers and the first stand are sufficient to form a new tertiary scale layer, about 10 μm thick, immediately on the hot steel surface, the behaviour of which may be ductile or brittle, depending on its temperature and thickness. This interface affects the frictional conditions, resulting in changes in the required roll forces, torques, and power consumption, as well as the overall roll wear and surface quality.

Understanding the formation and behaviour of the scale interface is important when examining the tribological phenomena that take place.

Three types of iron oxide phases make up the scale on the steel surface. These are, with increasing oxygen content, wüstite, FeO, magnetite, Fe_3O_4 and haematite, Fe_2O_3. Usually a scale with all three types of oxide phases is present on the steel surface with wüstite being closest to the steel matrix followed by the intermediate magnetite layer and the outermost haematite layer.

The most important attributes of the scale, at least when metalworking is considered, are its hardness and yield strength, since these indicate whether the oxides are abrasive. Loung and Heijkoop (1981) have reported room temperature hardness values of 460 H_v for FeO, 540 H_v for Fe_3O_4, and 1050 H_v for Fe_2O_3. Funke et al. (1978) analysed data obtained by Hirano and Ura (1970) and Stevens et al. (1971) and concluded that the hardness of the oxides is temperature dependent. They found the hardness of magnetite exceeding the hardness of cementite (which was the only carbide present in the roll material investigated) at all temperatures. Lundberg and Gustavsson (1994) have reported hardness values at 900°C for FeO, Fe_3O_4 and Fe_2O_3. These were 105, 366 and 516 H_v, respectively.

Blazevic (1983a, 1983b, 1985) and Ginzburg (1989) discussed scaling on steels during re-heating, roughing, finishing and coiling. The occurrence of surface defects caused by the scale and its location on the strip surface, including their causes and remedies, has also been described. The role of the tertiary scale was considered by Blazevic (1996).

According to Blazevic, the scale layer that enters the finishing train may be considered either thin and hot or thick and cold. In the first case of thin, hot scale, the scale fractures along fine lines as it is being compressed and elongated during deformation in the first stand. Hot metal is then extruded partially through the fine fractures as the deformation proceeds. At the same time, the hot metal deforms in the rolling direction, resulting in a simultaneous roughening and smoothing of the steel surface. In the second case of cold, thick scale, the scale is less plastic and therefore fractures severely upon elongation. The fractured scale is depressed into the steel surface, while hot metal extrudes outwards, causing a rough surface that is present even after pickling. The reason is that the metal that extruded upwards in the early stands will be over-pickled and will leave a mirror image of the prior roughness. This image remains although the cold rolling process creates an elongated and reduced image on the final product.

Scale formation has a very significant effect on friction and hence on the quality of the rolled surfaces as well as on the commercial value of the product. El-Kalay and Sparling (1968) were among the first to investigate the effect of scale on frictional conditions in hot rolling of low carbon steel. Different conditions were studied in a laboratory: light, medium and heavy scaling with both smooth and rough rolls at various velocities. Load and torque functions, according to Sims' equations, were calculated for these conditions. It was hypothesized that the scale acts as a poor lubricant and that its effect on the frictional conditions varies along the arc of contact as it fractures. It was found that the presence of scale could reduce the roll loads by as much as 25%. A thick scale reduced the loads more than a thin scale since the thick scale breaks up into islands that transmit the load from the rolls to the strip. The islands become separated as the strip is elongated. Hot metal then extrudes between the islands and sticks to the rolls while the sliding islands move further apart and promote tensions applied to the sticking portion, thereby reducing the load. It was also found that thin scale promotes sliding friction with smooth rolls, but sticking friction with rough rolls. The load functions increased with temperature in rolling with rough rolls, but decreased with temperature for smooth rolls.

Roberts (1983) used the data of El-Kalay and Sparling (1968) to empirically model the coefficient of friction in terms of scale thickness, roll roughness and temperature. The model predicts an increase in the coefficient of friction with increasing roll surface roughness and decrease in scale thickness or increased temperature.

Li and Sellars (1996) found that sticking friction takes place in hot forging of scaled low carbon steel, but a certain degree of forward slipping, indicating partly or completely sliding friction, occurs in the rolling of the same material. Comments, similar to Blazevic's (1996), were made on the break-up of the scale. They found a limited number of cracks on specimens with thin scale. A scale layer can follow a similar reduction and elongation as the steel only if its hot strength is equal to or lower than that of the hot steel.

Schunke et al. (1988) presented a hypothesis on the effect of partial oxygen pressure on friction coefficients at room temperature, although additional information for temperatures below 600°C was presented for various Fe alloys. While analysing data obtained by other researchers, they found that the coefficient of friction during sliding was dependent on the partial pressure of oxygen as well as the sliding length. Generally, the coefficient of friction decreased with increased oxygen pressure and temperature, as these cause an oxide layer to grow more rapidly on the surface. The drop in friction was explained as follows: the oxide particles are fragmented when deformed and become further oxidized and compacted onto the metal surfaces where they form islands in the next cycle. When these islands grow in area, a large portion of the shearing is at these islands, causing the total contact area to be reduced. Friction is then lowered because of the brittle nature of the oxide particles that are being sheared. Shaw et al. (1995) determined fracture energies of oxide-metal and oxide-silicide interfaces. It was concluded that the fracture energy depends primarily on interfacial bond strength, although roughness of the interface, microstructure of the compounds, and porosity also have some effect.

An up-to-date exposition of the role of the layer of oxide in the hot rolling process was given by Krzyzanowski and Beynon (2002).

The effect of the layer of scale was investigated using an AISI 1018 steel, containing 0.18% C and 0.71% Mn as well as an HSLA steel, containing 0.067% C and 0.0764% Nb. The samples 12.65 mm thickness and 50.8 mm width were heated under closely controlled conditions; hence, the scale growth was also well controlled.

The scale index, $\overline{\gamma}$, is defined in terms of the ratio of the growth rate of the scale and the time

$$\overline{\gamma} = k_{\mathrm{p}}\sqrt{t} \tag{5.68}$$

where the growth rate is given in terms of the activation energy for scale formation, Q_{scale}, the universal gas constant and the absolute temperature:

$$k_{\mathrm{p}} = k_{\mathrm{e}} \exp\left(\frac{-Q_{\mathrm{scale}}}{RT}\right) \tag{5.69}$$

where k_{e} is a constant.

A finite-element code, Elroll, whose predictions have shown to yield results in excellent agreement with both laboratory experiments and industrial data is used to determine the coefficient of friction, in an inverse manner. The output parameters are the temperature profiles in the sample and the work roll, roll pressure, roll separating force and torque, as well as the forward slip, defined as the relative difference in roll surface/strip exit velocity. Two parameters – the coefficients of heat transfer and friction – may be chosen at will. The heat transfer coefficient, to be discussed in Section 5.6, has been determined in previous experiments and a value of 10–30 kW/m^{2}K has been established, depending on pressure, contact time and scale thickness. The coefficient of friction is then chosen such that the calculated and measured values of the roll force, the roll torque and the forward slip agree as closely as possible. Following Wankhede and Samarasekera (1997) and Chen et al. (1993), the heat transfer coefficient, α is modelled as solely pressure dependent. Their predictions may over-estimate the coefficient of heat transfer under laboratory conditions, especially in rolling of highly scaled steel. The coefficient of heat transfer is therefore described empirically as:

$$\alpha = \frac{p}{3} - 40 \tag{5.70}$$

where p is the average roll pressure in MPa, taken as the ratio of the roll separating force and the projected contact area. This results in a coefficient of heat transfer that ranges between 10 and 30 kW/m^2, values that apply strictly to the laboratory mill.

5.5.5.2 The effect of the scale thickness on friction

The effect of scale thickness on the frictional conditions can be seen in Figure 5.32, in which the coefficient of friction is plotted against the temperature for a range of scale thickness. The reduction and the velocity are kept constant at 25% and 170 mm/s, respectively. It is evident that the coefficient of friction has its highest value for the thinnest scale layer of 0.015 mm, ranging between 0.35 at 825°C and the lower value 0.30 at 1050°C. The scale thickness is then increased, first to 0.29 and then to 1.01, followed by 1.59 mm. This results in a reduction of the coefficient of friction at all temperatures, to values ranging between 0.22 and 0.30 for a scale thickness of 0.29 mm. A scale thickness of 1.01 mm results in a variation in the coefficient of friction from 0.19 to 0.24. The lowest values are seen for the thickest scale, yielding a coefficient of friction between 0.195 and 0.215. It is realized, of course, that the thickness of the scale under industrial conditions is much below these values.

The scale thickness appears to have a significant effect on the frictional conditions. In analysing the experimental data, the gain in thickness was taken into account along with changes in the heat transfer coefficient due to the insulating effect the scale provides.

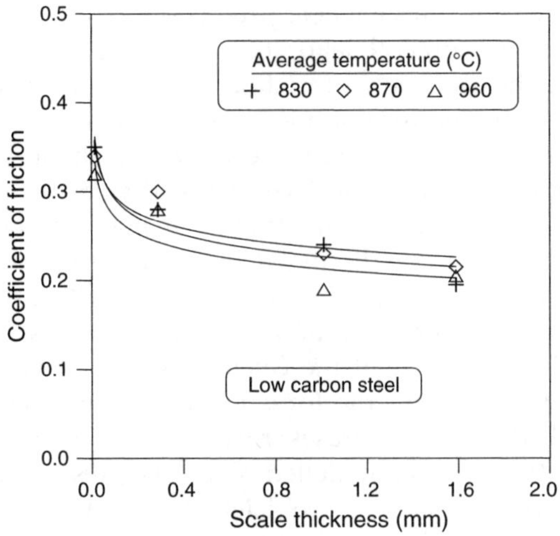

Figure 5.32 The dependence of the coefficient of friction on the temperature and thickness of the layer of scale (Munther and Lenard, 1999).

5.6 HEAT TRANSFER

The transfer of thermal energy at the contacting surfaces is affected by the same independent variables and parameters that affect frictional resistance. These have been discussed above in some detail and their interconnections were shown in Figure 5.1. In an approach similar to that followed when discussing the coefficient of friction or the friction factor, it is necessary to be realistic and limit attention to those parameters that may be measured, identified, determined or at the very least, may be assumed with some reasonable degree of confidence. In the present context, these are the areas in contact, the normal and shear forces on the contacting surfaces, their relative velocities and their bulk and surface temperatures. While the thickness of oxide layers and the use of lubricants must have a significant effect on the transfer of heat, in most instances only their presence or their absence are considered. Surface roughness, a significant parameter affecting friction, is not considered here, largely because its effect on the heat transfer has not been established in detail in the technical literature.[12] The relationship of the temperatures of the surfaces in contact is governed by the heat transfer coefficient, α, defined as the ratio of the heat flux – the amount of heat transferred per unit area and unit time – and

[12] At a recent conference in Vienna, Austria (Second World Tribology Congress, 2001) the present author heard the statement, made by a prominent researcher: "I am now working on the development of an asperity-based heat transfer model". Risking the display of some ignorance or inadequate literature search, I have not yet seen the publication of the study.

the difference of the temperatures of the hot and the cold surfaces. The usual manner of mathematical representation is to give the heat flux, $\dot{q}$, in terms of α and the average temperatures of the hot and the cold surfaces:

$$\dot{q} = \alpha \left(T_{\text{hot}} - T_{\text{cold}}\right) \tag{5.71}$$

where α is the heat transfer coefficient. The deficiencies of the formulation are apparent immediately. It is usually assumed in eq. 5.71 that the surface temperatures are uniform across the contacting surfaces, which, of course, is not correct. The heat transfer coefficient is also taken as a constant and in all likelihood, it varies with time as well as location and surface parameters. Nevertheless, eq. 5.71 works in the sense of "for all practical purposes" and is used almost universally. The thermal boundary conditions at the interface are usually formulated in terms of the heat transfer coefficient.

Relatively little work has been reported regarding procedures that lead to an estimate of the interface heat transfer coefficient in bulk forming processes. There are essentially two approaches by which the heat transfer coefficient may be estimated. One of these is the inverse technique, used to estimate the coefficient of friction, in which one chooses α such that calculated and measured temperature distributions – or, at least, the average surface temperatures – will agree closely. The other is to use the experimentally established time–temperature profiles to estimate the temperatures of the two contacting surfaces and use the definition of the heat transfer coefficient as the ratio of the heat flux and the temperature difference of the surfaces. Naturally, both of the methods have limitations. In the former, success depends on the quality, accuracy and rigour of both the measurements that are to match the predictions of a model and those of the model itself. The latter is also dependent on the measurements in addition to the technique of determining the surface temperatures and hence, their difference.

A detailed discussion of previous studies on the heat transfer coefficient was presented by Lenard et al. (1999). While the information given is important, only a summary will be reproduced here, in the form of Table 5.1.

Chen et al. (1993) present a relationship of the heat transfer coefficient and the interfacial pressure in the form:

$$\alpha = 0.695p - 34.4 \tag{5.72}$$

where p is the pressure in MPa and α is the coefficient of heat transfer in kW/m^2C.

Karagiozis (1986) and Pietrzyk and Lenard (1988, 1991) hot rolled carbon steel slabs, instrumented with several embedded thermocouples. The values of the coefficient at the interface are given in the last column of Table 5.2.

A comparison of the predictions of Chen et al. (1993) and the numbers in Table 5.2 indicates some difficulties. The average roll pressure in the experiment using the 19 mm thick strip, reduced by 21% at 4 rpm is estimated to

Table 5.1 Heat transfer coefficients

Reference	Material	α (kW/Km^2)	Comments
Chen et al. (1993)	aluminum	10–54	α varies along the arc of contact
Stevens et al. (1971)	steel	38.7	watercooled rolls
Silvonen et al. (1987)	steel	70	Obtained on a production mill
Bryant and Chiu (1982)	steel	7000	Obtained on a production mill
Harding (1976)	steel	2.055 at 700°C and 5.1 at 1100°C	Research mill

Table 5.2 Heat transfer coefficients, obtained on a laboratory mill (Karagiozis, 1986; Pietrzyk et al., 1994)

% red.	h_{entry} mm	rpm	α (kW/m^2 K)
7	15	4	12.78
7	15	10	15.35
6	15.4	3	15.27
10	15.5	4	10.85
21	19	4	12.78
21	19	10	13.99
19	19	4	9.8
20	19	4	12.31
11	30.9	4	13.06
20	38	4	15.93
19	38	4	11.79
20	38	10	20.76
18	20.1	12	13.74
24	18.3	12	9.6

be 162 MPa. Equation 5.72 predicts a heat transfer coefficient of 78 kW/m^{2}K while the calculations, based on the experimental data, give 12.78 kW/m^{2}K. It is apparent that eq. 5.72 doesn't include all of the significant variables to be useful in a general case; the relative speed of the work roll or the time of contact, the temperature and the thickness of the layer of scale should also be accounted for.

5.6.1 Estimating the heat transfer coefficient on a laboratory rolling mill

It is possible to estimate the coefficient of heat transfer with reasonable accuracy, as long as temperature data concerning the rolled strip are available.

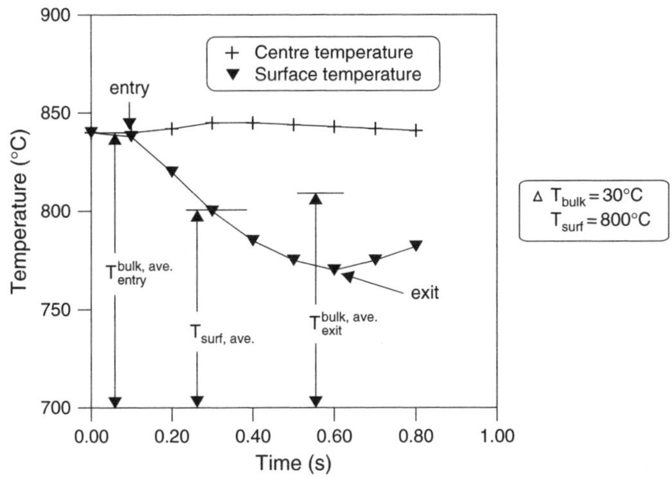

Figure 5.33 The temperature–time profile during hot rolling of a low carbon steel slab.

Such data are shown in Figure 5.33. The dependence of the temperature at the centre of the strip and near its surface as a function of the elapsed time is shown in the figure. A fairly thick strip of 38 mm entry thickness was reduced by 20% at a roll speed of 52 mm/s (4 rpm) in a laboratory rolling mill. Two thermocouples were embedded in the tail end of the strip to a depth of 25 mm. One of the thermocouples was located at the centre of the strip while the other was 2.5 mm from the surface.[13]

The suggested approach to estimate the coefficient of heat transfer is as follows. The heat flux has been defined above as the work done (W) on the rolled strip per unit surface area (A) and time (Δt). A simple model may then be written, giving $\dot{q}$ as:

$$\dot{q} = \frac{W}{A\Delta t} \tag{5.73}$$

which, realizing that $\Delta T_{\text{bulk}} = W/(V\rho c)$, may be re-written in terms of the temperature drop of the strip (ΔT_{bulk}) as:

$$\dot{q} = \frac{\rho c_{\text{p}} \Delta T_{\text{bulk}} h_{\text{ave}}}{2\Delta t} \tag{5.74}$$

where ρ is the density in kg/m^3, c_{p} is the specific heat in J/K kg, and $h_{\text{ave}} = (h_{\text{entry}} + h_{\text{exit}})/2$ is the average thickness of the strip in the pass. The volume

[13] Attempts to place a thermocouple closer to the surface were not successful as the stress concentration caused by the hole and the hard sheathing of the thermocouple caused cracking.

of the material in the deformation zone is $V = Lwh_{ave}$ and the contact area is $A = 2Lw$ where w is the width of the rolled strip. All of the information, necessary to calculate the heat flux, is available from Figure 5.33. The density of steel may be taken as 7850 kg/m^3; the specific heat is 625 J/K kg. The average thickness is 34.2 mm and the temperature loss of the strip is estimated as 30 K. The time elapsed from the entry to exit is read off the figure as 0.5 s. The heat flux is then obtained as 5.034×10^6 J/m^2 s. Now using eq. 5.71, the original definition of the heat transfer coefficient, the average surface temperature of the strip (800°C) and the average roll surface temperature as 100°C, the heat transfer coefficient is obtained as 7191 W/m^2 K.

The limitations of the approach just described need to be clearly understood. There are essentially three difficulties. The first is the use of average temperatures, based on the data obtained from only two thermocouples. The other is the estimate of the time elapsed from entry to exit, based on the reaction of the thermocouple near the surface. The third is the use of the data from the thermocouple, located 2.5 mm from the surface, as the temperature of the surface, ignoring the changes to the surface. Nevertheless, the magnitude calculated is quite realistic.

5.6.2 Measuring the surface temperature of the roll

Tiley and Lenard (2003) conducted hot rolling tests, using low carbon steel strips of 5.08 mm thickness. The rolling mill was instrumented with two optical pyrometers which allowed the monitoring of the entry and the exit surface temperatures of the strip. Force and torque transducers measured the roll separating forces and the roll torques. A shaft encoder measured the rolling speed. Eight thermocouples were embedded in the work roll, such that their tips were 0.5 mm from the roll surface, abutting 0.5 mm copper washers. The data collected were used to determine the interface heat transfer coefficient under three separate surface conditions:

1. The strips were rolled with the scales on and no lubricants were used;
2. The strips were descaled before the pass and no lubricant was used; and
3. The strips were not descaled and a mineral seal oil was used.

The results are shown in Figure 5.34, plotting the experimentally determined interfacial heat transfer coefficient against the interfacial pressure for the three conditions listed above. All three surface conditions cause the heat transfer coefficient to increase with the increasing pressure, albeit at very different rates. The magnitudes of the coefficients are of the same order of magnitude as those shown in Tables 1 and 2, but they are considerably lower than the predictions of eq. 5.72.

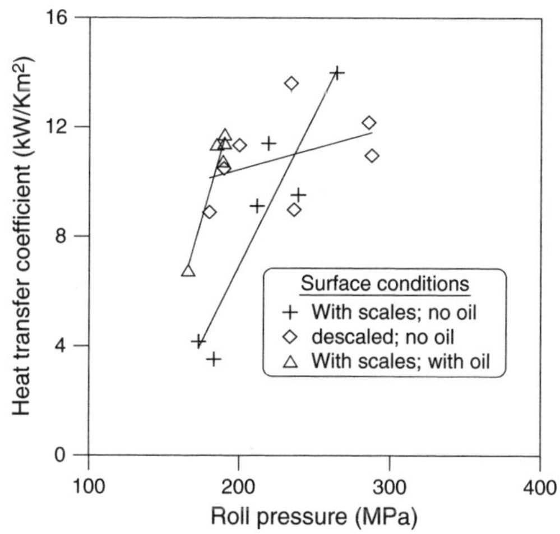

Figure 5.34 The heat transfer coefficient as a function of the interfacial pressure, on a laboratory size rolling mill (Tiley and Lenard, 2003).

5.6.3 Hot rolling in industry – the heat transfer coefficient on production mills

Even though direct measurements of the heat transfer coefficient under industrial conditions are rare, there is a consensus among the researchers and users that it appears to be significantly larger than values obtained in the laboratory. The difficulties in conducting trials using a full scale strip mill are probably impossible to overcome and this necessitates the use of inverse calculations.

Calculations were performed using data obtained from several hot strip mills. In the first instance, the heat transfer coefficient that best matched the temperature of the surface of the transfer bar before entry to the finishing train and after exit from the last stand was 50000 W/m^{2}K. In the second instance, when strip surface temperatures at the entry to each stand were available, the heat transfer coefficient varied from a low of 75000 W/m^{2}K at the first stand to 88000 W/m^{2}K at the last.

It is emphasized here that these numbers depend, in a very significant manner, on the data available from mill logs. Traditionally, these include the surface temperature of the strip after the rougher and before coiling but they don't provide stand-to-stand temperature data.

5.7 ROLL WEAR

Czichos' (1993) estimates that nearly 20% of the energy generated in the industrialized world is consumed by friction and that the losses form a significant portion of the gross national product. While he estimates 1–2% of the GNP is lost because of friction and wear, recall that Rabinowicz (1982) gave the much higher figure of 6%.

The recent review "Tribology in Materials Processing" by Batchelor and Stachowiak (1995) underlines these concerns, and suggests that the costs begin when the ore is extracted from the ground. They define wear and friction, and hence tribology, as chaotic processes in which predictions are not possible. They state categorically that an analytical approach to wear is impossible. There appears to be agreement with this view in the technical literature. For example, a few years earlier, Barber (1991), considering a tribological system, wrote that accurate pointwise simulation of such a system is inconceivable at the present time. The author continued to describe a caricature of a tribological research paper, commenting on the complexities of the physical system and the need for assumptions in the mathematical model. The opinions of Barber should be taken very seriously. He is absolutely correct in writing that predictions of models, without adequate experimental data supporting those predictions, are of little value. One possible addition to that sentence may be to request experimenters to use statistical methods to compare their measurements to the predictions of models. In this way, the accuracy and the consistency of the models' assumptions may be determined with confidence.

In spite of these comments, the relation given by Roberts (1983) to estimate the change in the radius of a work roll is found to be useful. The ratio of the change in the roll radius, ΔR, and the rolled length, λ, is given by:

$$\frac{\Delta R}{\lambda} = \frac{\overline{K}\mu L^2 r \sigma_{fm} \exp\left[\frac{\mu L}{h_{entry}(2-r)}\right]}{D^2 \sigma_{roll}} \tag{5.75}$$

where $\overline{K}$ is the wear constant, L is the contact length, r is the reduction in decimals and σ_{fm} and σ_{roll} are the flow strength of the strip and the yield strength of the roll, respectively. While the wear constant is not easy to determine exactly (Roberts, 1983, gives some further data on K), the formula gives realistic numbers for the loss of roll radius. Letting $K = 8 \times 10^{-5}$, the coefficient of friction equal to 0.4, the roll diameter equal to 400 mm, considering 40% reduction of an initially 10 mm thick strip whose flow strength is 250 MPa, that of the roll to be 600 MPa, the loss of the roll radius after 100 strips of 1000 m length each is estimated to be 5.4 mm, a reasonable number.

Batchelor and Stachowiak (1995) also discuss the mechanisms of friction and wear. Mechanisms of wear, shown in Figure 5.35, include abrasive, fatigue, erosive, cavitational and adhesive wear. Abrasive wear is caused by the ploughing action between the contacting asperities. Erosive wear is the result of impact of solid or liquid particles. Repeated contact causes fatigue wear and liquid droplet erosion causes cavitational wear.

Fitzpatrick and Lenard (2001) define the three phenomena that occur between two contacting surfaces that control the wear process, regardless of what mechanisms are causing the wear. These are:

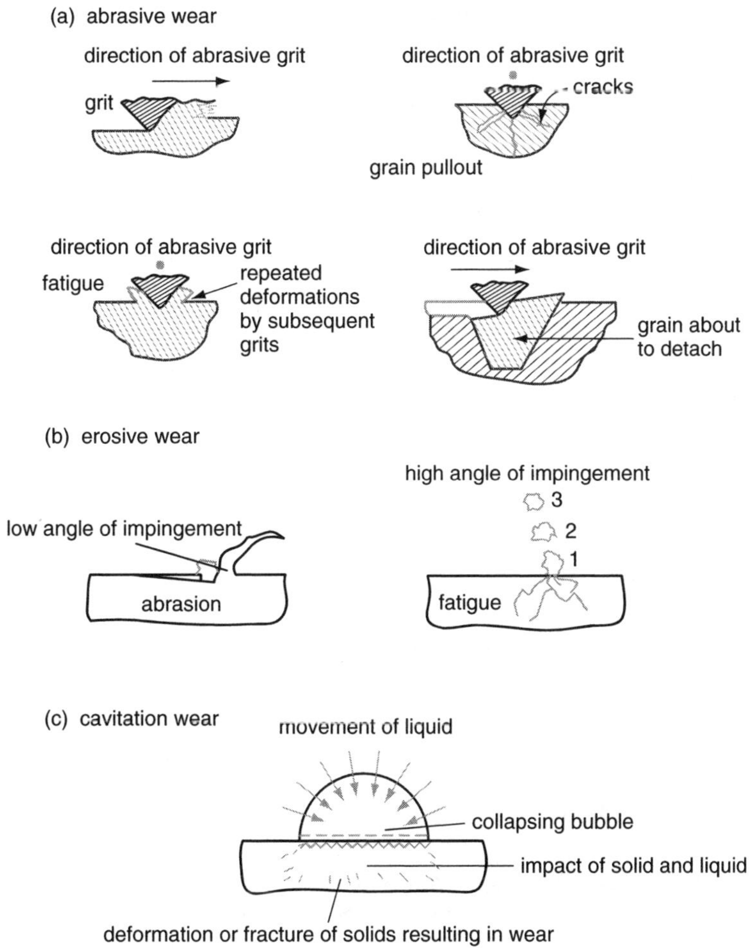

Figure 5.35 Mechanisms of wear (Batchelor and Stachowiak, 1995).

1. the chemical and physical interactions of the surface with lubricants and other constituents of the environment;
2. the transmission of forces at the interface through asperities and loose wear particles; and
3. the response of a given pair of solid materials to the forces at the surface.

These phenomena are not independent and changes have a dramatic effect on wear and wear rates. Adhesive wear results when the contacting surfaces form bonds between the asperities. Fatigue wear is caused by repeated application of the loads. Abrasive wear is observed when hard particles come in contact with the surface under load. Tribochemical reactions at the surfaces cause chemical wear. Hard, solid particles, causing impact, result in erosion. Similar to this is impact wear, occurring when the two surfaces come in contact under impact conditions. Finally, fretting wear is found when the contact surfaces experience oscillation with small displacements in the tangential direction. Czichos (1993) presented a well-thought out overview of wear mechanisms.

Previous studies show that roll wear rates are highest at temperatures of 850–950°C, precisely the temperatures used in the finishing stands of hot strip mills. Roll wear is also a function of the specific load, sliding length, and abrasive and corrosive particles in the cooling water. The low speeds of the roughing stands cause most of the wear and slippage as a result of too low friction also causes excessive roll wear. The parameters are many and the complex problem can be caused by either excessive or diminishing friction.

The fairly recent introduction of tool steel rolls gave a significant impetus to research on wear during both hot and cold rolling of steels. The papers presented at the 37th Mechanical Working and Steel Processing Conference agree that the change resulted in very significant drop of the rates of roll wear (Arnaud, 1995; Barzan, 1995; Hashimoto et al., 1995; Hill and Kerr, 1995 and Auzas et al., 1995). More recent works emphasized the improvements (Medovar et al., 2000; Gaspard et al., 2000 and Saltavets et al., 2001). Tool steels rolls, once implemented correctly, do provide benefits that offset their higher costs. The impact of lubricant interactions with these new roll chemistries have not been fully explored (Nelson, 2006, Private communication, R&D Department, Dofasco inc.).

5.8 CONCLUSIONS

The independent variables that affect surface interactions – friction, lubrication, heat transfer and roll wear – in the process of flat rolling have been identified and are classified below, according to the parameters of the process and the three components of the metal rolling system: the rolling mill, the rolled metal and their interface. In the present context, surface interactions refer to the transfer of mechanical and thermal energies at the contact and are characterized numerically by the coefficients of friction and heat transfer.

- **The rolling mill**
  1. The roll material and its diameter;
  2. Surface roughness and its direction;
  3. Surface hardness.
- **The rolled metal**
  1. The resistance to deformation;
  2. Surface roughness and its direction;
  3. Surface hardness.

- **The interface**
  1. Lubricant/emulsion viscosity;
  2. Flow rate;
  3. Pressure and temperature sensitivity;
  4. Density.
- **The process**
  1. The rolling speed;
  2. Reduction;
  3. Temperature.

It is recognized, of course, that the above list differs in a most significant manner, from the one given in Figure 5.1 as it is much reduced and much less comprehensive, driven by the need to retain only the most important parameters. The interactions of these parameters with the coefficients of friction are presented in Figures 5.36–5.37. The first versions of these figures were published by Lenard (2000). They are updated here, making use of recently accumulated experience.

The first step in constructing Figures 5.36 and 5.37 is the decision concerning the most important independent variables. Once the metal to be rolled, its chemical composition, dimensions and surface roughness are selected, the lubricant or the emulsion and their flow rates are prescribed, the rolling mill is chosen (bringing with it the roll dimensions and geometry, its surface hardness and roughness), the remaining decisions concern the reduction per pass and the speed of rolling. These two are then considered in Figures 5.36 and 5.37, respectively.

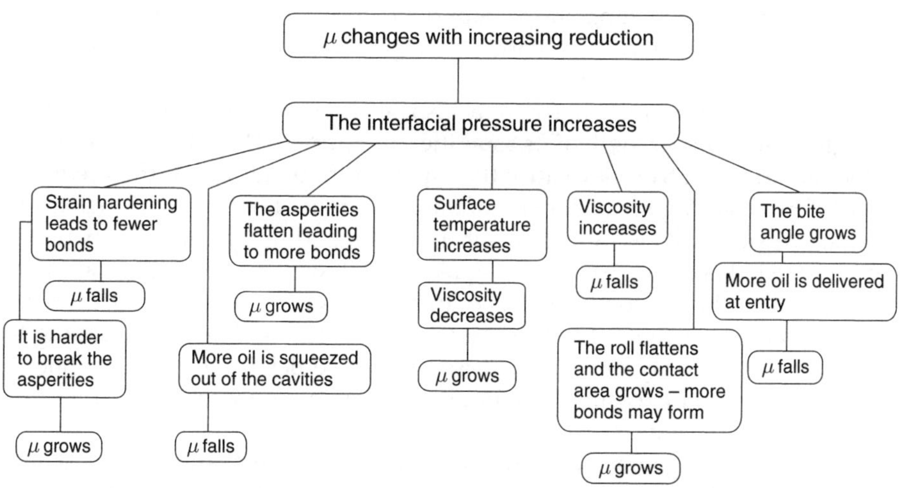

Figure 5.36 The effect of increasing reduction on the coefficient of friction.

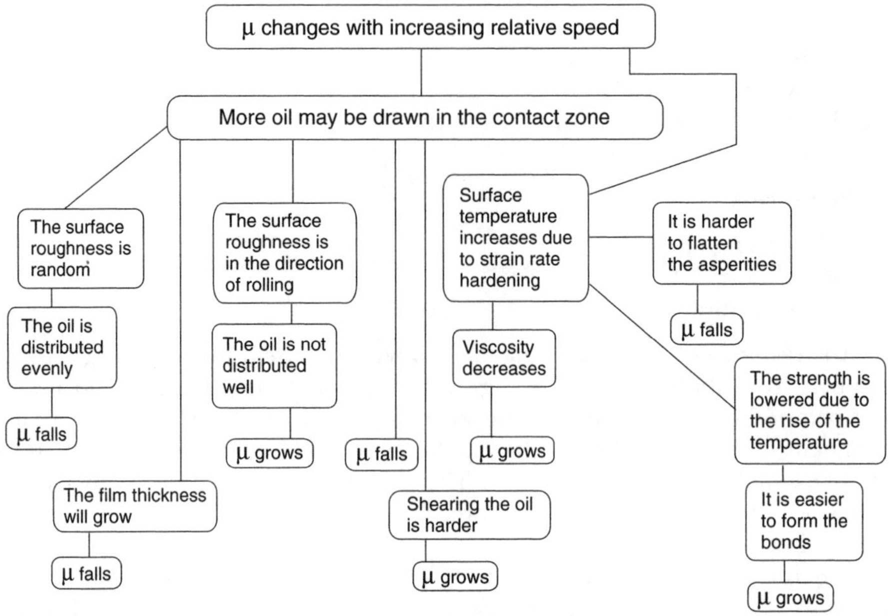

Figure 5.37 The effect of increasing relative velocity on the coefficient of friction.

It has been shown above that when the reduction is increased, the coefficient of friction drops in most cases. Increasing the reduction brings with it several changes, the first of which is the correspondingly increasing roll pressure, which in turn increases the stresses within the rolled strip. The metal may then experience strain hardening and, compared to a softer or less strain hardening material, it may be harder to flatten its asperities. Hence, keeping all other parameters identical, the harder metal will cause the coefficient of friction to drop. A contradictory mechanism may also be observed here: when the metal's strength grows, the strength of its asperities also grows. The relative movement of the roll and the strip may cause the asperities to break which then requires more effort, resulting in an increase of the coefficient of friction. The increasing normal pressure will also affect the lubricant trapped in the valleys in between the asperities. The oil is likely to be squeezed out, wetting the nearby surfaces, causing a drop in the coefficient. The lubricant's viscosity will be affected by two contradictory events. The lubricant's temperature will increase leading to a drop of the viscosity and an increase of the coefficient of friction while at the same time the increasing pressure will increase the viscosity, leading to a drop of the coefficient.

Changes of the geometry of the pass will also cause changes. The contact area will grow because the roll will flatten and more adhesive bonds may develop. As well, the bite angle will increase, delivering more oil to the contact zone. The mechanisms that cause a drop of the coefficient of friction appear to overwhelm the others in most instances.

Similar arguments may be made when the effects of the increasing relative velocity on the coefficient of friction are examined, shown in Figure 5.37. In preparing the figure it was assumed that no starvation is present.

It is generally agreed that when the relative motion between the roll and the rolled metal increases, more oil is available to be drawn in to the contact zone. The nature of the roll's surface roughness will determine if the lubricant is spread evenly on the contacting surfaces. If yes, lower coefficient of friction will result. Further, more oil will result in thicker lubricant films and lower coefficients. A limit may be reached here when the contact zone is saturated and the coefficient will not fall any more. The increasing speed may also contribute to increasing coefficient of friction, as long as the hydrodynamic condition is reached. The temperature may rise due to increasing strain rate hardening; the increasing speeds will increase the shear stress needed to shear the lubricant; the metal may experience strain rate hardening, and this in turn may make it harder to flatten the asperities. There is a contradictory phenomenon here as well: the increasing temperature will lower the lubricant's viscosity and the coefficient of friction will fall. The increasing temperature may also cause some softening of the rolled metal; however, this is not expected to be a very significant contributor. The time rates at which these changes occur are not known at this stage.

The heat transfer coefficient is also affected by the reduction and the relative velocity between the roll and the rolled strip. Increasing the reduction appears to have caused the heat transfer coefficient to increase, implying either a higher heat flux or a lower temperature difference. The true area of contact as well as the time of contact would increase with the reduction, lowering the coefficient. The temperature difference would also decrease and that is expected to overwhelm the other effects, resulting in an increase of the heat transfer coefficient.

Increasing relative velocities would decrease the time of contact and the temperature difference would increase as there would be less time for the heat to be transferred. The coefficient would therefore rise.

It is possible to make a few recommendations regarding the coefficients of friction and heat transfer, to be used in predictive modelling of the flat rolling process. While the numbers given below cannot replace values arrived at by independent experimentation, they nevertheless should aid in improving the quality of predictions when used in the mathematical models of the flat rolling process.

5.8.1 Heat transfer coefficient

When hot rolling steel in the laboratory, using relatively small rolling mills, values of 4–20 kW/m^2 K appear to be the correct magnitudes. If modelling hot rolling of steel under industrial conditions, values of 50–120 kW/m^2 K are more useful. In both cases, the layer of scale is an important parameter. Scale is an insulator, so its presence slows the cooling of the surface of the rolled metals. Since it is difficult to determine the heat transfer coefficient in experiments, use of the inverse method is recommended. The heat transfer coefficient during cold rolling of steel varies from a low of about 20 kW/m^2 K up to 40 kW/m^2 K. Somewhat higher magnitudes, by about 15%, are appropriate during cold rolling of aluminum strips.

5.8.2 The coefficient of friction

5.8.2.1 Cold rolling

When analysing the flat rolling process of cold rolling steel strips without lubrication, the magnitude of the coefficient of friction is likely to be in the range of 0.15–0.4. If efficient lubricants or emulsions are used, the coefficient drops to 0.05–0.15; use the lower values when the thickness of the oil film is high and the higher values when the roll's surface roughness is high. While rolling soft aluminum strips, the coefficient appears to be approximately 20% higher; when harder alloys are rolled, the coefficient is about 10% higher than that for steels. In general terms, increasing viscosity, speed and reduction cause a drop of the coefficient.

5.8.2.2 Hot rolling

While the suggestion by early researchers, indicating that sticking friction exists during the hot rolling process, has often been shown to be incorrect, the coefficient of friction is found to be significantly higher in the process. The constant existence of a layer of scale is one of most important contributors that affect the magnitude of the coefficient. The adhesion of the oxide layer to the work roll may also have an effect on the tribological conditions at the contact.

Values of the coefficient range from about 0.2–0.45 in lubricated hot rolling of steels. In contradiction to the experience found during the cold rolling process, increasing reductions appear to cause larger coefficients, due to the softer scale on the steel's surface. Higher velocities and thicker layers of scale cause a reduction of the coefficient. While hot rolling aluminum, values about 10% higher are recommended, caused by the accumulating oxide coating on the surface of the work rolls.

5.8.3 Roll wear

Roll wear was discussed briefly. Roberts' formula, predicting the loss of roll material, was shown to yield realistic numbers. The mechanisms of wear were discussed. There appears to be a general agreement that the use of tool steel rolls reduces the rate of wear of the rolls.

5.8.4 What is still missing

While several attempts at obtaining functional relationships for various surface interactions as functions of some of the independent variables have been reviewed above, a general equation of the form:

$$\text{Surface Interaction} = f(\text{load, speed, temperature, strength, roughness, viscosity} \ldots)$$

has not yet been presented. Its availability would ease modelling in a most significant manner.

CHAPTER 6

Applications and Sensitivity Studies

Abstract

Concepts, ideas and expressions concerning the flat rolling process were developed, derived or simply stated in the previous five chapters. Many of these equations were discussed and their predictions were analysed in their respective chapters. In what follows, a selection of them is analysed further by examining their sensitivity to some of the independent parameters in addition to their applicability in some specific situations. The first section deals with the sensitivity of the predicted magnitudes of the roll separating forces and the roll torques by the mathematical models considered in Chapter 3, to various independent parameters such as the coefficient of friction, the reduction, the strain-hardening coefficient and the entry thickness. This is followed by a comparison of the measured powers required for reducing a strip, calculated following the recommendations of Avitzur and Roberts. The roll pressure distributions, as obtained by the friction hill approach and by letting the friction factor vary along the contact, are compared. Several relations dealing with the metallurgical phenomena during rolling are examined next: these are the statically recrystallized grain size and the critical strain.

Chapter 4 contains a listing of several constitutive relations, for both hot and cold deformation of steels of various chemical compositions. Two of these are chosen for a comparison of their predictions: the hyperbolic sine stress–strain rate equation and Shida's equations, both developed to model hot deformation of low carbon steels.

6.1 THE SENSITIVITY OF THE PREDICTIONS OF THE FLAT ROLLING MODELS

The importance of the ability to predict the rolling variables prior to designing the draft schedule has been emphasized before. These predictions, which are usually performed using off-line models of the process, are dependent on several known and not-well known parameters.

6.1.1 The sensitivity of the roll separating force and the roll torque to the coefficient of friction and the reduction

It has been argued above that one of the less-well understood but arguably one of the most important parameters of the flat rolling process is the coefficient of friction. In this section, the dependence of the roll separating force and the roll torque on the coefficient is examined. In the calculations, the predictions of two models – that of the empirical model and the refined 1D model – are compared at various reductions as functions of the coefficient of friction; recall that both models are capable of reliable predictions. Cold rolling of low carbon steel strips is considered. The true stress – true strain relation of the steel is $\sigma = 150\ (1+234\varepsilon)^{0.251}$MPa; the entry thickness is 1 mm and the roll radius is 125 mm. Neither model accounts for the speed of rolling; not a significant problem since the metal's resistance to deformation is independent of the rate of strain. In a situation where the strain rate affects the metals' strength, a constitutive relation reflecting that dependence needs to be used. The rate-dependence of the roll separating force will then be demonstrated through the stress–strain relation.

The dependence of the roll separating force on the reduction and the coefficient of friction is demonstrated in Figure 6.1. The coefficient of friction is plotted on the abscissa and the force on the ordinate.

As expected, the forces increase as the reduction and the coefficient of friction increase. The unexpected observation concerns the extremely steep rise of the roll separating force, as predicted by the 1D model; which, it is to be recalled, uses the friction hill idea to obtain the roll pressure distribution,

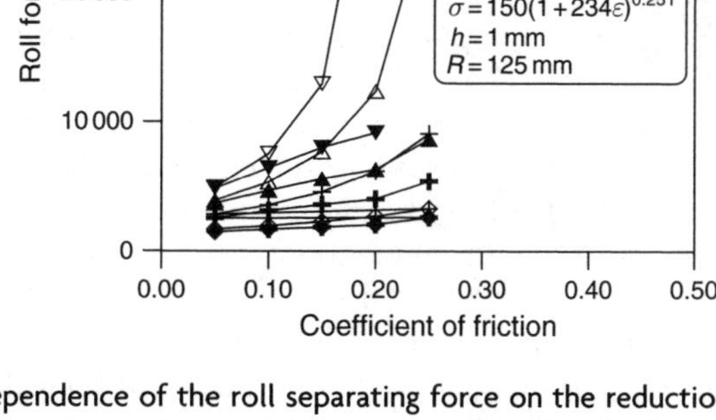

Figure 6.1 The dependence of the roll separating force on the reduction and the coefficient of friction; the predictions of the empirical model of Schey and that of the refined 1D model are shown.

the integration of which over the contact yielding the force. At a reduction of 30–40% and a coefficient of friction of 0.2 and 0.25, reasonable process parameters if rolling with no lubricants and a fairly rough roll surface are considered, the 1D model's predictions are approximately three times that of the empirical model. This rise, which is deemed quite unreasonable, is directly attributable to the use of the friction hill.

Similar deductions may be made by examining Figure 6.2, showing the dependence of the roll torque (for both rolls) on the same two parameters as in Figure 6.1. The rise of the torque, as indicated by the 1D model, is still too steep.

Based on the above, it is unclear at this stage which of these two models is to be recommended for off-line use, without some more calibration. Recalling Figure 3.6, however, in which the predictive capabilities of three models were compared, the 1D model is expected to be the most reliable. As well, it is appropriate to recall what should be expected of a model's predictive capabilities, mentioned already in Chapter 3. The two requirements are the accuracy and the consistency of the predictions. Of these two, consistency, shown by a low standard deviation of the difference between the measurements and the calculations, is the more important since the data may always be adjusted to be accurate. A model whose accuracy is good only sometimes is essentially useless. These concepts were demonstrated by Murthy and Lenard (1982), by comparing the mean and the standard deviation of the differences of the predictions and the experimental data of several models of flat rolling. In general, increased rigour resulted in decreasing standard deviations.

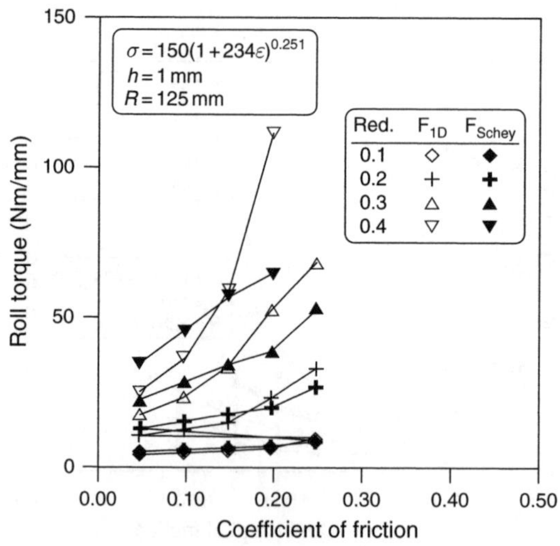

Figure 6.2 The dependence of the roll torque on the reduction and the coefficient of friction; the predictions of the empirical model of Schey and that of the refined 1D model are shown.

6.1.2 The sensitivity of the roll separating force and the roll torque to the strain-hardening co-efficient

Figures 6.3 and 6.4 show the dependence of the forces and the torques on the strain-hardening coefficient and the reduction, respectively. It is understood that while changing the strain-hardening coefficient leaves the yield strength unchanged, it raises the ultimate strength and therefore it raises the average flow strength of the rolled metal in the pass. This is clearly demonstrated in Figures 6.3 and 6.4. As expected, both the forces and the torques rise with increasing hardening.

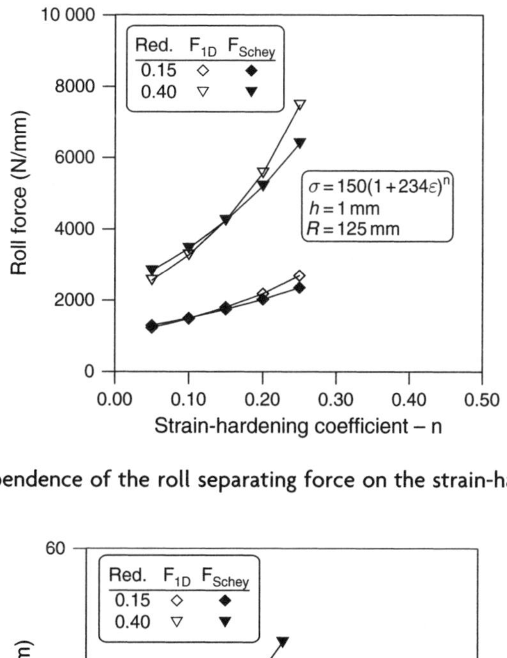

Figure 6.3 The dependence of the roll separating force on the strain-hardening coefficient.

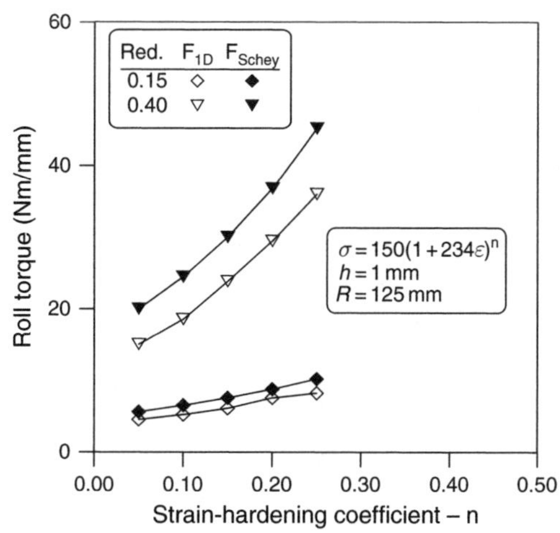

Figure 6.4 The dependence of the roll torque on the strain-hardening coefficient.

6.1.3 The dependence of the roll separating force and the roll torque on the entry thickness

The changing thickness in subsequent rolling passes affects the roll separating forces and the torques. These effects are illustrated in Figures 6.5 and 6.6, showing the roll forces and the roll torques, respectively. As above, the calculations

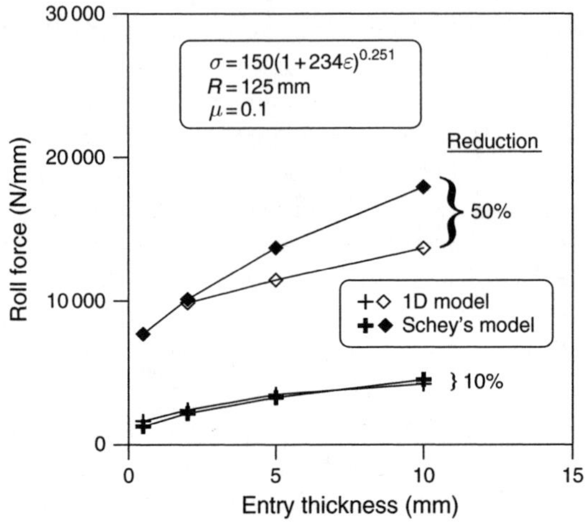

Figure 6.5 The dependence of the roll separating force on the entry thickness at high and low reductions.

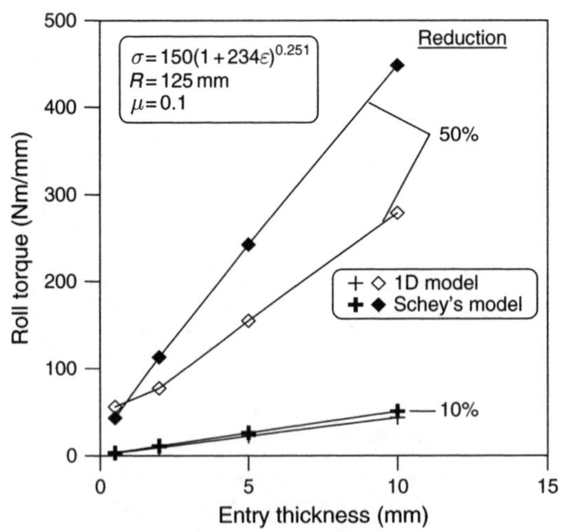

Figure 6.6 The dependence of the roll torque on the entry thickness at high and low reductions.

by the 1D model and the empirical model are contrasted. The same low carbon steel and rolling mill are used and the coefficient of friction is taken to be 0.1.

Figure 6.5 shows how the roll separating forces are affected by the increasing thickness at entry, at low and at high reductions.

6.2 A COMPARISON OF THE PREDICTIONS OF POWER, REQUIRED FOR PLASTIC DEFORMATION OF THE STRIP

Three independent models were presented in Chapter 3, each developed to estimate the power required to drive the rolling mill. This power included two parts: the power to produce permanent deformation of the rolled strip and the power to overcome friction losses in the drive system. The power needed to reduce the strip, as calculated by each of the three models is compared to the measurements in Figure 6.7, for two nominal reductions of 15 and 50%[1]. The data used have been developed by McConnell and Lenard (2000) and have been mentioned above. Briefly, low carbon steel strips were cold rolled in rolls of 250 mm diameter, using various lubricants in the process.

In the figure, the experimental data on the power was obtained by using the measured torque and the rolling speed. The coefficients of friction are needed in both the 1D model and the one presented by Roberts (1978); these

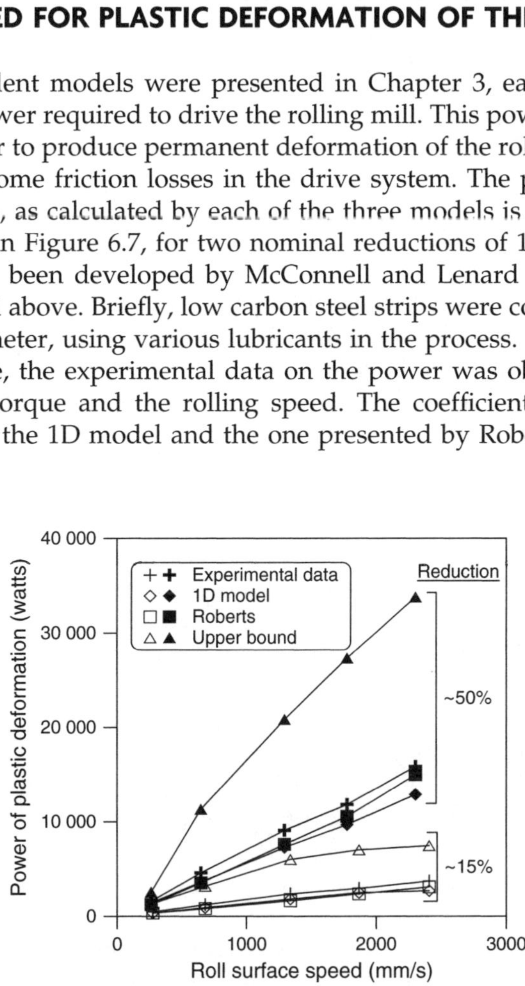

Figure 6.7 A comparison of the powers needed to cause permanent plastic deformation, measured and predicted by various approaches.

[1] Since friction losses in the bearings would be the same, they are not considered in Figure 6.7.

were obtained by inverse calculations, using the 1D model. The friction factor, needed in the calculations, was obtained using eq. 3.61.

Both the 1D and the Roberts models are remarkably close to the experimental data in their predictions. As expected, the upper bound prediction is conservative, yielding numbers much higher than the others.

6.3 THE ROLL PRESSURE DISTRIBUTION

While rarely used in simple models, the variation of the coefficient of friction or the friction factor, from the entry to the exit in the roll gap, is well acknowledged. The predicted roll pressure distributions using a constant coefficient of friction and a variable friction factor are compared in Figure 6.8 . The process parameters used to obtain experimentally the roll pressure distribution, by Lu et al. (2002) are employed in the calculations. These measurements were referred to above (see Section 3.7.2.2); briefly, the tests involved measuring the interfacial normal and shear stresses using pins and transducers embedded in the work roll. The shapes of the distribution curves are quite different. The saddle point, resulting from the friction hill and the use of the 1D model with $\mu = 0.34$, is not realistic while the rounded top, resulting from the smooth variation of the friction factor, is expected to be close to reality[2].

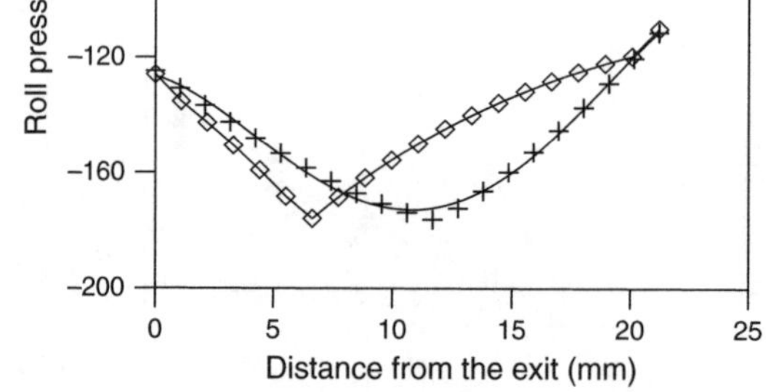

Figure 6.8 A comparison of the roll pressure distributions, as calculated by the 1D model using a constant coefficient of friction of 0.34 and by the model, using $m = m(\phi)$.

[2]See also Figure 3.11 where the measurements of Lu et al. (2002) are compared to the predictions of the model using $m = m(\phi)$.

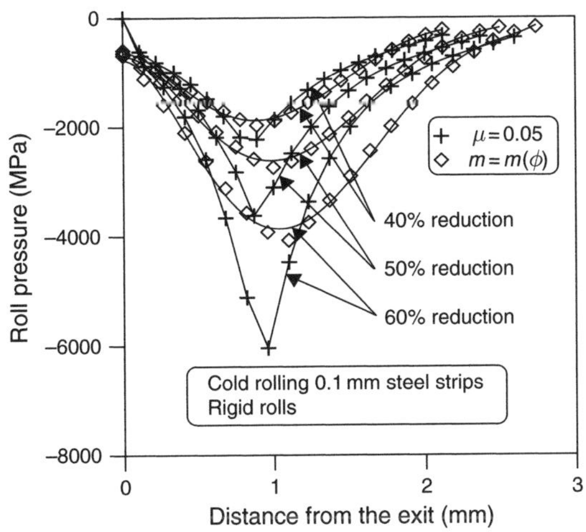

Figure 6.9 The pressure distribution as a function of the reduction. The distribution using the friction hill method becomes unrealistic at high reductions. The model, using the variable friction factor, predicts distributions which appear to be closer to expected experimental data.

The most significant shortcoming of the models which use the friction hill approach is their inability to yield realistic predictions when modelling passes in which large reductions of thin, hard strips are considered and the roll diameter to strip thickness ratios are much larger than unity. The reasons for this are clearly demonstrated in Figure 6.9. Cold reduction of 0.1 mm low carbon steel strips is considered with work rolls of 250 mm diameter and rolling to progressively larger reductions is modelled. In both models, rigid rolls are used. In the 1D model, the coefficient of friction is taken to be 0.05, a realistic number when a light lubricant is employed in the pass. At the lower reduction of 40%, both models lead to comparable distributions of the roll pressure. As the reduction is increased, the failure of the friction hill model becomes evident. The top of the saddle point rises to unrealistic magnitudes.

6.4 THE STATICALLY RECRYSTALLIZED GRAIN SIZE

Several empirical equations, relating the statically recrystallized grain size in a rolling pass, were presented in Chapter 3[3]. Their predictions are compared in Figures 6.10 and 6.11, as functions of the strain (Figure 6.10) and as functions of the temperature (Figure 6.11). In the computations, the models of Sellars (1990), Roberts (1983), Laasraoui and Jonas (1991a,b), Choquet et al. (1990) and Hodgson and Gibbs (1992) are used. In Figure 6.10, the initial austenite grain

[3] See equations 3.75–3.76.

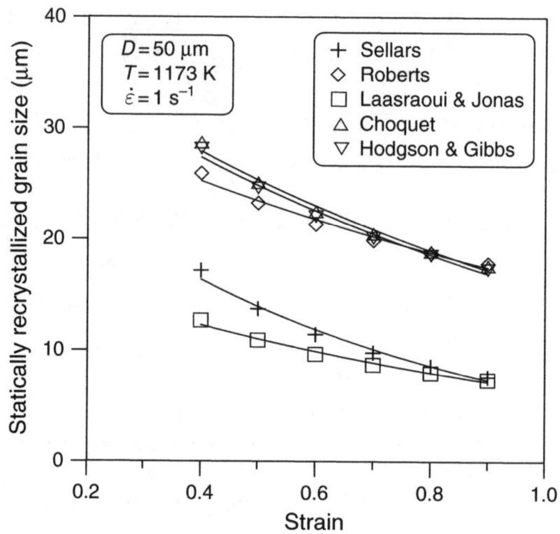

Figure 6.10 The statically recrystallized grain sizes as a function of the strain; predicted by Sellars (1990), Roberts (1983), Laasraoui and Jonas (1991a and 1991b), Choquet et al. (1990) and Hodgson and Gibbs (1992).

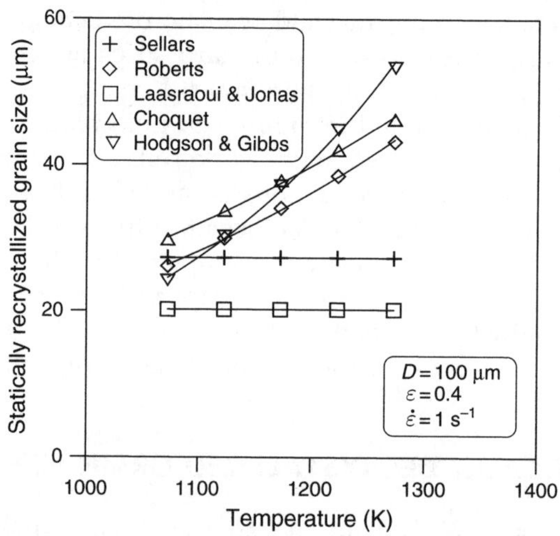

Figure 6.11 The statically recrystallized grain sizes as a function of the temperature; predicted by Sellars (1990), Roberts (1983), Laasraoui and Jonas (1991a,b), Choquet et al. (1990) and Hodgson and Gibbs (1992).

diameter is assumed to be 50 μm, the temperature is 1173 K and the rate of strain in the pass is 1 s^{-1}.

As expected, the predictions indicate that increasing strains cause the grain sizes to drop. The predictions are bunched into two groups, one giving numbers approximately twice the other. The predictions of Sellars (1990) and Laasraoui and Jonas (1991a,b) are close to one-another and are much smaller than those due to the other three researchers.

The dependence of the statically recrystallized grain sizes on the temperature of deformation is illustrated in Figure 6.11. The relations from the same studies as above, are used here; the strain is taken as 0.4 and the initial austenite grain size is assumed to be 100 μm. The expectations are that the grain diameters would grow with the temperatures and the predictions of Roberts (1983), Choquet et al. (1990) and Hodgson and Gibbs (1992) indicate this dependence. The two other equations, due to Sellars (1990) and Laasraoui and Jonas (1991a,b) do not include the temperature as an independent variable.

6.5 THE CRITICAL STRAIN

The shapes of high temperature stress–strain curves were discussed in Chapter 4, and the strain corresponding to the plateau of the curve was identified as the peak strain, corresponding to the peak stress. At that strain, the rate of hardening equals the rate of softening and just before that, the process of dynamic recrystallization has begun. The strain at which this restoration phenomenon becomes active has been identified as the critical strain. Equation 3.80 may be used to predict the magnitude of the critical strain, as a function of the Zener-Hollomon parameter (which is defined in terms of the strain rate, the temperature and the activation energy) and the austenite grain size.

The dependence of the critical strain on the temperature, the rate of strain and the austenite grain size is demonstrated in Figure 6.12. Two strain rates

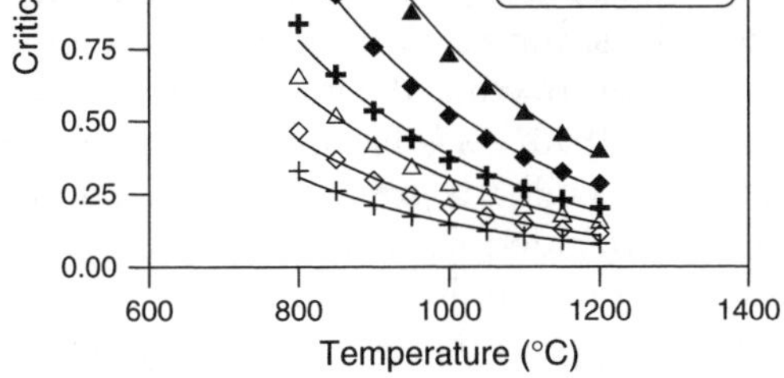

Figure 6.12 The critical strain, required for the initiation of dynamic recrystallization.

(0.1 s^{-1} and 50 s^{-1}), three grain sizes (25, 50 and 100 μm) and the constants and exponents of Sellars (1990) are used in the calculations. The process of dynamic recrystallization will begin when sufficient amount of mechanical and thermal energy has been given to the deforming material. This is indicated clearly in the figure. Increasing temperatures, decreasing strain rates and grain sizes reduce the magnitude of the critical strain.

6.6 THE HOT STRENGTH OF STEELS – SHIDA'S EQUATIONS

The shape of the stress–strain curves at high temperature was mentioned in Chapter 4, Section 4.7.2 and the metallurgical events occurring as the deformation is proceeding were discussed. The peak stress, the strain corresponding to the peak stress, the competing rates of the hardening and the restoration mechanisms and the attendant microstructures were described. In the Conclusions of Chapter 4, the recommendation was made: if no independent testing program to determine the material's resistance to deformation at high temperatures is possible, use Shida's equations in modelling the hot, flat rolling process. In what follows, Shida's equations will be examined. In the first instance, the shape of the stress–strain curve will be compared to the expected configuration. This will be followed by the estimated rise of the temperature as a hot compression process is proceeding. Finally, the ability of the relations to model the stresses in the two-phase, ferrite–austenite, region will be considered.

6.6.1 The shape of the stress–strain curve, as predicted by Shida

Figure 6.13 shows the shapes of the curves, at various rates of strain. These should be examined in comparison to Figures 4.13 and 4.14, which show high temperature stress–strain curves for a Nb-V steel in Figure 4.13 and in Figure 4.14 a schematic diagram of the expected curve is indicated.

The initial shapes of the curves are as expected, as the stress rises with strain and the slopes begin to drop slowly, indicating the competing rates of hardening and dynamic recovery. The peak stresses are also reached, but, not at the expected strains. The strains, corresponding to the peak stresses are expected to increase as the strain rates increase and the times available for the metallurgical phenomena drop; while the magnitudes of the strains are close to those predicted by Sellars (1990) – see Figure 6.12 – the growth is not reflected in Shida's curves. Further, while the grain size is not an independent variable in Shida's relations, it is included in Sellars' equation, so to some extent, the comparison is again one of the apples and oranges type.

The rise of the temperature of the sample during the compression process has been mentioned already and the need to correct for it was emphasized. This temperature rise is plotted against the strain in Figure 6.14, for the steel

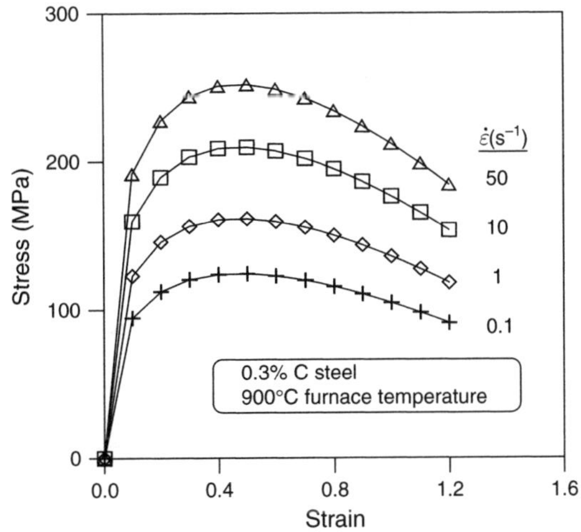

Figure 6.13 The shape of Shida's stress–strain curve.

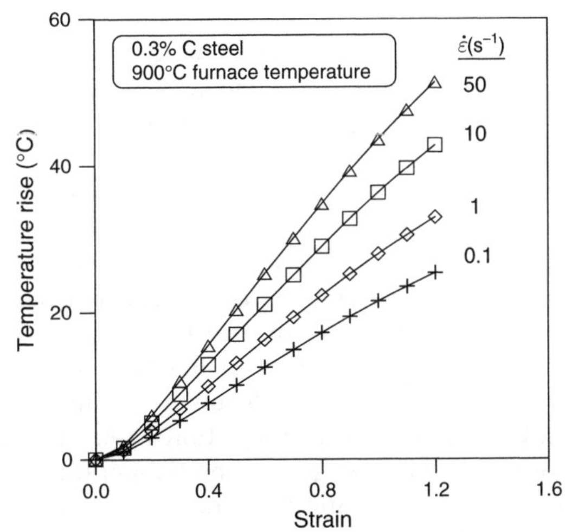

Figure 6.14 The rise of temperature during the compression process.

dealt with above. The initial temperature, rates of strain and the carbon content are also the same.

The importance of conducting isothermal tests, or to correct for the temperature rise during the application of the loads is emphasized while examining the details of Figure 6.14. At a strain rate of $50\,s^{-1}$ and a strain of 1.2, the rise is calculated to be approximately 45°C. The steel's flow strength at 900°C, at that

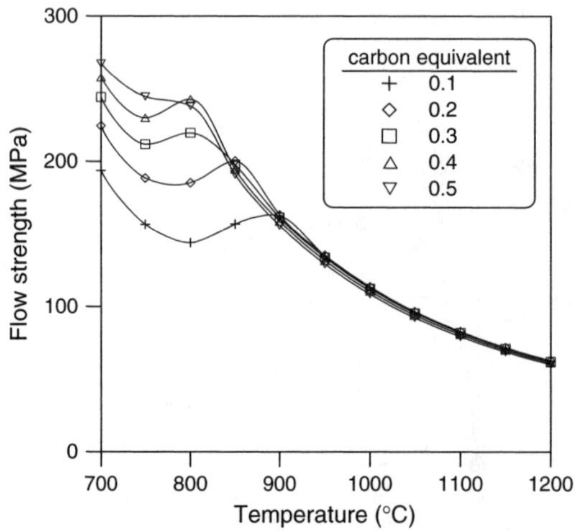

Figure 6.15 The flow strength, calculated by Shida's formula, for various carbon contents, showing the steel's behaviour in the two-phase region (Lenard et al., 1999, reproduced with permission).

strain is 184 MPa. At the higher temperature of 945°C the strength is estimated to be 159 MPa, a nearly 14% difference. Using the wrong magnitude for the flow strength in a set-up model of the hot strip mill would lead to incorrect settings.

Shida's equations correctly predict the steel's behaviour in the two-phase, austenite–ferrite, region as well. When the temperature is decreased, the deformation resistance of the steel is expected to increase. When the temperature indicating the appearance of the first ferrite grains is reached, the Ar_3 temperature, the strength is expected to fall with further temperature drop, since the strength of the ferrite is lower than that of the austenite. This phenomenon continues while all of the austenite transforms to ferrite and beyond the temperature indicating the end of the transformation, at Ar_1, the strength increases again. The dependence of the strength on the temperature is indicated properly in Figure 6.15. As observed, the influence of the carbon is present only at lower temperatures, in the two-phase region. The strength of the austenite appears to be independent of the carbon content.

6.7 CONCLUSIONS

The sensitivity of the predictive abilities of the mathematical models, presented in previous Chapters, was examined.

CHAPTER 7

Temper Rolling

Abstract

The temper rolling process is discussed in this chapter. The objectives of the process are listed – the most important of which is the suppression of the yield point extension – and their effects on the rolled metal are described. Several mathematical models, predicting the dependent variables, are critically reviewed. These include the empirical models, analyses based on extreme roll flattening as well as the finite-element studies and the application of artificial intelligence. The predictive abilities of the mathematical models are presented. The need to include an explicit and accurate account of roll deformation in the models, as pointed out in the studies, is emphasized. A schematic diagram of a temper mill is shown.

7.1 THE TEMPER ROLLING PROCESS

Temper rolling is a particular form of flat rolling. Its primary purpose is to suppress the yield point extension which, if present, would create Lüder's lines, a form of surface defect, shown in Figure 7.1. The presence of this defect in subsequent sheet metal operations – such as deep drawing, stretch forming and their combinations – would have very deleterious effects on the resulting products.

The temper rolling process subjects the flat product to a very low reduction of thickness, typically 0.5–5%. Other possible reasons for a temper pass include:

- Production of the required metallurgical properties;
- Production of the required surface finish;
- Production of the required flatness;
- Creation of magnetic properties; and
- Correction of surface flaws and shape defects.

The difficulties with temper rolling include the creation of non-uniform residual stresses in addition to the possibility of pre-existing non-uniform flatness of the starting product. Both of these may cause further processing difficulties.

Essentially, mathematical modelling of the temper rolling process needs to account for the same phenomena that were included in the traditional models

Figure 7.1 Lüder's lines (Wiklund and Sandberg, 2002).

of flat rolling. Pawelski (2000) presents a list of the differences, which, however, require additional attention. These are:

- The nearly equal elastic and plastic regions of the deforming strip, caused by the very small reductions per pass;
- The pronounced flattening of the work roll, which, if neglected or not accounted for carefully, would introduce large errors in the predictions of the roll separating forces; and
- The order of magnitude of the thickness reduction is comparable to that of the surface roughness.

7.2 THE MECHANISM OF PLASTIC YIELDING

Specialized textbooks dealing with plastic deformation of metals discuss the transformation of the recoverable elastic behaviour to that of permanent plastic equilibrium and flow. The two most often used yield and flow criteria, the maximum shear stress and the maximum distortion energy theorems, (developed by Tresca and Huber–Mises, respectively) form the mathematical bases of this change of the response of the metal.[1] Very briefly, the metal will enter the plastic deformation mode when and where the elastic stresses first satisfy either one of the criteria.[2] The elastic and the plastic regions will be separated by the elastic–plastic boundary, the location and the shape of which become extra unknowns to be determined.

[1] Please refer to the last section of Chapter 1 where further reading material on plastic deformation is listed.

[2] Neither criterion is a law of nature. When one of them is assumed to govern the metal's behaviour, one may write of a "Tresca material" or a "Huber–Mises material".

The mechanism of plastic yielding in a cold rolling pass has been discussed extensively by Johnson and Bentall (1969). They consider the rolling of thin strips and the longitudinal stresses acting on them. These include the tensile stresses as a result of the compression by the rolls and the elongation caused by them in addition to the compressive stresses imparted by the interfacial shear stresses, which always act towards the centre of the roll gap. When progressively thinner strips are rolled, the compressive longitudinal and normal stresses would subject them to hydrostatic compression in the deformation zone. This then implies that attempting further reduction of the metal would instead flatten the rolls more and would not result in any change of the strip's thickness; a limiting reduction is reached.[3] The just described phenomenon is, of course, critically dependent on the frictional conditions at the roll/strip contact.

The authors identify a neutral region – note that several researchers identify a neutral point, not a region – near the centre of the arc of contact where there is no relative motion between the roll and the strip.

In Johnson and Bentall's analysis (1969), two aspects are not considered: that of the bending of the work rolls and the possibility that the rolls may touch outside the width of the strip.[4]

In their treatment, the authors show that there are two regions of slip, one near the entry and another near the exit. Elsewhere, there is no relative motion between the rolls and the strip. Further analysis indicates that yielding will occur on a plane perpendicular to the direction of rolling and will be restricted to the entry and exit regions. They predict the presence of a state of plastic equilibrium in the neutral region with no change in thickness.

7.3 THE EFFECTS OF TEMPER ROLLING

7.3.1 Yield strength variation

Cold working the steel results in increased strength and decreased ductility. Roberts (1988) presents data on the effect of the elongation during a temper pass on the yield strength, showing that the yield strength decreases for reductions of less than ½% but when the reduction is increased beyond that, the metal's strength grows very significantly.

Fang et al. (2002) studied the effect of temper rolling on several mechanical attributes of two C–Mn steels. They found that the lower yield strength increases with the equivalent strain, according to the relations:

$$\sigma_y = -5.8\,\varepsilon_{eq} + 328.9 \tag{7.1}$$

[3] The limiting rollable thickness was discussed in Section 2.4.1.

[4] Johnson and Bentall (1969) acknowledge that the two phenomena just mentioned may affect the results of a more advanced analysis. Most other researchers simply ignore these two possibilities.

for the steel with 0.135% C and

$$\sigma_y = -9.4\,\varepsilon_{eq} + 309.2 \tag{7.2}$$

for the steel containing 0.019% C. In both equations above, the stress is in MPa and the strains are in %. Fang et al. (2002) also found some drop of the yield strength at low strains, as did Roberts (1988). The tensile strength of both steels increased monotonically with the strain. The uniform elongation, the yield point elongation and the strain hardening exponent dropped for both steels as the equivalent strain increased.

If the temper pass is performed above the ambient temperature, the strip hardness increases with the temperature for the same elongation. Considering tinplate rolling, increasing the strip's entry temperature from 21 to 149 °C, 1% elongation causes an increase in hardness of two points on the Rockwell scale. For 2.5% elongation, the same temperature rise will cause an increase of four points.

However, increasing the temperature of the strip results in a decrease of the ductility. The same temperature rise will cause a drop in the ductility of about 1–7%.

7.4 MATHEMATICAL MODELS OF THE TEMPER ROLLING PROCESS

7.4.1 The Fleck and Johnson models

Fleck and Johnson (1987) analyse cold rolling of thin foils and take careful account of the deformation of the work roll and the frictional conditions in the zone of contact. They also make use of the "planes remain planes" assumption and consider rolling an isotropic, elastic-perfectly plastic thin strip. The elastic and contained plastic deformation of the strip in a direction parallel to a line connecting the roll centres are ignored. Their analysis leads to similar results to that of Johnson and Bentall (1969). They also find that there is a neutral region in the contact zone where contained plastic flow occurs with no change in thickness. Plastic deformation is predicted to occur only at the entry and exit regions. The authors include some caution in the application of their model in the Conclusions and it is illuminating to quote their comments exactly:

> *It is thought that the new model provides a physical picture of the foil rolling process which is qualitatively correct. We express caution with regard to the quantitative results, as the location of plastic deformation in the roll bite and the rolling loads and torques are sensitive to the model chosen for deformation of the rolls. A more realistic treatment of the rolls is required in order to determine the accuracy of the present results.*

Fleck et al. (1992) refined the work of Fleck and Johnson (1987) by introducing an advanced treatment of roll flattening in the analysis, treating the rolls as elastic half-spaces. They retained the previous assumptions of having an elastic perfectly-plastic strip, homogeneous compression in the roll bite and a constant coefficient of friction. The general conclusions haven't changed in that plastic deformation still occurred near the entry and exit regions, separated by a neutral region. They determined the coefficient of friction by inverse analysis, equating the predicted and measured roll forces. When this value was used to calculate the roll torque, a 23% difference resulted between it and the measured torque. The authors attributed this difference to the use of a constant coefficient of friction. Another quotation from their Conclusions appears to be appropriate here:

> *There seems to be no point in refining the rolling model until more is known about the nature of friction in the roll bite.*

Roberts (1988) quotes Hundy (1955) who wrote that in the temper rolling process homogeneous compression of the rolled strips defined as uniform deformation across the thickness of the work piece – doesn't occur, opposing the "planes remain planes" assumption of Fleck and Johnson. Only some portions of the metal deform sufficiently to enter the plastic range.

It is known that the traditional mathematical models of the flat rolling process fail when applied at reductions experienced by the temper rolled strip. The causes for the inability to provide reasonable predictions of the roll separating forces include:

- the small reductions;
- the unexpected high values of the coefficient of friction;
- the need to include roll flattening in the model; and
- the lack of published experimental data to which a model's predictions may be compared.

Fleck and Johnson (1987) further conclude that the conventional models analysing the cold rolling process fail when the thickness of the rolled strip is less than 100 μm.

7.4.2 Roberts' model

An approximate model has been given by Roberts (1988) and is reproduced below. The model gives the magnitude of the roll separating force (P_r) in terms of the minimum pressure required to deform the strip (σ_p, MPa or lb/in^2), the thickness of the strip at the entry (h_{entry}, mm or in), the reduction (r), the coefficient of friction (μ) and the length of the arc of contact (L):

$$P_r = \sigma_p h_{entry}(1-r)\,\frac{\exp\left[\dfrac{\mu L}{h_{entry}(1-r)} - 1\right]}{\mu} \qquad (7.3)$$

with the length of the arc of contact given by:

$$L = 0.5\left[{Dr\mu}/{2} + \sqrt{\left({Dr\mu}/{2}\right)^2 + 2Dh_{\text{entry}}r} \right] \tag{7.4}$$

where D is the roll diameter. The minimum pressure required to deform the strip is prescribed in terms of the metal's yield strength (σ) in the form:

$$\sigma_p = 1.155\left[\sigma + a\log_{10}(1000\dot{\varepsilon})\right] - 0.5\left(\sigma_{\text{entry}} + \sigma_{\text{exit}}\right) \tag{7.5}$$

where the effect of the stresses at the entry and exit are taken into account. As recommended by Roberts, the material constant a is to be taken as 7500 lb/in^2 or 52 MPa.

Assuming that the stresses at the entry and exit are equal, the torque to cause the deformation of the strip is given by:

$$M = 0.5Dh_{\text{entry}}r\left(\sigma_{\text{fm}} - \sigma_a\right)\left(1 + {\mu L}/{h_{\text{entry}}}\right) \tag{7.6}$$

where σ_{fm} is the average, constrained dynamic yield strength of the metal and σ_a is the average tensile stress in the strip. Roberts (1978) shows that the predictions of eq. 7.3 compare quite well to measurements, taken when a low carbon steel strip was temper rolled.

Temperature rise in a temper rolled strip or sheet may be estimated by assuming that all of the work done on the rolled metal is converted into heat. As before, the rise of the temperature is then obtained from:

$$\Delta T = \text{power}/(\text{mass flow} \times \text{specific heat}) \tag{7.7}$$

The power is obtained from the roll torque, for both rolls, in terms of the roll velocity

$$P = Mv_r/R \tag{7.8}$$

where the roll surface velocity, v_r, is given by $v_r = 2\pi R\,\text{rpm}/60$, the mass flow is obtained by assuming mass conservation

$$\text{Mass flow} = \text{entry velocity} \times \text{entry thickness} \times \text{width} \times \text{density} \tag{7.9}$$

and the specific heat is given for steels as 500–650 J/kg K.

The temperature rise of the rolled strip, as calculated by the above formula, is an average value. It is important to realize that the temperature is not uniform across the strip. The strains experienced by the strip are the highest near the contact surface (Lenard, 2003) and thus indicates that the rise of temperature there, due to plastic work, is the largest. Counteracting this rise of the temperature near the surface is the cooling effect of the work roll and of the lubricant, if wet temper rolling is performed.

7.4.3 The model of Fuchshumer and Schlacher (2000)

Fuchshumer and Schlacher (2000) considered temper rolling as the last possibility to exert an influence on the strip by rolling. They developed a mathematical model for an industrial temper mill, the schematic diagram of which is given in Figure 7.2.

The authors also acknowledge that the conventional models don't apply here. They list the important parameters of the process: the forward and backward tensions, entry and exit thickness, material parameters, roll velocity, slip conditions at the roll/strip interface, the roll force and mill dynamics. The deformation of the rolls is accounted for by Jortner et al's (1960) analysis. The rolling regimes characterized by a central region of contained plastic flow, considered by Fleck and Johnson (1987), are not included here. The coefficient of friction is taken to remain constant in the roll gap. The assumptions in the derivation of the model are:

- Plane strain flow is present;
- Planes remain planes; and
- The transition between elastic and plastic zones occurs abruptly.

The result is a multi-input, multi-output system which is then used to control the process. The authors include the results of a typical example, showing the shape of the contact arc and the corresponding roll pressure distribution. The roll shape appears to include an indented portion but the flat portion, predicted by the Fleck and Johnson model, is not observed. The roll pressure distribution indicates the traditional friction hill model, with a sharp point at the pressure peak.

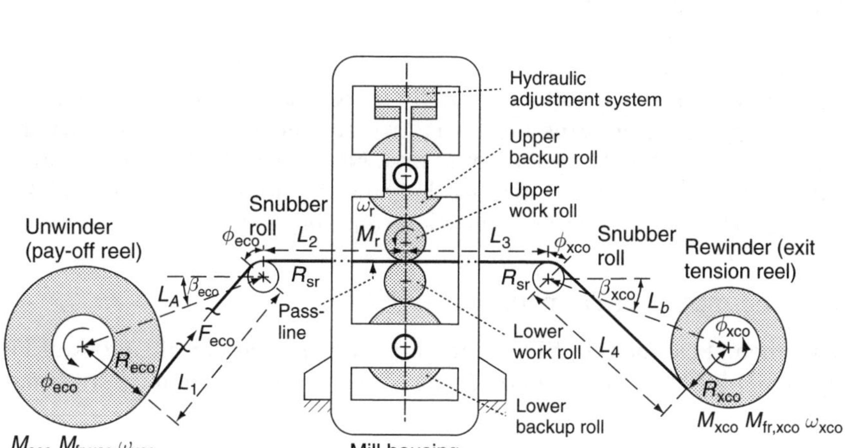

Figure 7.2 A temper rolling mill (Fuchshumer and Schlacher, 2000).

7.4.4 The Gratacos and Onno model (1994)

Gratacos and Onno (1994) agree with the conclusions of Fleck and Johnson that the classical models cannot predict the rolling parameters in the temper rolling process. They attribute the difficulties to convergence problems and/or unrealistic asymptotic behaviour, which occurs when the strip entry thickness to roll diameter ratio is much less than unity.

Gratacos et al. (1992) review the roll deformation models; these are – Hitchcock's formula, influence functions, FE modeling, Grimble's approach and a contact mechanics technique, and comment that when modelling the rolling of thin strips, foils and temper rolling, the manner of coupling the deformation of the work roll and that of the strip is the most important step. They apply their model to temper rolling and find their predictions of the roll separating force acceptable.

Gratacos and Onno (1994) employed two different 2D models to analyse flat rolling process. They considered both thick and thin strips and applied their model to temper rolling as well. One of the models was a full finite-element formulation and the other was a slab/finite-element approach. In both models, the elastic deformation of the work roll was calculated by the finite-element approach. The rolled metal's behaviour was described by the Prandtl–Reuss elastic–plastic relations and Tresca or Coulomb friction was employed.

The difficulties listed by the authors include the lack of precise evaluation of the sliding velocity between the non-circular deformed roll and the strip and the very high computational times.[5]

Considering 32% reduction of a 0.05 mm thick sheet, the central portion of the roll profile was found to be flat, as predicted by the Fleck and Johnson models. The roll pressure distribution was rounded on top. Both models yielded similar results. The temper rolling process was modelled next. The entry thickness was 0.4 mm, the reduction was 1.5% and the friction factor was assumed to be 0.7. Calculations of the equivalent strains in the deformation zone indicated that the deformation was not homogeneous and that the "slab hypothesis seems also insufficient locally under the flat part of the roll".

The shape of the deformed roll was strongly dependent on the entry thickness, which was taken to be 1.4 mm at first. No flat roll contour was found. As the thickness was decreased, flatness began to be noted at an entry thickness of 0.6 mm. The roll pressure curves were rounded at the top.

7.4.5 The model of Domanti et al. (1994)

The mathematical model of Domanti et al. (1994) builds on the work of Fleck et al. (1992) by keeping the general ideas and adding several improvements.

[5] The work of Gratacos and Onno was completed in 1994. It is very likely that computational times, using the high speed computers available in 2006, would greatly reduce these times.

The existence of a region within the roll gap where the roll contour is essentially flat and the strip is in a state of contained plastic flow is retained. The additional work includes a material model that accounts for the effects of the strain rate and the temperature on the rolled metal's resistance to deformation. A rare comparison of the predictions to experimental data is also presented. The roll force and the torque, as predicted by the model using a roll which remained circular under load, differed from the measurements in a most significant manner. When the non-circular roll profile was used, however, the predictions and the measurements were quite close.

7.4.6 The Chandra and Dixit model (2004)

A rigid-plastic finite-element model was used to study the temper rolling process. Roll deformation was analysed by assuming the roll to be an elastic half-space, as was also done by Fleck et al. (1992). The results indicated that the deformation in the roll gap is not homogeneous. The roll is found to flatten in the central zone where a rigid – actually elastic – region is found. Further, the authors report that under some conditions "the roll shape takes concave shape locally". Comparison to experimental data is also included in the study. However, one of the references quoted (Shida and Awazuhara, 1973) gave roll force and torque data for cold rolling, not temper rolling.

Since elastic behaviour appears to be a significant contributor to the roll force and torque in temper rolling, conclusions, using a rigid-plastic material may not indicate the correctness or otherwise of the predictions.

7.4.7 The models of Wiklund (1996a, 1996b, 1999, 2002)

Wiklund and Sandberg (2002) reviewed and summarized their studies of the temper rolling process in a state-of-the-art review. They described the application of several models and discussed their advantages and their abilities to predict the roll force and the contact length. They also considered the applicability of these models for online use and concentrated on two different models.

The first model is based on a finite-element approach, using an implicit Lagrangian code and four-node elements. Both the strip and the work roll are modelled. Both strain and strain rate are included in the material model.

While mostly rounded roll pressure distributions were obtained, in one instance, a double peak was observed, similar to the pressure peak at entry in the Fleck and Johnson (1987) and Fleck et al., (1992) models. When the strip thickness was decreased, the roll profile demonstrated the flat portion, again as postulated by Fleck and Johnson (1987) and Fleck et al., (1992). In these computations, surprisingly low coefficients of friction were used, in the range of 0.1–0.2. Homogeneous compression of the rolled strips are shown in Figures 15.9a and 15.9b of Wiklund and Sandberg (2002).

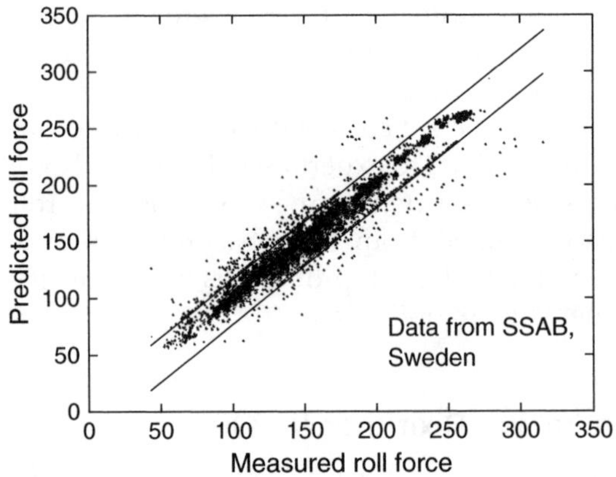

Figure 7.3 A comparison of the calculated and the measured roll separating forces (Wiklund and Sandberg, 2002).

The next model combined the FE method and neural networks, and a rare comparison of measurements and predictions was also included. The approach was shown to be very successful; see Figure 7.3 where the calculated and the measured roll separating forces are compared.

An interesting concept is introduced by Wiklund and Sandberg (2002); this is the "flattening risk factor", α^*, defined in terms of the contact length and the strip thickness at entry:

$$\alpha^* = \frac{L}{h_{entry}} = \varepsilon\sqrt{\frac{R'}{\Delta h}} \tag{7.10}$$

where $L = \sqrt{R'\Delta h}$ is the contact length, R' is the roll radius as calculated by Hitchcock's formula and h_{entry} is the entry thickness. The authors predict that severe roll flattening is expected when the risk factor exceeds 10.

In conclusion, Wiklund and Sandberg (2002) write that good predictive abilities were obtained by the use of neural and hybrid modelling. The use of cylindrical roll deformation models was found to be valid when the steel strips are thicker than 0.4 mm. Non-circular roll deformation models were found to be necessary below that thickness, leading to flat contact regions in the roll gap.

7.4.8 The model of Liu and Lee (2001)

Reasonable predictive abilities were demonstrated by the model, in which the preliminary displacement principle of Kragelsky et al. (1982) was applied. The authors present an argument for the need for a physically based model of the temper rolling process, in light of the significant differences between it and the conventional cold rolling process. Further, the authors question the existence of the flat contact region in the roll gap, stating that:

> *. . . the force of normal temper rolling is usually not large enough to build a central flat region because the soft annealed material is rolled and the reduction is slight.*

According to the model, the strip in the roll gap has both plastic and elastic regions and the main contact part of the strip is elastic. Hence, the friction in the deformation region is mostly governed by the contact between two elastic bodies. The authors make use of the work of Kragelsky et al. (1982) who gave two expressions for the friction stress in the elastic contact region, depending on the preliminary displacement, δ, and the limiting preliminary displacement, $[\delta]$, given by the following relations:

$$\delta = \int_{\phi_n}^{\theta} \left[\frac{h_{np}}{h(\theta)} - 1 \right] a(\theta)\, d\theta \qquad (7.11)$$

where h_{np} is the strip thickness at the neutral point and

$$[\delta] = \frac{2-\nu}{2(1-\nu)} \mu R_{max} \varepsilon \qquad (7.12)$$

In eq. 7.12, ν is the surface roughness coefficient, μ is the coefficient of friction and R_{max} is the maximum height of the work roll asperities. The elastic deformation of the work roll is given by $a(\theta)$, evaluated using influence functions

$$a(\theta) = \int_0^{\phi} U(\theta - t) p(t)\, dt + R \qquad (7.13)$$

where $U(\theta - t)$ is the influence function (Grimble et al., 1978), $p(t)$ is the roll pressure and R is the undeformed roll radius. The symbol ε indicates the relative approach of the contacting bodies, to be taken as unity. With these definitions, the interfacial shear stress is given by:

$$\tau = \mu p \quad \text{for} \quad \delta \text{ larger than } [\delta] \quad \text{and} \quad \tau = \mu p \left\{ 1 - \left[1 - \frac{\delta}{[\delta]} \right]^{(2\nu+1)/2} \right\} \text{ otherwise.} \qquad (7.14)$$

The calculations yield the roll profile, the roll pressure and shear stress distributions, all of which are compared to Grimble's predictions. A rounded pressure distribution is found but no flat portions of the deformed roll are located.

7.4.9 The studies of Sutcliffe and Rayner (1998)

Experimental verification of the predictions of a neutral region in which no reduction occurs (Fleck and Johnson, 1987; Fleck et al., 1992) has been given by Sutcliffe and Rayner (1998). Plasticine strips were rolled in elastomer rolls with steel cores. The roll diameters were 24 and 50 mm. Chalk was used as the lubricant and the ring compression method was used to determine the coefficient of friction.[6] Low rolling speeds were used (0.008 m/s). The rolls were stopped and separated fast after a certain part of the plasticine was rolled and the partially rolled strip profile was carefully measured.

The authors conclude that when thin strips are rolled, a clearly noticeable reduction near the inlet is observed. As well, a central region exists which is relatively flat, confirming two of the Fleck and Johnson predictions. There are two more predictions of the theory, however, which are not confirmed: the measured profiles don't show plastic reduction at the exit and the measured roll loads are almost an order of magnitude lower than predicted.

7.4.10 The model of Pawelski (2000)

In order to read Pawelski's (2000) statement in context, that the magnitude of the thickness change in a temper pass is comparable to that of the surface roughness, the examination of some numbers is helpful. Consider the entry thickness to be 0.25 mm, as in Pawelski's Figure 9 and the reduction to be 5%. The corresponding thickness change is 0.0125 mm or 12.5 μm. The surface roughness of a steel strip, ready for temper rolling may be 1–2 μm and the combined roll/strip roughness may be somewhat larger. If, however, a reduction of 1% in the pass is considered, the two numbers are very much closer, making Pawelski's idea very interesting and appropriate.

In his model of temper rolling he accounts for the roll deformation, considering it to be an elastic, semi-infinite space. This deformation, $U(s)$, supported at a distance R below the surface and loaded by a unit line load, is written as:

$$U(s) = -\frac{1+\nu}{\pi E}\left\{1 + (1-\nu)\ln\left(\frac{s}{R}\right)^2\right\} \tag{7.15}$$

where E is the elastic modulus of the roll, ν is Poisson's ratio and s is the distance between the load and the deforming roll. The vertical displacement of the roll, due to a pressure, $p(x)$, can then be calculated by:

$$u(x) = \int_{-\infty}^{\infty} p(\xi)\, U(x-\xi)\, d\xi \tag{7.16}$$

Pawelski then connects the pressure with the fractional area of contact of the surface asperities, using a slip-line field approach. Further, he takes the characteristics of the roll pressure and strip thickness distribution as was done

[6] The ring compression test was discussed in Chapter 5, Tribology; see Section 5.3.1.3.

by Fleck and Johnson (1987). Rounded roll pressure distributions are obtained. The roll gap profiles are similar to those of Fleck and Johnson. Coefficient of friction values varying from a low of 0.09 to 0.25 are determined.

7.5 COMMENTS FROM INDUSTRY

The demand for low speed temper rolling mills will grow in the next few years as it is expected that more and more strip processing centres will consider the installation of a temper mill in their strip processing lines.

Limited by the line speed, the temper rolling speed will be very low compared to the conventional temper mills of the strip producers. Since the friction and strain rate effects on the yield stress at very low speeds are somehow different with the high speed temper rolling process, temper rolling models need to be further developed under these special conditions (Levick, 2006, Quad Engineering; Private Communication).

7.6 CONCLUSIONS

The temper rolling process was described and the differences between it and the conventional cold rolling process were presented. The objectives of the process were listed; these include the effect of the process on mechanical and metallurgical attributes. Several mathematical models were described, including empirical models, 1D models, finite-element models as well as the use of artificial intelligence in predicting the rolling variables. Only in very rare instances were the predictions of the models compared to experimental data.

It appears that the efforts to model the mechanics of the temper rolling process are not quite finished and in the opinion of the present writer, the assumptions and arbitrary decisions made by the analysts are to blame. These include:

- The "planes remain planes" assumption. This step assures that the stresses across the thickness will remain constants and that ordinary differential equations result when forces are balanced on a slab of the deforming material. Integration of these equations results in the usual friction hill which has been shown not to represent actual conditions. Several researchers commented on the inapplicability of the "planes remain planes" assumption;
- The coefficient of friction is usually assumed to be constant along the arc of contact. Experimental evidence exists that this is not the case;
- The roll deformation models are realistic. However, they need to be coupled to the interfacial friction stresses, which, when non-constant frictional coefficients are used, may yield different flattened roll profiles than have been demonstrated so far.

CHAPTER 8

Severe Plastic Deformation – Accumulative Roll Bonding

Abstract

A brief discussion of processes that apply severe plastic deformation to a work piece in order to create small grains and thereby increase the strength is followed by a detailed description of one of these methods: that of accumulative roll bonding. The process is presented first, followed by a detailed discussion of a set of experiments. In that process ultra low carbon steel strips containing 0.002% C were rolled at 500°C. Strips of 32 layers were created. The mechanical attributes after rolling and cooling were examined and the development of edge cracking was monitored. The metal's yield and tensile strengths increased by 200–300% while the ductility dropped from a pre-rolled value of 75 to 4%. The rolling process was stopped when cracking of the edges became pronounced. The shear strength of the bond was about 60% of the yield strength in shear. The accumulation of the retained strain after dynamic recovery caused cracking at the edges. A potential industrial application of the accumulative roll bonding process, that of the creation of tailor rolled blanks, is discussed.

8.1 INTRODUCTION

The interest in bulk nanostructured materials, processed by methods of severe plastic deformation, is justified by the unique physical and mechanical properties of the resulting products. The advantage of these over other processes is concerned with overcoming the difficulties connected with residual porosity in compacted samples, impurities from ball milling, processing of large-scale billets, and the practical application of the resulting materials.

Methods of severe plastic deformation create ultra fine-grained structures with prevailing high-angle grain boundaries. They should also be able to create uniform nanostructures within the whole volume of a sample to provide stable properties of the processed materials and they should not suffer mechanical damage when exposed to large plastic deformations.

Most of the information in this chapter was provided by Dr Krallics of the Budapest University of Technology and Economics. The experimental portion is based on Krallics and Lenard (2004).

8.2 MANUFACTURING METHODS OF SEVERE PLASTIC DEFORMATION (SPD)

There are several different SPD methods: high-pressure torsion (HPT), equal channel angular pressing (ECAP), accumulative roll bonding (ARB), multiple forging (MF) and repetitive corrugations and straightening (RCS). These methods can be divided into four main groups: methods based on torsion and compression, methods using the extrusion process, methods based on rolling, and methods based on forging. In what follows, a review of these methods and possible industrial applications are given.

8.2.1 High-pressure torsion

The Bridgman anvil type device (Bridgman, 1952), in which an ingot is held between two anvils and strained in torsion under the applied pressure of several GPa magnitude, was the first equipment utilizing this method. The approach has since been used by several researchers. The lower holder rotates and the surface friction forces deform the ingot by shear. During the process, the sample experiences quasi-hydrostatic compression. As a result, in spite of the very large strains, the deformed sample is not destroyed. The samples, processed by severe torsional straining, are usually of a disk shape, from 10 to 20 mm in diameter and 0.2–0.5 mm in thickness. While significant changes of the microstructure are observable after 1/2 rotation, several rotations, however, are required to produce a homogeneous nanostructure. In spite of the very large deformation in which true strains of more than 100 may be reached without defects, the process is used mostly in laboratory experiments and to the best of the authors' knowledge, there are no industrial applications. Recent investigations show that severe torsional straining can be used successfully not only for the refinement of the microstructure but also for the consolidation of powders (Valiev, 1996; Alexandrov, 1998).

8.2.2 Equal channel angular pressing (ECAP)

Segal and co-workers developed the method of ECAP in the beginning of the 1980s, creating the deformation of massive billets via pure shear (Segal et al., 1981, 1984). The goal of the method was to introduce intense plastic strain into the materials without changing their cross-sectional area. In the early 1990s, the method was further developed and applied as an SPD method for the processing of structures with submicron and nanometer grain sizes (Valiev et al., 1991; Furukawa et al., 1996). Since the cross-sectional dimensions of the sample remain unchanged with a single passage through the die, the sample may be pressed repeatedly through the die in order to achieve very high total strains. The overall shearing characteristics within the crystalline sample may be changed by rotation between the individual pressings. It is possible to define three distinct processing routes: one in which the sample is not rotated between repeated pressings, another in which the sample is rotated by 90° between each pressing and the third in which the sample is rotated by 180° between each pressing. A further possibility may be introduced when it is noted that the second route may be undertaken either by rotating the sample by 90° in alternate directions between each individual pressing, or by rotating the sample by 90° in the same direction between each individual pressing. To obtain a desired microsructure using ECAP, about 8–10 passes are usually required. For each pass, the pressed billet must be removed from the die and re-inserted for the next pass, often after re-heating in a separate furnace. This makes the process inefficient and difficult to control. A new ECAP method using a rotary die was developed to make ECAP more industrially viable (Nishida et al., 2001). Most of the experiments so far used work pieces with diameters of 15–20 mm, but the results of Horita et al. (2001) demonstrate the feasibility of scaling ECAP to large sizes (40 mm) for use in industrial applications. An important variable is the length/diameter ratio of the specimen. In current practice that ratio is 6–7. Implementations of the ECAP in industry require labour-intensive handling of the work pieces between process steps.

8.2.3 Cyclic extrusion-compression

Very large deformations are imposed by the cyclic extrusion–compression method, which, as the name indicates, combines the extrusion and compression processes. The sample is placed in a two-piece sectional die consisting of an upper and lower chamber of equal diameters. The chambers are connected by a constriction whose diameter is smaller than that of the dies. The deformation proceeds by the cyclic flow of the metal from one chamber to the other. Compression occurs simultaneously with the extrusion so that the sample is restored to its initial diameter. It has been found (Reichert et al., 2001) that strain localization in the long-range shear bands crossing the whole volume of the samples is the main deformation mechanism. As a result of mutually crossing shear bands and microbands, nearly equiaxial sub-grains are formed, creating a homogeneous structure.

8.2.4 Multiple forging

Another method for the formation of nanostructures in bulk billets is multiple forging (Valiahmetov, 1990; Imayev et al., 1992). The process of multiple forging is usually associated with dynamic recrystallization. The principle of multiple forging assumes multiple repeats of free forging operations: setting-drawing while changing the axis of the applied load. The homogeneity of strain provided by multiple forging is lower than in the case of ECA pressing and torsional straining. However, the method allows one to obtain a nanostructured state in rather brittle materials because processing starts at elevated temperatures. As well, the specific loads on the tooling are low. The choice of appropriate temperature and strain rate regimes of the deformation leads to a minimal grain size.

8.2.5 Continuous confined strip shearing

The continuous confined strip shearing process is based on equal channel angular pressing. The process is designed to apply simple shear to the metal strip in a continuous mode. A specially designed feeding roll with grooves on its surface is used, delivering the power required to feed the metal strip through the ECAP channel at a given speed. The experimental results indicate that ECAP can be used as a means not only for enhancing the tensile strength but also for controlling the texture of the strips suitable for subsequent sheet forming applications.

8.2.6 Repetitive corrugation and straightening (RCS)

In this process, a work piece is repeatedly bent and straightened without significantly changing its cross-section. Large plastic strains are imparted to the material, which lead to the refinement of the microstructure. The RCS process can be easily adapted to large-scale industrial production.

8.2.7 Accumulative roll-bonding (ARB)

Flat rolling is acknowledged as the most applicable deformation process for continuous production of bulk sheets (Saito et al., 1998). It is often stated that up to 90% of all metals are rolled at some point in the manufacturing process. The rolling process, however, has serious limitations.[1] One of these concerns the possible total reduction in thickness, i.e., the total strain achieved per pass, which is limited because of the resulting tensile straining and the attendant cracking at the edges.

Accumulative roll bonding, developed by Saito et al. (1998), the topic of the present section, is one of the techniques capable of creating the metallurgical and mechanical attributes, demanded of metals with very small grains, as listed above. While the major objective of the accumulative roll bonding process is to produce the very small grains within the rolled metal, another interrelated objective is to achieve this without damage and this requires the minimization of the development of the tensile cracking of the edges.

The process is simple. The roll surface is cleaned carefully and a strip of the metal is rolled to 50% reduction, usually without lubricants. After rolling, it is cut into two parts, cleaned very carefully and stacked, one on top of the

[1] The limitations of the flat rolling process have been discussed in Chapter 2.

other part, resulting in a strip whose dimensions are practically identical to the starting work piece. The stacked sheets are rolled again to 50% reduction and the two sheets are cold bonded during the rolling pass while creating bulk material. Hence, ARB is not only a deformation process but also a roll-bonding process. After the second pass, the process is repeated and continued until edge cracking is severe, such that the resulting product may not be usable any further. To achieve good, strong bonding, surface treatments such as degreasing with a non-greasy detergent and wire brushing, preferably using stainless steel brushes, of the sheet surface are done before stacking. Rolling at elevated temperature is advantageous for joining ability and workability, though too high temperatures may cause recrystallization and cancel the accumulated strain. Therefore, the rolling (roll bonding) in the accumulative roll-bonding process is preferably carried out at "warm" temperatures. Research indicates that the process may be repeated numerous times and while rolling strips of several layers, the occurrence of edge cracking, if not eliminated completely, is reduced in a significant manner.

Saito et al. (1998) rolled strips of fully annealed commercially pure aluminum of 1 mm thickness using the ARB process, with no lubrication. The strips were held in the furnace at 473 K for 300 s. The pre-rolled, pre-heated grain diameters were measured to be 37 μm. The reduction in each pass was 50% at a mean strain rate of $12\,s^{-1}$. No cracks were observed even after eight cycles. While the process created ultra-fine grains of 670 nm mean grain diameter after eight cycles, the grain diameter was already under 1 μm after the third rolling pass. After six cycles, the grain distribution was uniform. The tensile strength of the metal increased from about 90 MPa to nearly 300 MPa and the elongation decreased from about 40% to under 10% after eight cycles. Tsuji et al. (1999a,b) used the accumulative roll-bonding process to reduce the grain size of 5083 aluminum alloy from 18 μm to 280 nm, in five cycles of rolling. Testing at higher temperatures after the roll-bonding process indicated that the metal has become superplastic, elongating to nominal strains of 200–400%. Saito et al. (1999) rolled Ti added interstitial free steel strips at 773 K, employing the accumulative roll-bonding process. The pre-rolled average grain size was measured to be 27 μm. After five cycles, the grains decreased to less than 500 nm. The changes in tensile strength and elongation of the IF steel were given by Tsuji et al. (1999a,b), indicating that the strength increased from about 280 MPa to over 800 MPa after seven cycles of roll bonding and the elongation dropped from just under 60% to under 5%. The accumulative roll-bonding process as well as other techniques that create ultra-fine grains have been reviewed recently by Tsuji et al. (2002).

Park et al. (2001) used the ARB process to create grains of 0.4 μm in 6061 aluminum alloys after five passes, starting with a grain size of approximately 40 μm. The rolling passes were performed at 523, 573 and 623 K, at strain rates of $18\,s^{-1}$. The authors showed that no delamination of the rolled sheets was observed.

Lee et al. (2002a,b) examined the effect of the shear strain experienced by the samples in accumulative roll bonding. The authors write that during the rolling pass the effect of the interfacial conditions between the rolls and the rolled metal on the characteristics of the deformation process is most significant. They subjected commercially pure aluminum sheets to eight cycles of the ARB process, without any lubrication. A pin was inserted into the samples and the distortion of the pin was used to infer the amount of shearing in the passes. The distribution of the shear strain across the thickness of the sheets was found to correspond well to the grain size distribution.

Lee et al. (2002a,b) used 6061 aluminum alloy sheets and found that after eight cycles of ARB, the tensile strength increased from 120 to 350 MPa, while the elongation dropped from about 30% to a low of 5%. Xing et al. (2002), rolling AA3003 aluminum alloys, reduced the grains from a starting magnitude of 10.2 μm to 700–800 nm, in six cycles of the ARB process. The results and the conclusions were similar to those of Lee et al. (2002a,b).

The review indicates that the first concerns of the researchers are the changes in the metallurgical attributes of the multi-layered strips, demonstrating the very pronounced decrease of the grain diameters accompanying the accumulative roll-bonding process. The increasing tensile strength and the loss of ductility were also indicated for a number of materials, including an ultra-low carbon steel, containing 0.0031% C and several aluminum alloys. While the surface hardness, the strength of the bonds and the bending strength of the multi-layered strips following the process would contribute to the success of potential industrial applications, these have been treated less intensively in the technical literature. Further, the parameters of the successive rolling passes have not been given. Experiments were conducted to study these phenomena and in what follows, a detailed account of the work will be presented. These form the topics of the next section, in addition to a discussion of the potential industrial use of the multi-layered strips.

In what follows, the effect of the progressively increasing number of layers on the mechanical attributes of the multi-layered strips, after rolling and cooling is examined.[2] An ultra-low carbon steel, containing 0.002% C, somewhat less than contained in the steel of Tsuji et al. (1999a,b), is used. The parameters of the warm-rolling process are documented. The changes of the hardness, the yield and tensile strengths, the corresponding loss of ductility and the behaviour of the multi-layered strips in three-point bending are followed as the number of layers is increased. The strength of the bond is determined. The number of layers that may be bonded without producing edge cracking is indicated and the causes of the cracking of the edges are discussed. A suggestion for a potential industrial use of the multi-layered strips is presented.

[2] Most of the work, reported here, was performed during Dr Krallics' tenure as a Visiting Professor in the Department of Mechanical Engineering, University of Waterloo. Dr Krallics is an associate professor at the Budapest University of Technology and Economics.

8.3 A SET OF EXPERIMENTS

8.3.1 Material

An ultra-low carbon steel was used in the tests. The chemical composition of the steel is given in Table 8.1. The steel is comparable to that of Tsuji et al. (1999a,b) except for the lower carbon content. The grain structure of the as-received steel, obtained using a scanning electron microscope, is shown in Figure 8.1, indicating grain sizes of 25–35 μm.

The true stress–true strain curve of the metal, determined in a uniaxial tension test at 22°C, is $\sigma = 183.2\,(1+51.7\varepsilon)^{0.317}$ MPa.

8.3.2 Preparation and procedure

The ultra-low carbon steel strips, nominally 2.5 mm thick, were cut into samples of 25 mm width and 300 mm length. The surfaces of the strips were roughened by using a wire brush, removing as much of the layer of scale as possible and creating a somewhat random surface of roughness of 1.5–1.8 μm R_a. After brushing, the surfaces were cleaned using acetone. The strips were then joined on the roughened surfaces and while holding them in a vice to ascertain that they lie flat against one another, the leading and the trailing edges were spot welded.[3] The leading edge was tapered to ease entry to the roll gap.

Table 8.1 The chemical composition of the steel, (weight %)

C	Mn	P	S	Cu	Nb	Ti	Al	N	Si	Fe
0.002	0.133	0.01	0.0099	0.02	0.005	0.053	0.048	0.0067	0.009	Rest

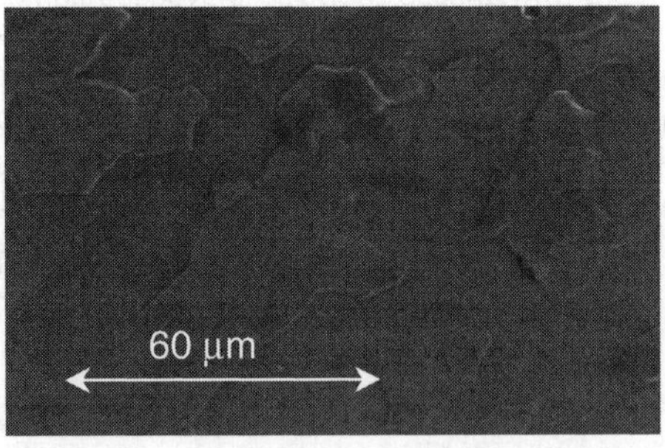

Figure 8.1 The grain structure of the ultra-low carbon steel, as received.

[3] In hindsight, welding the edges was not the best approach to keep the strips from sliding over one another during the rolling pass. A soft wire, wrapped around the strips would have been much better and is used in current practice.

The strips were then placed in a furnace, pre-heated to 515°C and held there for 10 minutes in air, before rolling. After the soaking period, the strips were rolled, without any lubrication, to a nominally 50% reduction, at a velocity of 0.39 m/s (50 rpm), creating strain rates of approximately $20\,s^{-1}$. Ten pairs of strips were prepared. Ten two-layered strips were rolled in the first pass, four layers in the second, eight in the third and so on. The rolled strips were visually inspected for the appearance of edge cracking and successful bonding. One of the strips was removed for mechanical testing. The remaining nine strips were cut into two samples of equal length and the procedure was repeated, rolling the four-layered strips, at the same temperature and the same rolling speed to the same nominal reduction. The experiments were stopped when the cracking of the edges became pronounced.

8.3.3 Equipment

All experiments were carried out on a 15 kW, two-high, STANAT laboratory mill with a four-speed transmission and tool steel work rolls of 150 mm diameter, hardened to Rc = 55 and having a surface roughness $R_a = 1.7\,\mu m$, obtained by sand blasting. The surface roughness is expected to be random and also helpful in drawing the strips into the roll bite. The mill is instrumented with two load cells positioned over the bearing blocks of the top roll. The data are collected and stored in a personal computer. The top speed of the mill is 1 m/s and the maximum roll force is 800 kN. The furnace is located beside the mill, so transfer of the strips for rolling caused minimal loss of heat and the entry temperature may be safely assumed to be very close to the furnace temperature.

8.4 RESULTS AND DISCUSSION

8.4.1 Process parameters

A typical experimental matrix and the observations for three sets of tests are shown in Table 8.2. The entry thickness and the width, the exit thickness and the width, the reduction per pass and the measured roll separating forces per unit width are given in the table. The number of layers and qualitative observations concerning the bonds and the appearance of cracking of the edges are also indicated.

In the first few passes the bonds are generally well formed, as long as the reduction the strips experience is above a certain limit, estimated to be approximately 50%. When the reduction is much below that level, bonding appears unsuccessful, as in the test where the reduction reached only 37.5%. Minor cracking of the edges appears after 16 layers have bonded well. When the 32-layered strip is rolled to a reduction, beyond 50%, bonding is acceptable but cracking of the edges is pronounced. When the reduction

Table 8.2 The experimental matrix and observation while roll bonding the ultra-low carbon steel

h_{in} (mm)	w_{in} (mm)	h_{out} (mm)	w_{out} (mm)	Reduction (%)	P_r (N/mm)	Layers	Comments
First set of experiments							
5.6	29.50	2.40	33.20	57.1	12024	2	Good bond
4.8	33.20	2.38	34.55	50.4	10614	4	Good bond
4.76	34.55	2.47	35.70	48.1	10901	8	Good bond
4.94	35.70	2.80	37.65	43.3	10716	16	Small cracks
5.6	37.65	2.55	40.00	54.4	12870	32	Large cracks; bonding
Second set of experiments							
5.6	29.50	2.42	33.20	56.8	12799	2	Good bond
4.84	33.20	2.45	34.65	49.4	10853	4	Good bond
4.90	34.65	2.43	36.90	50.4	11391	8	Good bond
4.86	36.90	2.80	37.65	42.4	10437	16	Small cracks
5.60	37.65	3.50	40.00	37.5	11703	32	No bonding; few cracks
Third set of experiments							
5.61	29.50	2.40	N/A	57.2	12024	2	Good bond
4.80	33.20	2.38	N/A	50.4	10614	4	Good bond
4.76	34.55	2.47	N/A	48.1	10901	8	Good bond
4.94	35.70	2.8	N/A	43.3	10716	16	Good bond
5.60	37.65	2.6	N/A	53.6	12870	32	Edge cracking; bonding

is less, only 37.5%, no bonding and, as expected, much less edge cracking is observed.

The similar magnitudes of the roll separating forces per pass are to be pointed out. As the following sections will indicate, the room-temperature strengths of the rolled strips depend on the number of layers and the amount of cold – or, more precisely, warm – rolling process they are subjected to. The cumulative effect of the repeated warm working and the accumulation of residual stresses after cooling are observed to be causing the increasing resistance to deformation. The similar magnitudes of the measured roll forces/pass in the warm rolling process are expected to be caused by the nearly ideally plastic behaviour of the ultra-low carbon steel, which at the rolling temperature of 500°C experiences dynamic recovery only. Using the simple, empirical model of the flat rolling process,[4] inverse calculations give the effective flow

[4] See Section 3.2, Chapter 3.

strength of the strip in each pass and, in general, a slowly increasing trend is noted, indicating some accumulation of strain.

8.4.2 Mechanical attributes at room temperature

8.4.2.1 Hardness

The hardness on the edges of the rolled samples was measured in the transverse direction, and the averages of the measurements are shown in Figure 8.2. The average Vickers hardness, obtained using a force of 200 g, is given on the ordinate and the number of layers rolled is shown on the abscissa. As expected, the hardness increases as the number of passes is increased. The hardness of the as-received steel, prior to rolling, is 111 Hv. At the end of the fifth pass, after rolling the 32-layered strip, the hardness has become 293 Hv, indicating significant hardening. It is observed that the largest increase in the hardness, nearly 100%, is created when the two-layered strip was rolled. The hardness increased in the subsequent passes, but at a progressively lower rate. It is probable that if the cracking of the edges did not limit the process, a limiting hardness would have been reached.

8.4.2.2 Yield, tensile strength and ductility

The yield strength, the tensile strength and the elongation have been determined in standard tensile tests. The results are shown in Figure 8.3, plotted against the number of layers contained in the rolled strips. The strengths change as a result of the cumulative effect of warm working, in a manner

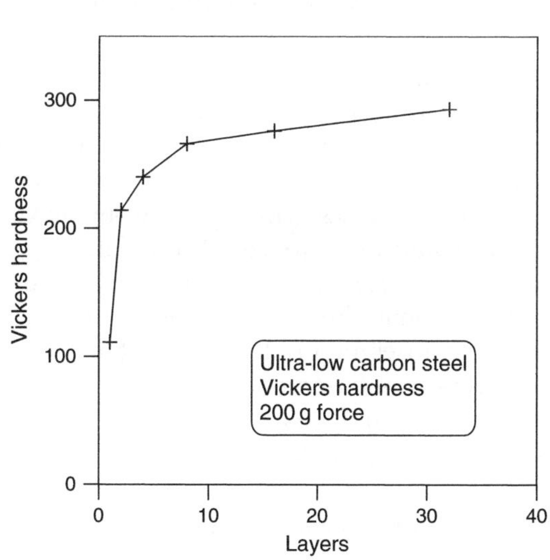

Figure 8.2 The hardness on the edges of the rolled samples, measured in the transverse direction. The averages are given.

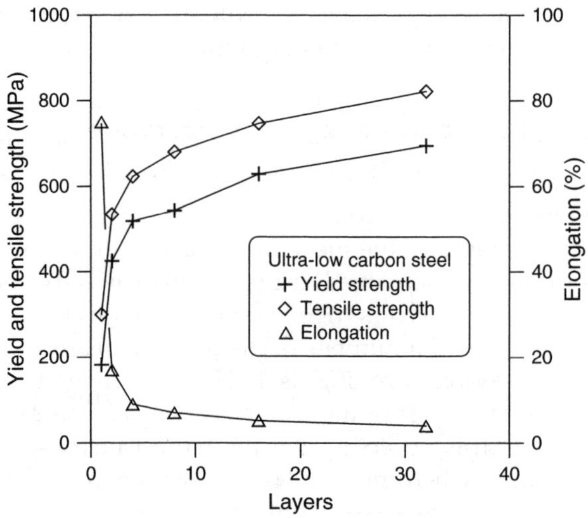

Figure 8.3 The yield strength, the tensile strength and the elongation.

similar to the hardness, increasing by approximately the same percentage. The major increase is again observed to occur in the first pass. The yield strength increases from a low of 183 MPa before the rolling process to a high of 695 MPa, after five cycles of rolling. The tensile strength increases from 300 to 822 MPa. At the same time, however, the ductility decreases from a high of nearly 75% to 4%, indicating a very pronounced loss of formability.

8.4.2.3 The bending strength

Three-point bending tests were performed in order to observe the behaviour of the multi-layered strips in a potential sheet metal forming operation. These tests subject the sample to significant tensile and compressive stresses in their plane in addition to shear stresses which vary from maximum at the neutral axis to zero at the outermost surfaces. The results of the tests are given in Figure 8.4, plotting the force/sample width versus the vertical displacement. Up to the displacement shown in the figure, the tensile strains were not excessive and no fractures occured. Some, but not excessive, delamination of the bonded layers was observed. As the number of layers increased, fewer instances of delamination were noted, indicating that the bond strength increased after repeated rolling passes.

8.4.2.4 The cross-section of the roll-bonded strips

The cross-sections of the rolled strips are indicated in Figure 8.5a, showing two layers and in Figure 8.5b, showing the 32 layers. The interfaces are visible only slightly, indicating the possibility that the roll-bonding process was successful.

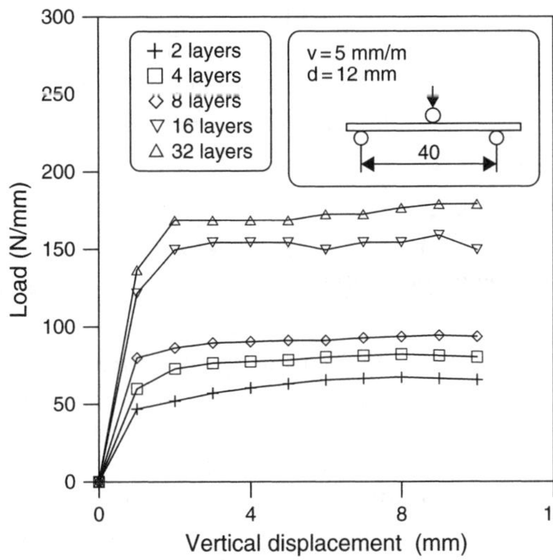

Figure 8.4 Three-point bending test results.

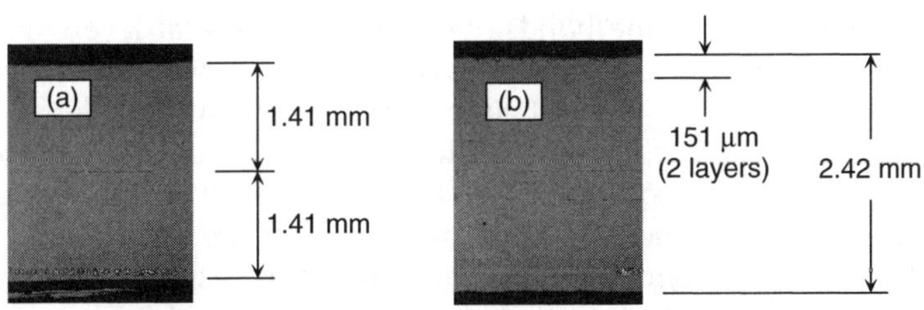

Figure 8.5 (a) Cross-section of a two-layered strip. (b) Cross-section of a 32-layered strip.

8.4.2.5 The strength of the bond

The bond strength was tested, following a test procedure, shown schematically in Figure 8.6. The figure indicates a four-layered strip and the strength of the bond in the middle, just formed, is to be tested. As shown, two narrow slots are milled at about 10 mm from each end of the sample, to carefully controlled depths. Tension tests, conducted at a speed of 1 mm/m, are then performed on an Instron tensile tester. Two tests, performed to test the bond in the middle of a four-layered and an eight-layered strip, indicated that the shear stress necessary to separate the bond, is in the order of 52–53 MPa, somewhat less than expected but still indicating that reasonably successful bonding was achieved. The third test, performed on a 32-layered strip, was designed to test the strength of the bond on the second layer from the surface. The sample broke as a result of the tension test, at a tensile stress of 730 MPa while the shear stress at the designated layer reached nearly 100 MPa which, however, did not separate.

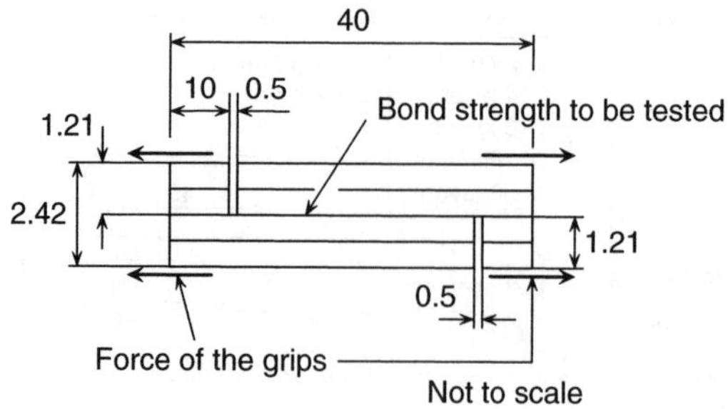

Figure 8.6 The test for the bond strength.

8.5 THE PHENOMENA AFFECTING THE BONDS

Several phenomena, mechanical and metallurgical in nature, are involved in the accumulative roll-bonding process. The drastic decrease of the grain size and the attendant changes of the mechanical properties are among the major features. The strength of the bonds, created when several layers are rolled, also contributes to the success or otherwise of the process. Further, the ability of the multi-layered strips to resist edge cracking is of interest.

The cumulative hardening and the loss of ductility during cold or warm rolling are well understood. As predicted by the Hall–Petch equation, the decreasing grain sizes and the increasing strength are clearly related. The loss of ductility associated with these changes has also been discussed in the technical literature. As mentioned above, the focus in this study is on the post-rolling, room-temperature mechanical attributes, the strength of the adhesive bonding in between the layers and the occurrence of cracking of the edges.

The roll-bonding process is a form of cold welding. In the process, two sheets, usually but not exclusively metals, are rolled and hence, bonded together. The strength of the bond depends on providing the appropriate conditions for adhesion of the materials: cleanliness, closeness and pressure. When contact is made, the phenomena there are best explained in terms of the adhesion hypothesis (Bowden and Tabor, 1950), which examines the origins of the resistance to relative motion in terms of adhesive bonds formed between the two contacting surfaces that are absolutely clean and are an interatomic distance apart. Bowden and Tabor (1973) credit the French scientist Desaguliers, living and working in the eighteenth century, with this idea and reproduce his account of an experiment with two lead balls which, when pressed and twisted together by hand, created what must have been adhesive bonds. The top ball held the bottom ball, a load of nearly 7.3 kg.

The parameters that influence the adhesion of metals are discussed in detail by Gilbreath (1967). He lists the material properties, the interfacial pressure, the duration of the contact, the temperature and the environment as those that affect the adhesion coefficients, defined as the ratio of the strength of the bond to the strength of the parent metal. The study, conducted in high vacuum, indicates that while adhesion is inversely proportional to hardness, it increases with increasing loads, the time of contact and the temperature. Further, even small amounts of oxygen or air decrease adhesion. In the present set of tests, these parameters were kept constant. Another parameter of importance is the roughness of the surfaces to be joined, also kept constant here.

The roughness of the surfaces, created manually by wire brushing, was measured to be in the order of $R_a = 1.5$–$1.8\,\mu m$. These would create large true areas of contact that would be expected to aid adhesion. Since the normal pressures are several times the metal's resistance to deformation, the major change of the true area of contact is expected to occur in the first pass. Subsequent passes, during which the rolled metal experiences pressures of similar magnitude, would likely not increase the true area of contact by any significant measure. As several rolling passes have indeed increased the bond strength, this is likely due to the increasing chemical affinity which would result in stronger interfacial adhesive bonds.

8.5.1 Cracking of the edges

The occurrence of edge cracking is indicated in Table 8.2 and it is observed that in most instances 16 layers of the steel were rolled successfully, while the edges did not crack much. Only in the last pass, when rolling the 32 layers, was cracking pronouced. The process was ended at that point. It is recalled that Saito et al. (1999) trimmed the cracked edges and continued to roll the multi-layered strips.

In the accumulative roll-bonding process, as followed in this work, the true strain experienced by the strips in each pass is near 0.7. The total true strain, that is, the sum of the strains per pass of the strips, is approximately 3.5–4. This corresponds to a reduction of over 97%, much larger than what can be achieved in one conventional pass, without edge cracking.

The maximum reduction obtainable in one pass of the cold rolling process is limited by the metal's ductility, the through-thickness and the transverse non-homogeneity and when lubricants are used, by the directionality of the roll's surface roughness. Since the rolling passes were performed dry, only the non-homogeneity of the deformation needs to be considered as the probable limiting mechanism. Non-homogeneity through the thickness may cause alligatoring. Since the ratio of the roll diameter and the strip thickness was quite large and the shape factor was significantly larger than unity, alligatoring was not expected nor was it observed. Transverse non-homogeneity may cause splitting of the rolled samples in the direction of rolling. This limit of workability was not observed either when the ultra-low carbon steel strips were rolled. It is worthwhile in this context to refer to a few unsuccessful accumulative roll-bonding tests using medium carbon steel strips. Following the procedure of the ultra-low carbon steel, the medium carbon steel strips split in the direction of rolling, no doubt due to lack of ductility in addition to transverse non-homogeneity.

The process was limited by the appearance of significant cracking of the edges. In a study concerning the workability of aluminum alloys in the hot rolling process (Duly et al., 1998), the occurence and the direction of edge cracking were identified as a result of the state of stress at that location. The same approach indicates that at the centre of the sample, the tensile stresses in the direction of rolling and the compressive stresses transverse to that direction cause the maximum shear stresses to occur in a direction of 45°. The majority of the cracks of the 32-layer strip, as shown in Figure 8.7, are in general, in that orientation. Also noted are several cracks in other directions, at various angles, and not 45°. The reasons for the orientation of the cracks lie in the complex stress distribution at the edge during the rolling passes and are considered beyond the scope of this study. The reasons for the ability of the strips to resist edge-cracking, however, may be explained by considering the ultra-low carbon steel's resistance to deformation at the rolling temperature of 500°C. At that temperature the true stress–true strain curves exhibit dynamic recovery and almost perfect ideally plastic behaviour. The dynamic recovery process, in which some of the stored internal energy is relieved by dislocation motion without affecting the size of the grains, allows the samples to recover some, but not all, of their original softness. Some of the strain is then retained and as the passes are repeated, these strains accumulate. When the accumulation

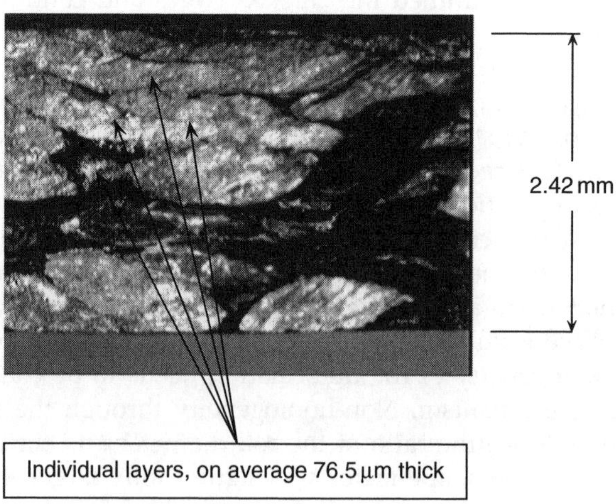

Figure 8.7 Cracking at the edges.

is sufficient to reach the limit of workability, cracking occurs at the most higly stressed location near the edges.

8.6 A POTENTIAL INDUSTRIAL APPLICATION: TAILORED BLANKS

Tailor welded blanks are made up of two sheets of unequal thickness which are welded to form a blank for subsequent sheet metal operations involving bending in one or two directions, such as in the deep drawing or the stretch forming processes. While welding techniques are well advanced and the interruption of material continuity can be accounted for in the design of the forming processes, the strength of the welds is often less than that of the parent metal (Worswick, 2002). The accumulative roll-bonding process may lead to blanks of uniform thickness but significantly different strength and formability from one portion of the blank to another. Limited number of tests have been performed – to be reported on at a later date – and the refinement of the technique is continuing, but in essence the procedure is as follows. The surface of a strip is roughened by a wire brush and cleaned with acetone, as above. Two strips are then placed on one another such that over half of the length the strip is made up of two layers. The strip is then warm rolled, and dried to a reduction of 50%. The end result is a strip made up of two bonded layers over about half the length of the sample, the remaining part being a single layer. The bonded portion has smaller grains, increased strength and reduced ductility. The unrolled portion's mechanical attributes have not changed. Tests performed so far allow some cautious optimism that the removal of the welding process and the attendant discontinuity may result in improved formability.

8.7 A COMBINATION OF ECAP AND ARB[5]

An aluminum alloy, used in the automotive industry, is processed by equal channel angular pressing (ECAP) and repeated rolling. First, the metallurgical attributes caused by one pass of the ECAP process are examined. The alloy is then rolled in several passes and the changes of its attributes are monitored. The objective is to determine whether repeated applications of the rolling process are able to create grains of magnitude smaller than those that were produced by several passes of the ECAP process. 6082-T3 aluminum alloys

[5] This study was conducted by R. Bogár, in the Department of Mechanical Engineering, University of Waterloo.

were used, possessing an initial yield strength of 125 MPa and a tensile strength of 180 MPa. The ductility of the metal is 55%.

The samples were processed by ECAP and the effect of different processing routes on the development of the tensile strengths and that of the grain structure were investigated. While rotating the sample around its longitudinal axis after each of a total of eight passes by 180°, the tensile strength increased from 180 to 260 MPa (Krallics et al., 2004). The pre-ECAP grain size of 2.5 μm was reduced to 300–500 nm after one pass and did not decrease any further after seven subsequent passes (Krallics et al., 2002).

In what follows, the potential advantages of combining the rolling process with the ECAP are examined. Using the aluminum alloy, two processing steps are employed. First, the alloys are subjected to one pass through the equal channel angular pressing dies and their grain structures are examined. These are followed by repeated, unlubricated rolling passes and the influence of the combination of the processes on the resulting metallurgical attributes is monitored. All rolling experiments are conducted at a nominal roll speed of 170 mm/s. The equal channel angular pressing tests were performed using the press, described by Krallics et al. (2002).

Before the experiments, all samples were annealed at 420°C for one hour and allowed to cool with the furnace at a rate of approximately 1°C/s.

8.7.1 The ECAP process

The grain size of the aluminum sample, after the heat treatment but before the ECAP process was 2.5 μm for the 6082 alloy, see Figures 8.8a and 8.8b. The figures show the transverse (a) and the longitudinal sections (b) The grain boundaries are clearly visible in the transverse sections and the

(a) (b)

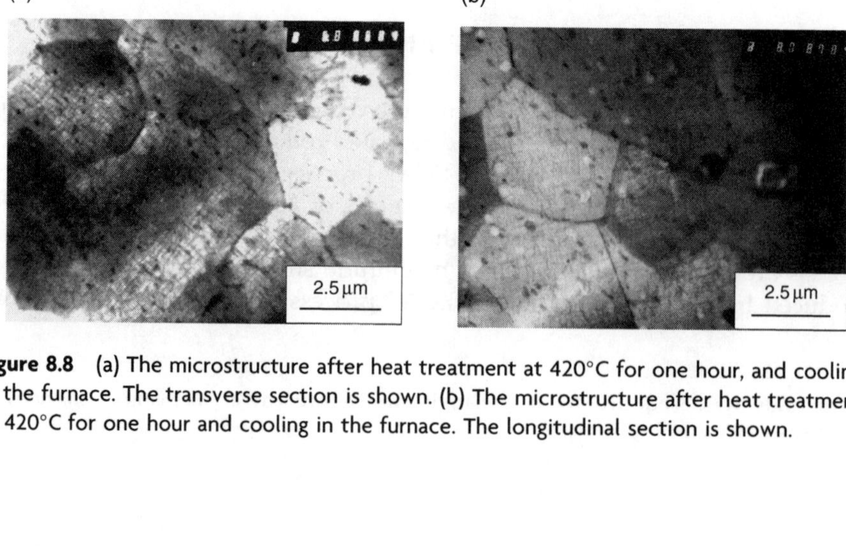

Figure 8.8 (a) The microstructure after heat treatment at 420°C for one hour, and cooling in the furnace. The transverse section is shown. (b) The microstructure after heat treatment at 420°C for one hour and cooling in the furnace. The longitudinal section is shown.

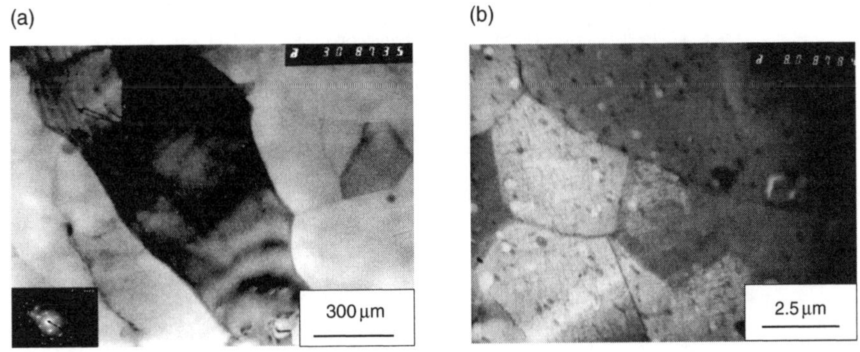

Figure 8.9 (a) The microstructure after one pass eg. ECAP; the transverse section is shown. (b) The microstructure after one pass eg. ECAP; the longitudinal section is shown.

effects of the prior extrusion process are also observable in the longitudinal directions.

After one ECAP pass the yield strength increased to 190 MPa, the tensile strength to 230 MPa and the ductility dropped to 45%.

The grain structure after one pass of the equal channel angular pressing process is shown in Figures 8.9a and 8.9b, taken in the transverse and the longitudinal directions, respectively. The diffraction patterns are also indicated in the lower left corners of the figures.

8.7.2 The rolling process

The samples were subjected to essentially flat rolling passes, even though the cross-sections were circular at the start. The circular shape changed to an almost completely flat cross-section by the end of three passes of 50% reduction each. During the passes, the roll separating forces per unit width were measured and they are reported in Figure 8.10 as a function of the effective strain. The actual width of the contact was measured before and after each pass and the average was used in the calculation of the specific roll force. In the figures, the specific force is plotted on the ordinate and the effective strain is given on the abscissa.

The deformation of the samples was far from homogeneous and as a result, the distribution of the strains was also highly non-homogeneous. The roll pressure was applied on a fairly small area, in contact with the work rolls, in the first pass. The contact area increased in each pass, until the work pieces became practically completely flat. It is interesting to note that in spite of the increasing contact area and the attendant increasing resistance to frictional forces, the roll force, after an initial increase in the first pass, decreases as the strains accumulate. When the nominal reduction per pass is increased to

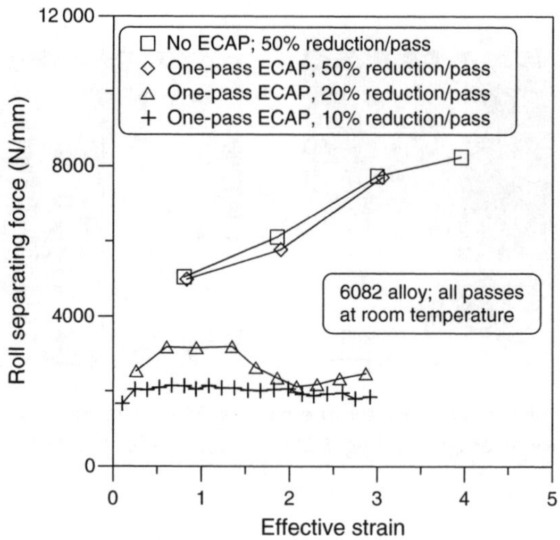

Figure 8.10 The roll separating force as a function of the effective strain. Rolling only and rolling after ECAP are shown.

50%, the picture changes in a significant manner. As observed in Figure 8.10, the roll forces increase with the effective strain, indicating the effect of the strain hardening of the metal. It is noted that the roll separating forces in four, 50% reduction passes with no ECAP (indicated by open squares) and three 50% reduction passes following one press through the ECAP die (indicated by diamonds) are practically identical.

In the rolling passes the transverse edges, especially near the centreline of the samples, experienced uniaxial tensile stresses and strains. The presence or the lack of tensile cracking there is indicative of the ductility of the sample. No cracking was observed in the first 50% reduction and some minor cracks were created after the second and the third passes.

8.7.3 The microstructure after ECAP and the rolling passes

The transmission electron microscope photographs shown in Figures 8.11–8.14 indicate the development of the microstructure as a result of the ECAP and the rolling processes. In Figures 8.11 and 8.12 the effect of one pass at 50% reduction is demonstrated, while in Figures 8.13 and 8.14 the effects of three passes are given. The microstructure and the diffraction patterns are given.

It is found that the most intense reduction of the grain size occurs in the first pass through the ECAP die. Subsequent rolling passes caused no significant further reduction.

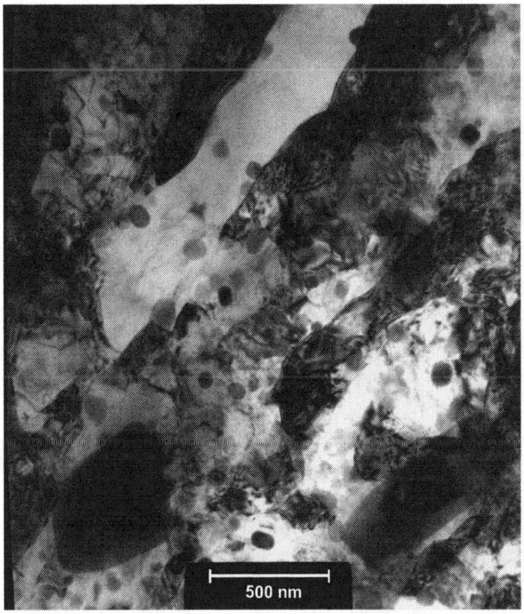

Figure 8.11 The microstructure after heat treatment at 420°C for one hour, cooling in the furnace and subjected to one pass of the ECAP process and a rolling pass of 50% reduction. The longitudinal section is shown.

Figure 8.12 The diffraction pattern after heat treatment at 420°C for one hour, cooling in the furnace and subjected to one pass of the ECAP process and a rolling pass of 50% reduction.

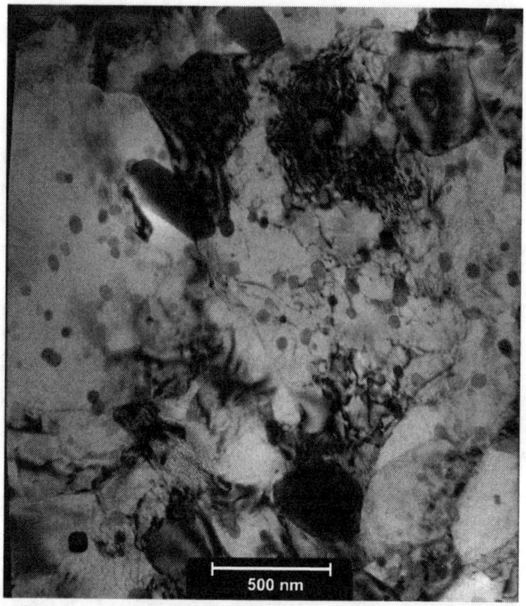

Figure 8.13 The microstructure after heat treatment at 420°C for one hour, cooling in the furnace and subjected to one pass of the ECAP process and three rolling passes of 50% reduction each. The longitudinal section is shown.

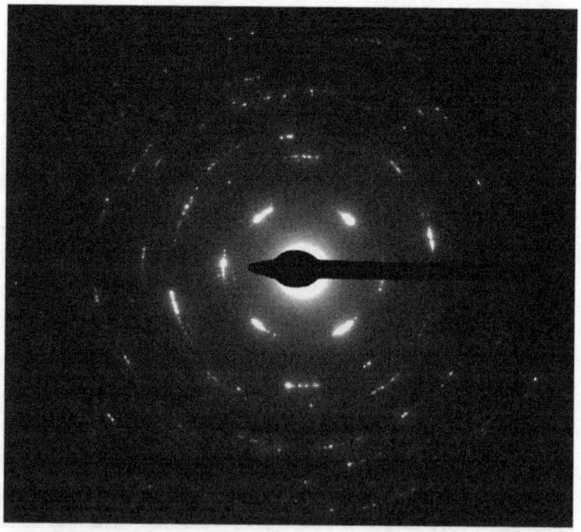

Figure 8.14 The diffraction pattern after heat treatment at 420°C for one hour, cooling in the furnace and subjected to one pass of the ECAP process and three rolling passes of 50% reduction each.

8.8 CONCLUSIONS

Ultra-low carbon steel strips were warm rolled, following the accumulative roll-bonding process. Strips made up of 32 layers were rolled and bonded successfully. The process was limited by the occurrence of cracking of the edges, caused by the state of stress at that location. The effect of cumulative warm working was monitored, and the hardness, the yield and the tensile strengths increased significantly as the process continued. The ductility decreased to very low levels, indicating that post-rolling sheet metal forming processes may have to be planned with care. The most pronounced changes of the mechanical attributes were observed to occur in the first rolling pass. The bond strength was also investigated in selected instances. The shear stress necessary to separate the centre bond was found to be about half of the metal's original yield strength in shear. The strength of the adhesive bonds near the edge appeared to be higher, affected by the number of rolling passes. Edge cracking was most likely initiated when the strains, retained after dynamic recovery, reached the limit of workability of the metal. A possible industrial application is discussed: that of the creation of tailored blanks of uniform thickness in which part of the blank is stronger and less ductile while the remainder's attributes are unchanged.

Combining the ECAP and the ARB processes resulted in sharply reduced grain sizes. Most of the reduction, however, occurred in the first ECAP pass and subsequent rolling passes contributed little to the formation of small grains.

CHAPTER 9

Roll Bonding

Abstract

6111 aluminum alloy strips were roll-bonded at warm and cold temperatures. The parameters that create successful bonds were determined. The shear strengths of the bonds were measured and found to increase when the temperature or the interfacial pressure is increased. Successful bonds, whose shear strength approached that of the parent metal, were created at room temperature only after the alloy was annealed. At warm temperatures the bond strength reached the strength of the parent metal and depended strongly on the entry temperature.

9.1 INTRODUCTION

Since components produced by cold pressure welding include automotive parts, bimetal products and household items, understanding the mechanisms and details of the process is of significant industrial importance. The solid-state joining technique can be used on a large number of materials, which may be the same, possessing identical attributes, or may be different, possessing widely varying mechanical and metallurgical properties. As mentioned by Bay (1986), materials that cannot be welded by traditional fusion often respond well to cold welding.

The cold-welding process causes bonding by adhesion[1] and as described by Bowden and Tabor (1950) this requires the surfaces to be clean and to be an interatomic distance apart. Considering the comment of Batchelor and Stachowiak (1995) that "surfaces are always contaminated" and Figure 3.1 (b) of Schey (1983), reproduced as Figure 5.2 in Chapter 5, showing the layers of oxides and adsorbed films on the surface of a metal, cleanliness of a surface is difficult to achieve without a controlled atmosphere and significant plastic deformation, large enough to break up the contaminants. Under industrial or

Based on "A Study of Warm and Cold Roll-Bonding of an Aluminum Alloy", Yan and Lenard, 2004.

[1] Adhesion is discussed in Chapter 5, Tribology.

laboratory conditions without the provision of protective environments, complete cleanliness is simply not achievable. The normal pressures are expected to be sufficiently large to satisfy the second criterion of the adhesion hypothesis, that the surfaces be close to one-another and that at least some new metal surface be created.

Wu et al. (1998) write that diffusion bonding and mechanical bonding are two types of solid-state bonding. They define the first as bonding that occurs in a considerable amount of time and it involves the application of temperature and pressure. Mechanical bonding, on the other hand, occurs practically instantaneously or over a very short time and depends, among others, on the forces of attraction between the atoms. In their experiments on several metals they find that the bond strength depended on the exponential of the temperature, implying that diffusion played a role.

It is the shear strength of the bond that determines the usefulness of the two-layered component in subsequent metal forming processes in which bending in two directions takes place, such as in deep drawing, stretch forming or a combination of the two. The process parameters affecting the bond strength involve the surface expansion and normal pressure, the surface roughness, the storage time between surface preparation and the welding process in addition to the time during which the normal pressure is applied (Bay, 1986). Gilbreath (1967) also includes the temperature and vibratory loading as two further parameters that may affect the strength of the bond. Kolmogorov and Zalazinsky (1998) add the strain at the interface to the list of parameters.

As shown by Bay (1986), the bond strength of a cold-welded Al–Al combination may approach the strength of the parent metals at high levels of surface expansion, defined as the increase in total bonding area, as compared to either the initial or the final area. As mentioned above, the large magnitude of surface expansion is required to cause the oxide layer to break up and to allow the fresh metal in between the cracks to make contact and thus, adhere to one-another. The pre-bonding preparation of the surface is also shown to affect the strength of the bond (Clemenson et al. 1986). Scratch brushing, using a brush with medium stiffness at high speeds was found to create the strongest bond while the normal pressure during brushing was found to have no effect.

A model of the cold-welding process was presented by Zhang and Bay (1997), making use of the observation that the strength of the weld between absolutely clean surfaces is approximately equal to the applied normal stress (Zhang and Bay, 1997; Bay, 1979). Kolmogorov and Zalazinsky (1998) base their model of the bond strength on the kinetic energy of micro-damage accumulation, resulting in the rupture of oxide films. They applied their model to the production of steel–aluminum wires. Manesh and Taheri (2005) used the upper bound theorem to examine the rolling of bimetal strips. Their model correctly predicted the measured peel strength as a function of the composite reduction.

Cold welding by rolling, that is, roll-bonding, is well suited to the creation of two-layered strips or plates. The rolling process is capable of producing the high interfacial pressures required to cause strong adhesion of the components. The process was studied by Hwang et al. (2000), presenting a mathematical model of rolling sandwich sheets. The model is an extension of an earlier study, and is based on the upper-bound method, using stream functions to define the velocity field. Mean contact pressures under 10 MPa magnitudes were considered. Experimental data, obtained while rolling aluminum and copper sheets were correctly predicted by the model. A mathematical model of roll-bonding, hot and cold, was presented by Tzou et al. (2002), deriving the stress field during the process of roll-bonding. They conclude that the important parameters needed to create strong bonds include the reduction, the friction factor at the interface and the tension, enlarging the bonding length during the bonding process. Zhang and Bay (1997) identified the threshold surface expansion, caused by plastic deformation, necessary to initiate cold welding. Weld efficiency, defined as the ratio of the strength of the weld and the strength of the base metal, was examined by Madaah-Hosseini and Kokabi (2002) during cold roll-bonding of an aluminum alloy. The strain hardening of the metal was included in their model which predicted the weld efficiency with good accuracy.

The expansion of the surfaces in contact is the result of the application of the normal pressures, or, in other words, the work done in the rolling process. It is hypothesized that the shear strength of the bond will approach that of the parent metal when sufficient amount of energy – the activation energy to initiate the bonding process – has been given the two components to be joined. This energy may be provided by heating and/or by mechanical means.

A discussion and examination of the process parameters that create bonds by rolling between two layers of an aluminum alloy, whose strengths are comparable to the strength of the original metal is given below. The independent parameters are the reduction/pass, the entry temperature and the rolling speed. The work done during the passes, necessary to create the bonds, is identified and a correlation between it and the activation energy of bond formation is demonstrated.

9.2 MATERIAL, EQUIPMENT, SAMPLE PREPARATION, PARAMETERS

9.2.1 Material

Strips of cold rolled aluminum alloy Al 6111, commonly used in the automotive industry, is experimented with in the tests. The chemical composition of the alloy, by weight %, is given below, in Table 9.1.

The stress–strain curve of the 6111 alloy, obtained in a uniaxial tension test, is closely approximated to the relation $\sigma = 150(1 + 157\varepsilon)^{0.245}$ MPa. The pre-rolling surface roughness of the Al 6111 strips was $R_a = 0.5\,\mu\text{m}$.

Table 9.1 The chemical compositions of the alloy, weight %

	Cu	Mn	Si	Zn	Mg	Al
6111	0.82	0.21	0.21	0.02	<0.01	rest

9.2.2 Equipment

The two-high experimental mill has 249.8 mm diameter by 150 mm long, D2 tool steel rolls, hardened to $Rc = 63$. The surface roughness is $R_a = 0.4\,\mu\text{m}$, and the roughness direction is random, created by manually finishing the surface using 100-grit carbide paper. The mill is driven by a 42 kW, constant torque, DC motor. Two force transducers are located under the bearing blocks of the lower work roll, measuring the roll separating force. Two torque transducers are placed in the drive spindles, leading to the roll torque. A shaft encoder monitors the rotational velocity of the roll. A digital data acquisition system is used to collect and process the results. The screwdown is governed by a hydraulic feedback control system.

9.2.3 Sample preparation

The samples for two-layered bonding are prepared, as follows. The strips of 300 mm length, 15 mm width and 0.97 mm in thickness are cut from a cold rolled sheet, parallel to the original rolling direction. All edges are deburred. The leading edges of the strips are tapered to ease entry to the roll bite. The surfaces are cleaned using acetone. The two pieces of the strip are stacked together and bound, near the leading edge, using a soft aluminum wire. No special surface preparation is done prior to performing the warm roll-bonding tests. When cold roll-bonding is performed, the surfaces to be bonded are roughened manually, using a stainless steel brush. The average surface roughness after brushing is in the range of $4 \sim 5\,\mu\text{m}$. The surface roughness is random. After brushing, the surfaces are cleaned using acetone.

9.3 PARAMETERS

The effects of the temperature, reduction, speed and annealing temperature on the shear strength of the bond were considered in the experiments. In one set of tests, the entry temperature was raised to the warm rolling level while in the next set, room temperatures were used. Prior to the tests at 22°C, the strips were annealed at various temperatures for 2 hours and cooled in air. The ranges of the parameters used in the tests were

- the reduction, varied from 40–70%;
- the entry temperature, varied from 22–280°C;

- the rolling speed, varied from 50–380 mm/s;
- annealing temperatures: 350–450°C.

9.4 TESTING OF THE SHEAR STRENGTH OF THE BOND

The shear strength was measured, following the approach of Bay (1979) by carefully cutting narrow slots approximately 1 mm wide on both sides of the roll-bonded samples so when pulling, the bonded areas should experience shearing only. The tests were conducted using an Instron electro-hydraulic testing system. Crosshead speed was 5 mm/m and temperature was 22°C. The bonding area after fracture was measured using an optical travelling microscope. It is noted that in several previous studies, the peel test was used to determine the strength of the bond; in that test, a normal stress is applied to the bonded surfaces. Since the bonded work pieces would eventually be formed further in sheet metal forming processes such as deep drawing or stretch forming, in which the stresses at the bonds would be created mostly by a shearing action, the shear test is deemed more relevant to evaluate the success or otherwise, of the bonding process.

9.5 RESULTS AND DISCUSSION

9.5.1 The roll force and the torque

The specific roll separating force and the roll torque, measured during warm roll-bonding, are shown in Figures 9.1a,b, respectively, as a function of the reduction and the entry temperatures.

As expected, the roll forces increase as the reductions increase. The roll torques reach a maximum at a reduction where the contact length, defined as $L = \sqrt{R'\Delta h}$, and hence, the moment arm, begin to drop. The entry temperatures and the speed of rolling are also shown. The 4-rpm roll speed or the 52-mm/s linear speed indicates that the contact time is in the order of a few hundred ms. The effect of the slightly increased entry temperature and the attendant drop of the metal's resistance to deformation are also observable.

9.5.2 The shear strength of the bond

9.5.2.1 The effect of the speed of rolling

The shear strength of the bond, obtained at an entry temperature of 280°C is presented in Figure 9.2 as a function of the rolling speed.

The data are given for two nominal reductions, 55 and 65%. The shear strength is observed to decrease sharply as the speed is increased leading to shorter times of contact, confirming the conclusions of Gilbreath (1967).

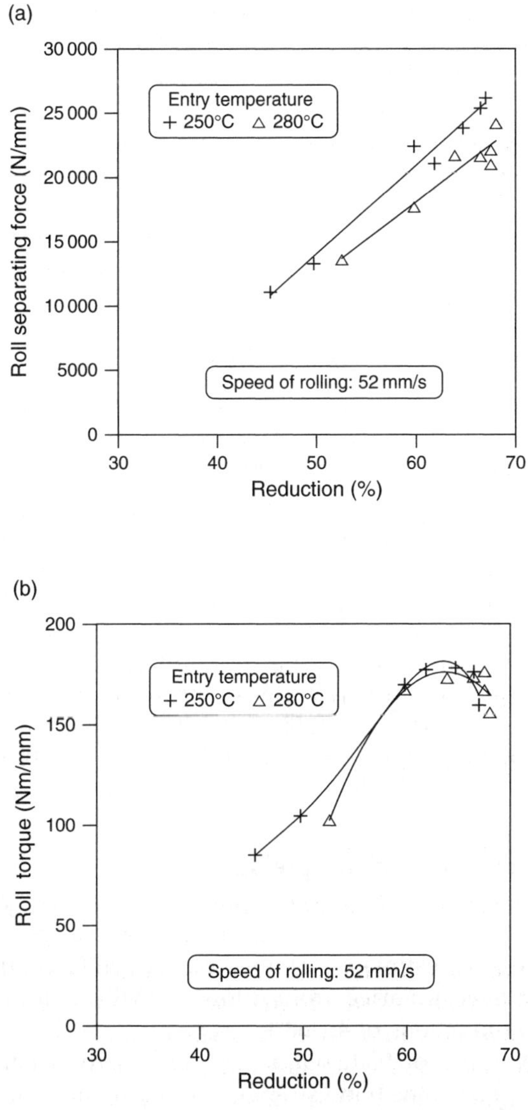

Figure 9.1 The roll force as a function of the reduction and the entry temperature, (b) The roll torque as a function of the reduction and the entry temperature.

Also observed is the effect of the reduction; at the higher reductions the bond strengths are higher. At the higher reduction, and at a low speed of 52 mm/s the time of contact is about 220 ms, sufficient to create a bond whose shear strength is near 230 MPa, approximately equivalent to the shear strength of the original material. Reducing the time of contact leads to sharply lower bond

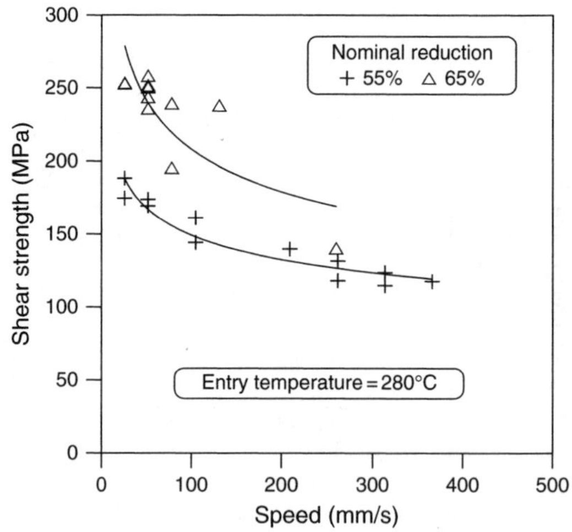

Figure 9.2 The shear strength of the bond, as a function of the rolling speed and the reduction.

strength; at the reduction of 55% and a speed of 350 mm/s the contact time is about 30 ms and the shear strength of the bond is reduced to 120 MPa, nearly half that at the lower speed. Since the bonding mechanism is expected to be adhesion and probably some limited amount of diffusion, the effect of the time of contact is expected and understandable.

9.5.2.2 The effect of the normal pressure

The results, shown in Figure 9.3, were produced at an entry temperature of 280°C and at a rolling speed of 52 mm/s, giving ample time for the bonds to form. The roll pressure, estimated by dividing the roll force/unit width by the projected contact length, varied from a low 950 MPa to a high of 2000 MPa, corresponding to reductions of 40–68%.

At the roll pressure of 950 MPa a very small drop of the rolling speed resulted in a steep rise of the bond strength, emphasizing again the importance of the time of contact in creating the adhesive bonds. The bond strength increased with the increasing normal pressures and the attendant growth of pure metal-to-metal contact, as expected. A maximum of the shear strength of the bond was reached at a pressure of approximately 1100 MPa and no further increase in the strength was obtained. When one surface approaches the other, the forces of attraction between the atoms increase until the attraction equals the repellent forces. Beyond that the adhesive forces and hence, the bond strength, remain stable. Note that the normal pressures are orders of magnitudes higher than those, considered by Hwang et al. (2000).

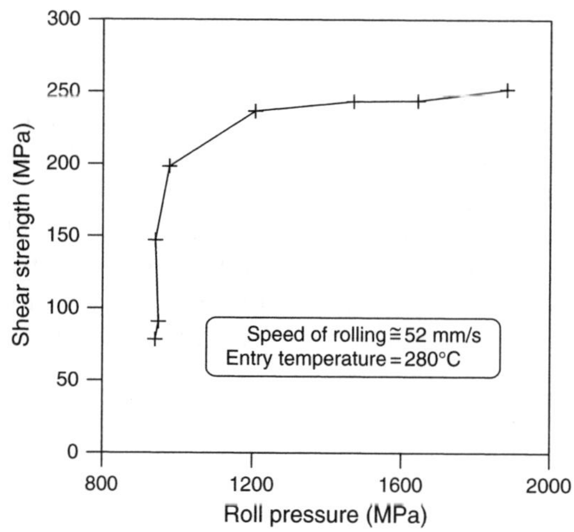

Figure 9.3 The shear strength of the bond as a function of the roll pressure.

9.5.2.3 The effect of the entry temperature – warm bonding

The relationship between shear strength of the bond and the entry temperature of the strips is shown in Figure 9.4, at a reduction of 66% and a nominal rolling speed of 88 mm/s.

It is observed that with increasing temperatures the shear strength of the bond also increases at a relatively fast rate at lower magnitudes. The rate decreased and a maximum strength was reached at 280°C with no significant

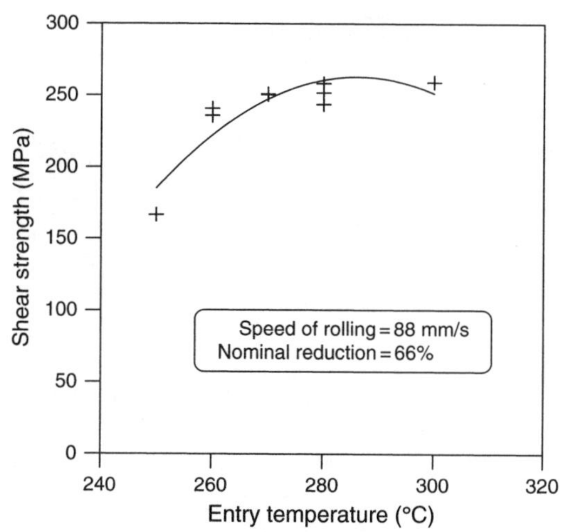

Figure 9.4 The shear strength of the bond as a function of the entry temperature.

change beyond that temperature. Bonding experiments were also conducted at 200°C entry temperature at a reduction of 66%. No successful bonds were created. The temperature affected the bonding properties greatly by influencing and accelerating the adhesion process and the break-up of the oxide layer. At higher temperatures, the strengths of the oxide layer, of the parent metal and the bond between the oxide and the strip all drop. As well, the rate sensitivity of all three increases. These phenomena enhance the possibility of the two surfaces conforming to one-another, leading to higher bond strength. While Peng et al. (1999) bonded aluminum and copper sheets at various temperatures, they also indicated that an optimum temperature to reach the maximum bond strength exists.

9.5.2.4 The effect of the entry temperature – cold bonding

Roll bonding experiments of the 6111 alloy at room temperatures were not successful at first. The two strips of aluminum adhered together but it was easy to pry them apart manually. Annealing the strips at various temperatures for 2 hours and air cooling before bonding, however, resulted in relatively strong bonds. The results are given in Figure 9.5, showing the bond strength in terms of the reduction and the annealing temperature. Annealing at the highest temperature of 450°C created the strongest bonds. When the reduction was high, nearly 72%, the bond strength reached 50 MPa, about 30% of the strength produced by warm bonding. Decreasing annealing temperatures caused sharply lower strength.

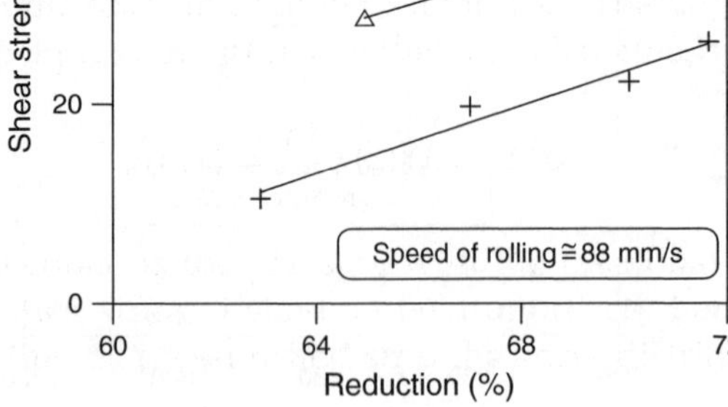

Figure 9.5 The shear strength of the bond as a function of the annealing temperature and the reduction.

The annealing process affected the microstructure of the aluminum strips. At lower temperatures, static recovery did not change the grain sizes but caused the dislocations to move in such a way that the effect of some or all of the prior cold working was removed and the material regained some lost softness. At the higher temperatures, static recrystallization created strain free, new grains and hence, the softening. Both processes enhanced the deformation of the asperities and allowed the formation of larger true areas of contact, leading to higher bond strength.

9.6 EXAMINATION OF THE INTERFACE

Using scanning electron microscopy, the appearance of the bonded surfaces, after separation, has been examined. In evaluating these surfaces, it is to be kept in mind that some damage during separation by the shear test must have affected their appearance. Results obtained during warm bonding are given first, followed by the observations in cold bonding. Following these, the side views of the warm and the cold bonds are shown.

9.6.1 Warm bonding

Three bonds of various strengths have been chosen for examination. In Figure 9.6 the surface of a strong bond is indicated. In Figure 9.7 a bond of medium strength is given while in Figure 9.8 a poor bond is presented. In each figure, the process parameters are also shown and while they are not identical, the differences are not expected to affect the surface appearance. The differences as a function of the bond strength are clearly observable. Figure 9.6, showing a strong bond, indicates significant permanent deformation of the

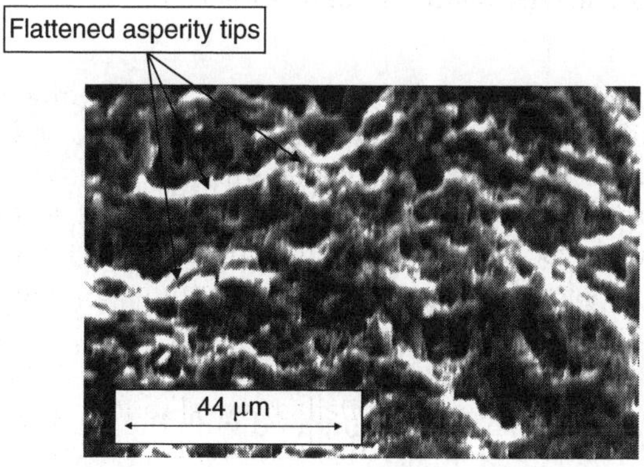

Figure 9.6 The bonded surface; entry temperature: 280°C; rolling speed: 52 mm/s; reduction: 67.5%; bond strength: 244 MPa.

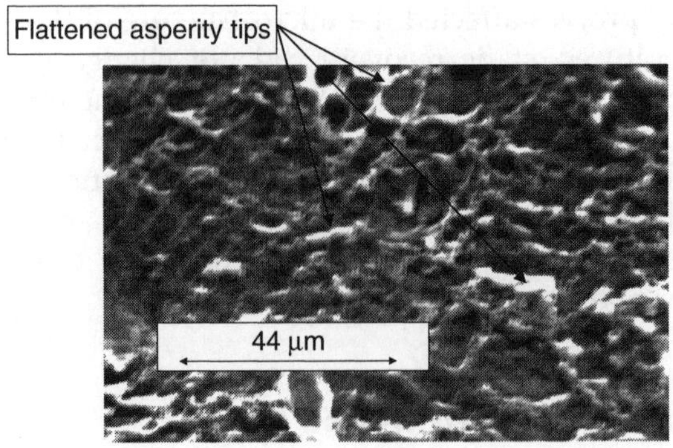

Figure 9.7 The bonded surface; entry temperature: 280°C; rolling speed: 78.5 mm/s; reduction: 66.5%; bond strength: 195 MPa.

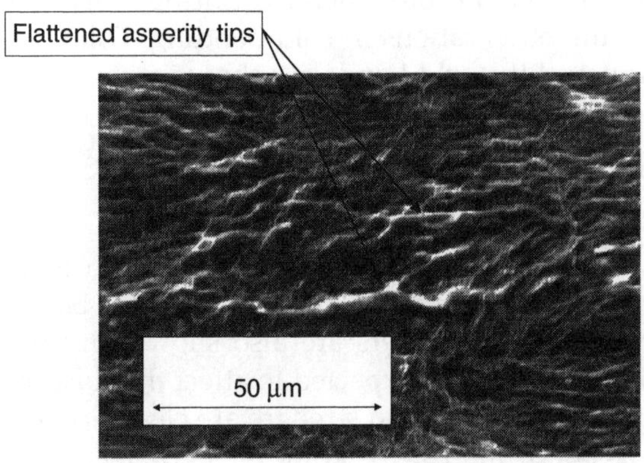

Figure 9.8 The bonded surface; entry temperature: 280°C; rolling speed: 73.9 mm/s; reduction: 48.5%; bond strength: 140 MPa.

asperity tops, giving rise to relatively large true areas of contact, resulting in a large number of adhesive bonds. Figure 9.7, demonstrating the average bond strength, indicates similar behaviour but the contact surfaces are somewhat smaller than those in Figure 9.6. The surfaces of the poorly bonded samples are shown in Figure 9.8, demonstrating much lower roughness and a noticeably smaller true area of contact. The appearance of the fracture surface suggests little permanent deformation of the asperity tops and that bonding only occurred at the top of the asperities.

Figures 9.6, 9.7 and 9.8 may be compared with the SEM photograph of Dick and Lenard (2005), obtained after the surfaces were separated by the peel test. The figures are similar in a general sense. The method of separation – shear versus the peel test, applying shear or normal stresses, respectively, to cause the separation – contributed to the appearance of the surfaces and it is expected that the peel test interfered less with the asperities.

9.6.2 Cold bonding

The surface of contact in cold bonding, after separation by the shear test, is shown in Figure 9.9. The bond strength was approximately 50 MPa, obtained at a reduction of 70.7%. The surface structure appears to be significantly different from that of the warm-bonded specimens. It is indicated that the bonds only occurred at the asperity tops. Also indicated is a possible relative movement of the strips during bonding, evidenced by asperity deformation in the direction of rolling, similar to the surfaces seen after rolling, using sand-blasted rolls (Dick and Lenard, 2005).

9.6.3 Side view of the bond

The side view of the bond of a two-layer warm-rolled blank is shown in Figure 9.10. The process parameters were: 280°C entry temperature, 52 mm/s rolling speed, 67.5 % reduction. The shear strength of the bond was 240 MPa. The large bonded areas are clearly evident as is the conformance of the surfaces to one-another.

The cross section of the bond of a cold-rolled specimen is shown in Figure 9.11, produced at 90 mm/s speed, a reduction of 70.7%, after annealing at 450°C, for 2 hours. The bond strength was nearly 50 MPa. The difference between the warm and the cold bonds is immediately noticeable. Much smaller contact area as well as much less conformance of the surfaces are noted when the process was conducted at room temperature. There are several clearly

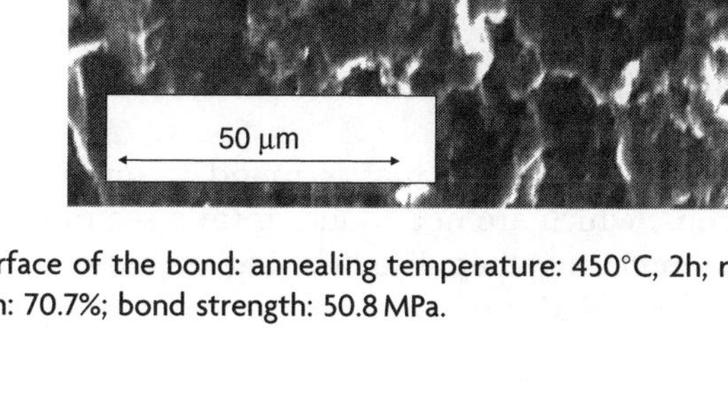

Figure 9.9 The surface of the bond: annealing temperature: 450°C, 2h; rolling speed: 90 mm/s; reduction: 70.7%; bond strength: 50.8 MPa.

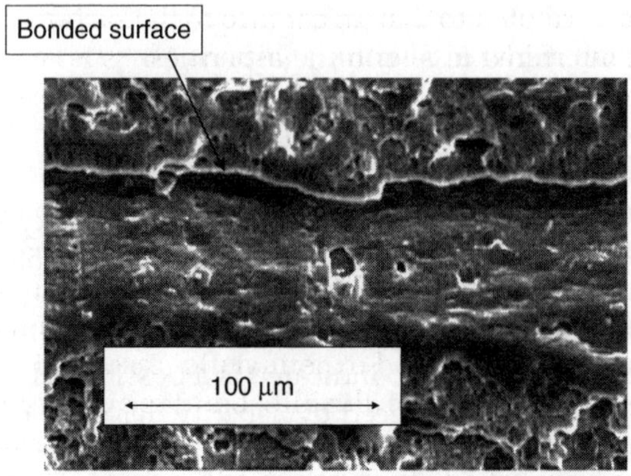

Figure 9.10 The cross section of warm-bonded specimen Entry temperature: 280°C; rolling speed: 52 mm/s; reduction: 67.5%.

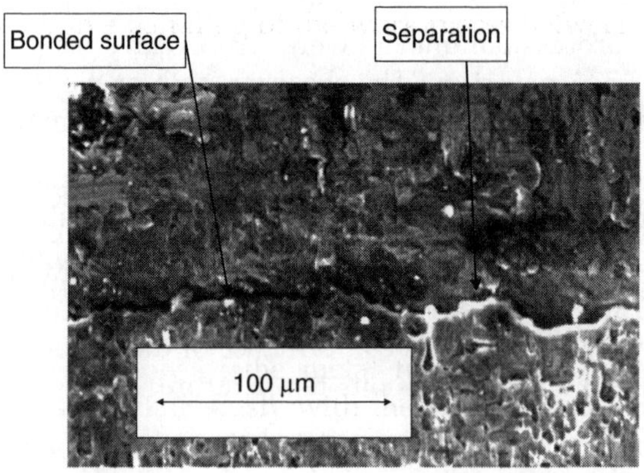

Figure 9.11 The cross section of cold-bonded specimen, Annealing temperature: 450°C, 2h; rolling speed: 90 mm/s; reduction: 70.7%.

visible gaps between the two layers. It is indicated again that bonding only happened at the asperity tips of two surfaces.

9.7 THE PHENOMENON OF BONDING

Metallic bonding, the mechanism that is found in aluminum, occurs when the valence electrons, which are not bound, form a sea of electrons (Callister, 2000), holding the metal together. The bonding energy of aluminum is given by Callister, (2000), as 324 kJ/mol.

The figure of Schey, referred to above (Schey, 1983) needs to be considered when the events, occurring during the roll-bonding process, are studied. Both surfaces are covered with contaminants in addition to a layer of aluminum oxide of approximately 1–2 μm thickness. The oxide is very hard; Rabinowicz gives its hardness as 2150 kg/mm^2 (Rabinowicz, 1965) but it is brittle and it breaks at tensile strains of approximately 0.0006–0.0008 (Callister, 2000). As the two strips placed on one-another enter the roll gap, the oxide layers are subjected to increasing normal pressures and when the strips deform in the direction of rolling, the oxides cannot follow. The layers then break-up into islands, exposing the new, clean, virgin metal surfaces, a prerequisite for the formation of adhesive bonds. This phenomenon is similar to the break-up of the iron oxides during hot rolling of steels, studied by Li and Sellars (1998) and Krzyzanowski and Beynon (2002), who concluded that the hot, new material, extruding in between the cracks in the oxides and contacting the colder roll surface, increases the coefficients of heat transfer and friction. The asperities of the freshly created aluminum surfaces in contact are now subjected to pressures in the range of 900–2000 MPa, see Figure 9.3. Calculations, using a finite element program designed to predict the variables during flat rolling of metals, indicate that the interfacial temperatures may reach as much as 180°C, softening the materials and enhancing the creation of the adhesive bonds (in the calculation of the temperature at the centre of the two-strip work piece, a heat transfer coefficient of 20 000 W/m^2 K, entry temperature of 22°C and a coefficient of friction of 0.25 were used, subjecting the strips to a reduction of 70% at a speed of 4 rpm.)

In order to identify the mechanisms that may be involved in the creation of the bonds, a brief summary of the results follows:

- No successful bonds were created when rolling the as-received metal at room temperatures;
- Bonding was successful when the strips were rolled at room temperatures only after they were annealed for two hours prior to rolling. The bond strength, which was a fraction of the strength of the parent metal, increased with increasing annealing temperatures;
- Higher entry temperatures resulted in higher bond strength;
- The time of contact in one pass was in the order of 30 – 220 ms.

Diffusion bonding needs time, pressure and temperature to proceed. It is hypothesized that in the present set of experiments, mechanical bonding, caused by the force of attraction of the atoms overwhelmed diffusion bonding. This is based mainly on the time available for diffusion, and when that is in the order of 200–300 ms, not much of mass transfer may have occurred, in spite of the large interfacial pressures. Annealing softened the metals and softened the adhesion in between the oxide layer and the parent metal, making it easier to fracture the aluminum oxide and creating larger true areas of contact. The softness allowed the metals to conform to each other well. The strength of the softened aluminum alloy was somewhat rate sensitive, partially accounting for the observation of the effect of the time of contact. As also indicated by Zhang and Bay (1997), when the clean, new metals make contact, mechanical bonding occurs.

The activation energy required for bond formation, a material property, may be defined in terms of the shear strength of the bond, and noting the strong temperature dependence, an Arrhenius type relation may be formulated:

$$\tau_{\mathrm{bond}} = \tau_0 \exp\left(-\frac{Q_\mathrm{b}}{RT}\right) \tag{9.1}$$

where the shear strength of the bond is designated by τ_{bond}, the shear strength of the original metal is τ_0, Q_b is the activation energy required for bond formation, R is the universal gas constant and T is the entry temperature in K. The activation energy needed to form bonds is plotted versus the energy input during the rolling pass, see Figure 9.12, where Q_b is given on the ordinate and the energy input for rolling – calculated using the measured torque, the speed of rolling and the time taken to travel through the roll gap – on the abscissa. Three distinct sets of points are shown, for room temperature and for 250 and 280°C entry temperatures. Bond formation appears to require the least amount of activation energy when the entry temperature is the highest, at 280°C, near the amount given by Callister (2000), 324 kJ/mol. While the figure indicates that as the energy input is increased – the reduction is increased – less energy is required to cause successful bonds, this change is not very pronounced and one may infer that the activation energy is not strongly dependent on the

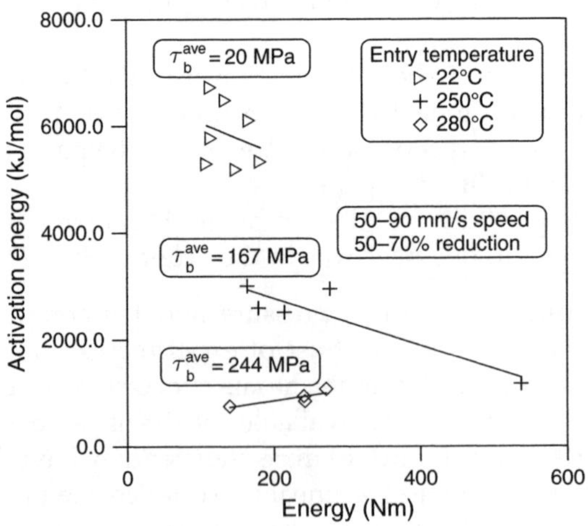

Figure 9.12 The activation energy of bonding as a function of the energy input and the temperature.

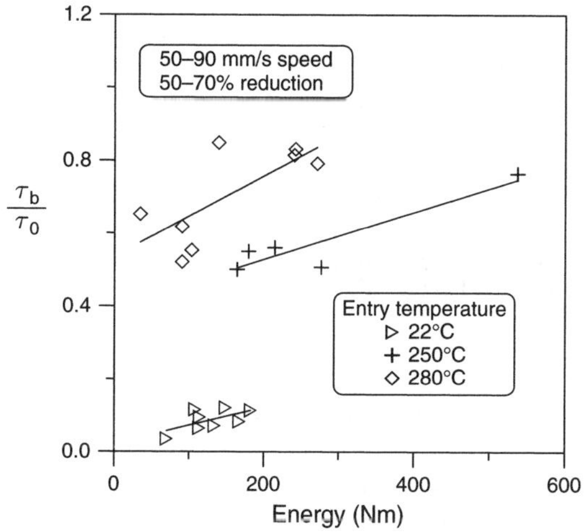

Figure 9.13 The ratio of the bond strength to the strength of the original material as a function of the energy input.

energy input. It is, however, strongly dependent on the temperature-induced structural changes, which are more pronounced at higher rolling temperatures.

A limitation of the physical meaning of equation (9.1) needs to be mentioned. The equation indicates that the bond strength reaches the shear strength of the metal when the temperature reaches infinity. The only conclusion that should be arrived at is that the ratio of the shear strength of the metal to the bond strength will approach unity as the temperature is increasing.

The energy input is related to the ratio of the shear strength of the bond to the shear strength of the original metal, in Figure 9.13. The process parameters are as given in Figure 9.12. The figure confirms earlier observations that the creation of stronger bonds requires higher effort. As before, the strongest bonds have been produced at the highest temperature and the weakest bonds at the lowest temperature.

9.8 CONCLUSIONS

- Strong bonds whose shear strengths are comparable to that of the parent metal can be created with the Al 6111 alloy, by warm roll bonding;
- Entry and annealing temperatures, the reduction and the rolling speed are the four important parameters of the roll bonding process. The most important factor is the temperature. Warm roll bonding creates higher bond shear strength than cold roll bonding. In cold bonding, it was possible to approach but not to reach the same strength as the parent material;
- Mechanical bonding was the dominant mechanism.

CHAPTER 10

Flexible Rolling

Abstract

Flexible rolling is a relatively new technology. Its objective is to produce sheet material for components which will be subjected to loads and stresses that vary with location. By controlling the roll gap during the pass, the product thus created possesses a varying thickness; thick where the loading is heavy, thin where it is not. As a direct competition to tailor-welded blanks, the material discontinuity is removed. After a brief review of the background, the results of a set of experiments on flexible rolling are presented.

Aluminum and steel strips are rolled such that the roll gap is changed during the pass, creating strips whose geometrical and mechanical attributes vary along their lengths. The resulting attributes and parameters of the process are related. The fracture strains of the strips are determined and compared to that of tailor-welded blanks. The rolling passes cause a significant increase in the metals' resistance to deformation at the expense of drastic reductions of the ductility. It is apparent that post-rolling annealing is necessary to restore sufficient formability to permit subsequent sheet metal forming processes to be successful.

10.1 INTRODUCTION

One of the most important current objectives of the automotive industry is to develop the technology to produce lightweight components. New materials are being introduced, among them the interstitial free steels, transformation induced plasticity steels, the high strength low alloy steels, bake hardenable steels, dual phase steels, martensitic and manganese–boron steels, having yield strengths that vary from a low of 200 up to 1250 MPa. The elongation of these steels decreases as the strength increases, from a high of nearly 40% to a low of 4–5%, affecting their subsequent applications (Flaxa and Shaw, 2003). A recent review indicates, however, that light construction steels with induced plasticity possess tensile strength in the order of 1000 MPa and total elongation of 60–70%. (Ehrhardt et al., 2004)[1]. New processes are also being examined in order to satisfy the dual objectives of reduced weight and increased strength.

[1] See Chapter 4 for a discussion of some of the recently developed steel alloys.

One of the methods combines advanced computing techniques to determine the exact load-carrying needs of a component with the application of tailor-welded blanks.

Tailor-welded blanks are made up of two sheets of unequal thickness which are welded to form a blank for subsequent sheet metal operations. Their importance in the sheet metal forming industry is indicated by a statement, extracted from the American Iron and Steel Institute's website (AISI.com, 2005):

> *Demand is expected to jump from 20 million tailor-welded blanks in 1999 to more than 90 million in 2005.*

As Worswick (2002) writes, the weld metal strength in welded aluminum alloy blanks is less than that of the parent metal which contrasts the behaviour of steel welds that tend to be overmatched. Chan et al. (2003) developed forming limit diagrams of tailor-welded blanks, made of cold rolled steel sheets, butt-welded by Nd:YAG lasers. They concluded that while the strengths of the tailor-welded blanks and the parent metals were similar, the formability of the blanks was lower and depended significantly on the thickness ratio. Higher thickness ratios resulted in lower formability. Min and Kang (2000), testing two steels, with 0.12 and 0.35% C contents, agree. As well, there is agreement that higher gauge ratios tend to push failure away from the weld into the more ductile thinner gauge material. Figure 10.1, reproduced from Worswick (2002), shows a typical weld cross-section of an aluminum alloy.

It is observed that the weld is not ideal and exhibits an undercut at the top surface and shows some weld metal porosity. The weld seams cause discontinuities and thus affect formability as well as the fatigue resistance of the subsequent components, produced using the tailor-welded blanks. Kampuš and Balič (2003) compare the deep drawing capabilities of tailor-welded blanks using laser and MIG welding and four steels of similar yield strengths. They comment that the use of laser welding of thin sheet metals is questionable

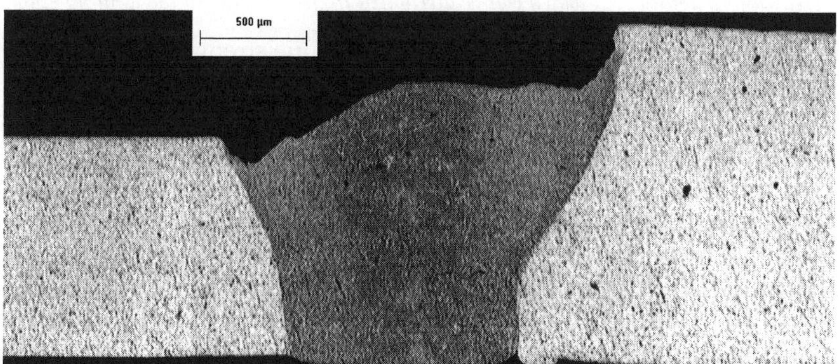

Figure 10.1 The cross-section of the weld in a tailor-welded blank (Worswick, 2002).

because of the high power requirements and hence, lasers are used less often. Using MIG welding, the microhardness in the weld of two of their metals increased to 400 Hv and fracture always appeared in the weld. When using laser welding, the hardness increased to 700 Hv. They found that drawing the laser-welded cups was not successful, even after annealing.

Venkat et al. (1997) evaluated the effect of autogenous CO_2 laser butt welding using 6111-T4 aluminum alloys on their elongation under uniaxial tension. The uniform elongation of the welds varied from a low of 8.3% to a high of 18.3%, compared to 24% of the base metal. Ahmetoglu et al. (1995) conducted deep drawing experiments using aluminum killed deep drawing quality steel sheets. They found that failure occurred at the flat bottom, parallel to the weld line. They attributed this to the non-uniform distribution of the deformation and concluded that the difficulties occasioned by the use of tailor-welded blanks in sheet metal forming operations make process planning more difficult and new design guidelines need to be developed.

It may be concluded that while tailor-welded blanks represent a major advance in the quest for weight reduction of automotive components, the discontinuity caused by the weld seam often causes difficulties.

Merklein and Geiger (2002) write about the need to develop new materials and processes to produce innovative lightweight construction, of special importance in the automotive industry. Kopp et al. (2002) and Kopp (1996) describe a fairly new technique to produce tailored blanks – that of flexible rolling. In that process, the roll gap is changed during the pass and depending on the desired final product, it may be increased or decreased, producing a work piece whose attributes – geometrical, mechanical and metallurgical – vary along the length. The authors also conducted limiting dome height tests and found that when the thickness transition is appropriate the heights were near that of the parent metal. They further comment that:

> *this is a great advantage compared to Tailor-welded Blanks, which are known for their decreased formability . . . . which can be attributed to the presence of the welding seam.*

While rarely reported, it is expected that the fatigue strength of the continuous material, rolled but not interrupted by welding, will be improved.

In the more-reduced portion the grain sizes are decreased, strength is increased and the ductility is decreased when compared to the less-reduced portion. The weld and the attendant discontinuity are eliminated. The manner and the rate of change of the roll gap will affect the stress concentration at the changeover.

The advantages and disadvantages of strip profile rolling and flexible rolling when compared to tailor-welded blanks are discussed by Kopp et al. (2005) in addition to a description of the behaviour of the rolled products in subsequent sheet metal forming processes. They conclude that both processes may offer cost savings over the welded blanks. It is worthwhile to quote directly from the website Mubea.com (Mubea.com, 2004) describing the first commercial application of flexible rolling.

> *Flexible Rolling is a new rolling process with considerable lightweight construction potential for the automotive industry. This technology is characterised by the fact that the roller gap is selectively altered during the rolling process by means of a control system, enabling the metal thickness to be adapted everywhere along the length of the sheet steel to suit the respective loads to which the component is to be subjected. Furthermore, Flexible Rolling shortens the process of manufacturing the sheet metal and enables it to be formed by further processing stages, such as deep drawing or hydroforming. The reduction will directly affect the steels' strength and ductility and as such, is expected to be one of the most important process parameters.*

A more recent check of Mubea.com shows how the product line on tailor-rolled sheet metals has evolved. Recently, Danieli of Italy commissioned a flexible rolling mill for Mubea.

Hirt et al. (2005) write that up to 50% thickness changes are now possible using Strip Profile Rolling. They also consider wrinkling during deep drawing of the tailor-rolled blanks and describe the die modifications that minimize the problems – use of an elastic blank holder is most beneficial.

In this chapter, the relation of the process and material parameters to the resulting formability of the tailor-rolled blanks is examined. Three materials are experimented with, an aluminum alloy, often used in the automotive industry, and two low carbon steels. The roll gap is changed as a step-function shortly after entry into the roll gap; however, the inertia of the screw-down system leads to a gradual change of the thickness. The reduction in the two parts of the sample is varied in a systematic manner, keeping the total reduction to 55–56%. The results of the two-stage passes are compared to those obtained in single-stage passes. The roll separating forces, the changes in the roll gap and the resulting changes to the rolled metals' resistance to deformation and strain at fracture are examined.

10.2 MATERIAL, EQUIPMENT, PROCEDURE, SAMPLE PREPARATION

10.2.1 Material

Two low carbon steel strips and an aluminum alloy were experimented with. The first set of steel samples was made of AISI 1030, cold drawn steel alloy and the other is of AISI 1008, deep drawing quality steel. The engineering stress–strain equation for the cold drawn AISI 1030 steel, obtained in a conventional, uniaxial tension test is $\sigma = 400\,(1 + 169\varepsilon)^{0.083}$ MPa and for the 1008, deep drawing quality (DQSK) it is $\sigma = 243\,(1 + 139.5\varepsilon)^{0.095}$ MPa. The strain at fracture of the 1030 steel, before rolling, is 25–27% and that of the

Table 10.1 The chemical compositions of the alloys, weight %

	Cu	Mn	Si	Zn	Mg (max)	C (max)	P (max)	S (max)	Fe	Al
6111	0.82	0.21	0.21	0.02	0.01	–	–	–	–	rest
1030	–	0.6–0.9	–	–	–	0.28–0.34	0.4	0.05	rest	–
1008	–	0.3–0.5	–	–	–	0.1	0.04	0.05	rest	–

1008 steel is, surprisingly, only slightly over 18%. The strength of the aluminum alloy, Al 6111-T4, an automotive grade, is well described by the relation $\sigma = 150\ (1 + 157\varepsilon)^{0.245}$ MPa. The aluminum alloy is identified as among those that provide the best forming properties along with very good weldability, demonstrating a fracture strain of nearly 45%, though in the present tests only 22% strains were obtained. The carbon steels are comparable to the materials used by Min and Kang (2000) and Chan et al. (2003).

10.2.2 Equipment

The two-high experimental mill, described above in Chapter 9, was used here as well.

10.2.3 Procedure

The rolls and the strip to be rolled are cleaned using acetone. The roll gaps for the single or the two stages of rolling and the roll speed are set and the roll pass is started. The digital data acquisition system begins the data collection after the roll force reaches a predetermined level and stops automatically when the rolled strip exits. All passes are performed without lubrication.

10.2.4 Sample preparation

The steel samples are 1.32 (AISI 1030) and 1.12 mm (AISI 1008) thick, 25 mm wide and 600 mm long. The aluminum samples are 1.02 mm thick, 25 mm wide and 600 mm long. The edges are deburred and the strips are cleaned using acetone. The long dimension is always in the direction of previous rolling passes.

10.3 RESULTS AND DISCUSSION

10.3.1 Roll separating forces and the roll gap

10.3.1.1 AISI 1030 steel, cold drawn

The variations of the roll force during single- and two-stage rolling passes are shown in Figure 10.2. The specific roll force is plotted on the ordinate while

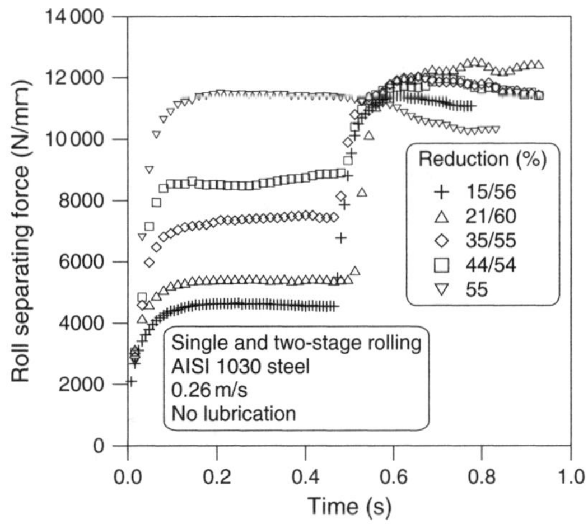

Figure 10.2 The roll forces in single and two-stage rolling; AISI 1030 alloy.

the time elapsed from the start of the pass is given on the abscissa. All tests are conducted at a roll surface speed of 0.26 m/s. The roll gap is changed at approximately the middle of the pass, 0.5 s after entry. No lubricants are used. The data from a single-stage and four two-stage experiments are indicated in the figure and the reductions/pass in each test are also presented. In all four tests the aim was to roll to progressively larger reduction in the first stage and keep the total reduction to as near 55% as possible so the reduction in the second stage was progressively smaller. The objective in the choice of these reductions was to test the metals' reaction to cold working when subjected to various sequences. The roll force obtained in a single-stage pass of 55% reduction is shown for comparison.

The roll force in the single-stage pass is 11370 N/mm, steady for most of the pass, dropping slightly near the exit. In the second pass of the two-stage reductions – all of which are approximately of similar total magnitudes – all of the forces rise beyond that value, with one exception. While the differences are not very large, of the order of approximately 1000 N/mm, about 10%, they are still noticeable and are not expected to be caused by experimental scatter.

It is hypothesized that tribological factors are at play. Since no lubricants are used, the interfacial tribological mechanisms include adhesion and ploughing. The rolls were ground and finished manually, creating shallow-angle asperities, making the contribution of adhesive forces overwhelm those of ploughing. Hence, as implied by the adhesion hypothesis, the creation and the rate of growth of adhesive bonds would control the transfer of energy and the frictional conditions at the contact surface.

The reductions in the first stages of the four two-stage tests are smaller than in the 55% single-pass experiment so the flattening of the asperities up

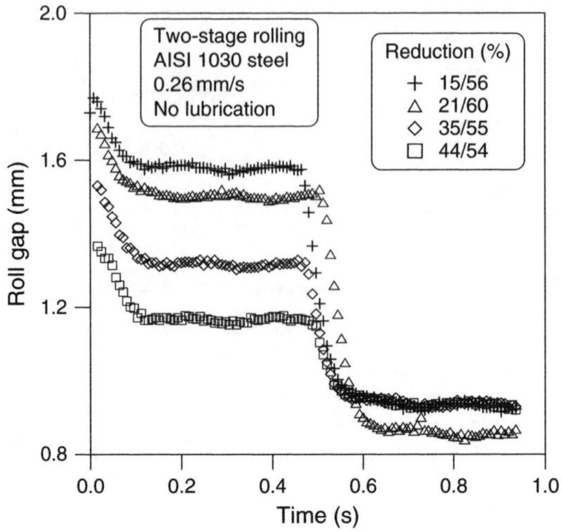

Figure 10.3 The roll gap changes during two-stage rolling; AISI 1030 alloy.

to the change in the roll gap is not as pronounced as in the second stage. It is the sudden change of the roll gap that is responsible for the change of tribological conditions. Interface events depend, among other mechanisms, on the size and the relative movement of the contacting surfaces. The roll gap change, completed in about 100 ms in a direction perpendicular to the direction of rolling, adds to the usual relative movement along the roll surface, increases the contact areas and adds to the creation of adhesive bonds; hence, the approximately 10% growth of the roll separating forces.

The changes of the roll gap during the two-stage passes, while rolling the 1030 steel alloy, are shown in Figure 10.3. The roll gap is plotted on the ordinate and as before, the time is given along the abscissa.

In addition to the changes of the roll gap during the two stages of rolling, the dynamic response of the mill is also observable in the figure. As soon as the strip enters the roll gap, the upper and the lower rolls separate and in about 100 ms the requested roll gap is reached. The command to change the distance between the rolls is according to a step-function and approximately another 100 ms are needed to reach the steady-state roll gap distance of the second pass. The 100 ms delays are contributed to by two events: the response of the servovalves with a frequency response of 100 Hz, and the inertia of the massive top roll and its connections to the hydraulic apparatus.

10.3.1.2 AISI 1008 steel, cold drawn

The roll separating forces obtained in a single-stage pass of 55% reduction and two two-stage passes of 14/39% and 20/33.6% reduction are given in Figure 10.4, measured when rolling the softer AISI 1008, deep drawing quality steel. Figures 10.2 and 10.4 are quite similar. The dynamic response of the

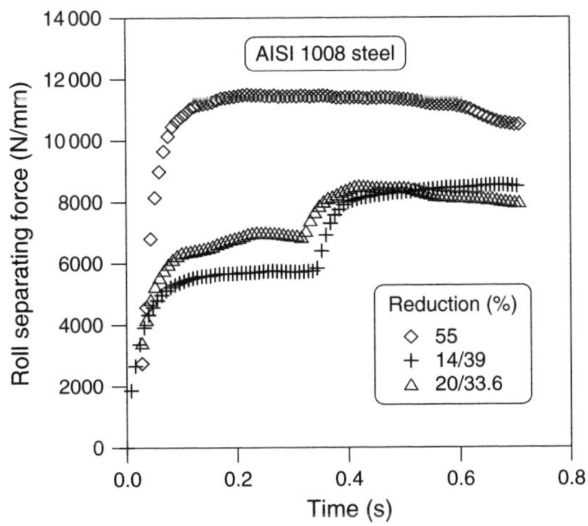

Figure 10.4 The roll forces in single and two-stage rolling; AISI 1008 alloy.

mill is also similar to the previously observed behaviour. As expected, the roll separating forces increase with the reduction and remain fairly steady as the process continues. At the time the roll gap is changed, the roll forces increase to approximately the same magnitude, as the total reduction in the two passes are nearly identical.

10.3.1.3 Al 6111 aluminum alloy

Figures 10.5 and 10.6 demonstrate the behaviour of the 6111 aluminum alloy when subjected to single- and two-stage rolling passes. The roll forces are shown in Figure 10.5 and the roll gap changes are given in Figure 10.6.

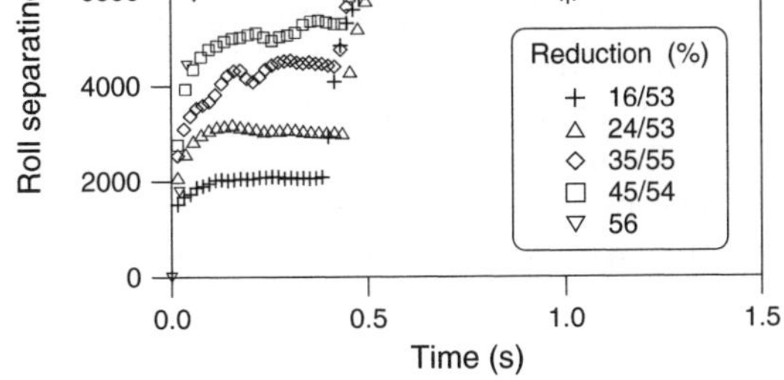

Figure 10.5 The roll forces in single and two-stage rolling; 6111-T4 alloy.

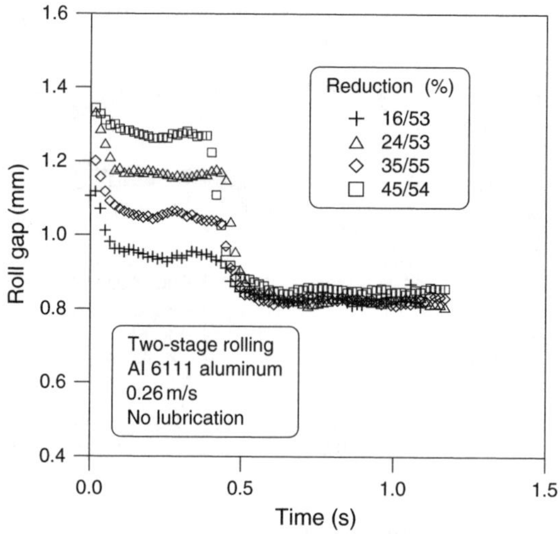

Figure 10.6 The roll gap changes during two-stage rolling; 6111-T4 alloy.

The reductions in the single- and double-stage passes are approximately the same, see Figure 10.5. The roll forces in the two-stage passes don't rise beyond the single-stage data, opposing the observations when the harder steel alloy was rolled. This is caused again by the flattening of the asperities. When the softer aluminum alloy enters the roll gap, its asperities flatten rapidly and even at the smaller reductions they reach their maximum deformation; hence, no changes in the tribological behaviour result. The dynamic response, see Figure 10.6, is as observed before.

10.4 PREDICTIONS OF A SIMPLE MODEL

The measured roll separating forces are realistic and consistent with earlier results. Using a simple 1D model of the flat rolling process, in which the elastic deformation of the roll is analysed using the theory of elasticity[2] (Roychoudhury and Lenard, 1984) and the stress–strain relations, given above, are employed, the forces are calculated and found to be quite close to those measured. The predictions and the measurements are compared in Table 10.2 for the 1030 steel. The coefficient of friction values used in the computations are also indicated. In the first five rows, the first passes are

[2] See Section 3.4.

Table 10.2 Comparison of the measurements with the predictions of a simple model; rolling of the 1030 steel strips

Reduction (%)	Roll force (N/mm) measured	Roll force (N/mm) calculated	Coefficient of friction
15 (single stage)	4580	4610	0.095
21 (single stage)	5400	5510	0.085
35 (single stage)	7400	7403	0.078
44 (single stage)	8800	8803	0.078
55 (single stage)	11370	11316	0.082
55 (second pass)	12000	11892	0.086

modelled, including the single pass, at 55% reduction. As mentioned above, the 55% reduction in the second passes caused the roll forces to increase due to increased frictional resistance and the predictions of the model indicate that this is, in fact, the case. The force measured in the second stage of the 35/55% two-pass rolling is approximately 12000 N/mm. Calculations demonstrate that the model predicts a force of 11892 N/mm, when using a coefficient of friction of 0.086, a 5% increase over the value needed to predict the force on the single-stage pass. Admittedly this is not much of a change, but it is not taken to be insignificant. When designing pass schedules for flexible rolling, the changes of tribological conditions, however small, need to be taken into account.

Table 10.3 gives similar results for the 1008 steel. The magnitudes of the coefficient of friction are realistic and as published elsewhere (Lenard, 2001). As expected, they decrease as the reduction is increased. As well, their magnitudes are higher than those found in the case of the 1030 steel, showing the applicability of the adhesion hypothesis and the enhanced ease of the formation of adhesive bonds with the softer steel. The predictions of the model for the Al 6111 aluminum alloy are given in Table 10.4 and the conclusions are as for the two steel alloys. The aluminum alloy is the softest of the three metals and the formation of adhesive bonds occurs easiest.

Table 10.3 Comparison of the measurements with the predictions of a simple model; rolling of the 1008 steel strips

Reduction (%)	Roll force (N/mm) measured	Roll force (N/mm) calculated	Coefficient of friction
35.6	6788	6897	0.15
56.8	8935	8962	0.105

Table 10.4 Comparison of the measurements with the predictions of a simple model; flexible rolling of the aluminum strips

Reduction (%)	Roll force (N/mm) measured	Roll force (N/mm) calculated	Coefficient of friction
16	2000	2023	0.04
24	3150	3110	0.068
35	4500	4557	0.075
45	5100	5141	0.06
56	7380	7388	0.07

10.5 STRAIN AT FRACTURE

The longitudinal strains at fracture in a conventional tension test are a reasonable indication of the metals' formability. These were determined using the 6111 aluminum and the AISI 1008 steels, as a function of the processing parameters, and are shown in Figures 10.7–10.9. The loss of ductility as a result of cold working is very pronounced in the aluminum alloy, see Figure 10.7.

The fracture strain, approximately 22% in the as-received metal, dropped to about 1% after two-stage rolling to a total of 59% reduction, indicating the impossibility of further plastic forming. The AISI 1008 steel shows similar behaviour, as demonstrated in Figure 10.8. The strain at fracture before rolling is almost 19%. When rolled to 58.5% in a single pass or to 52.5% in two passes, the strains reduce to about 1%, as with the aluminum. Some, but not all,

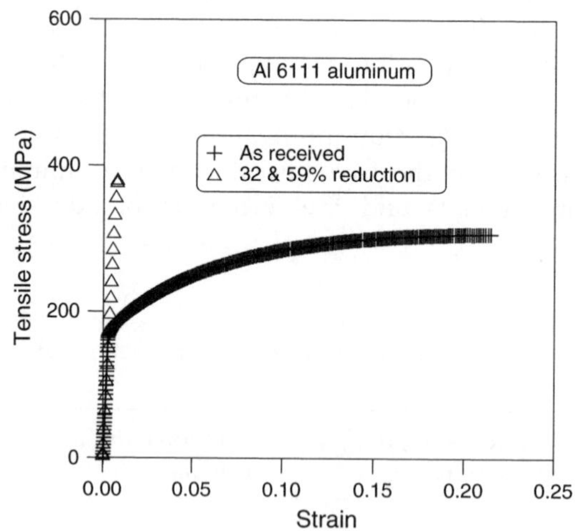

Figure 10.7 The stress–strain curves of the 6111-T4 alloy.

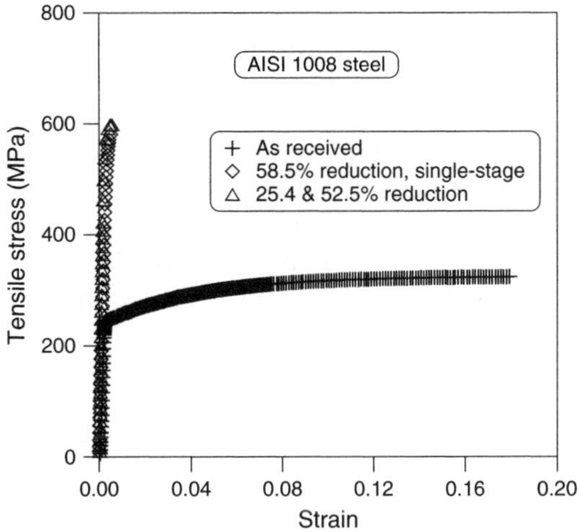

Figure 10.8 The stress–strain curves of the AISI 1008 steel.

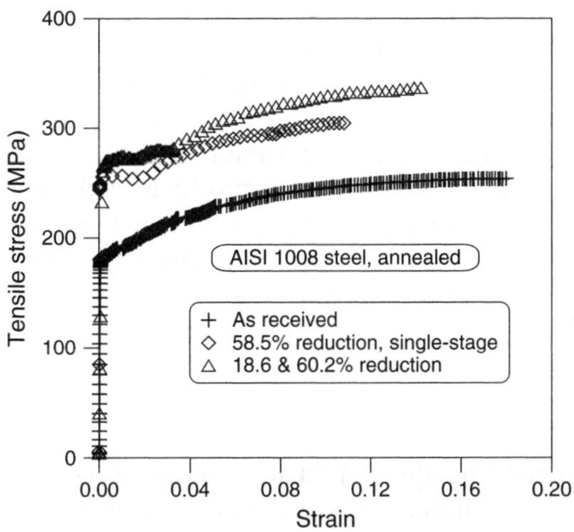

Figure 10.9 The stress–strain curves of the AISI 1008 steel, annealed.

of this loss is regained after annealing, indicated in Figure 10.9, confirming the conclusions of Dyment (2004) whose experiments in tube bending and annealing showed significant recapture of ductility.

It is evident that careful annealing before subsequent processing is needed. This would increase the formability, albeit at the expense of the increased strength.

The ductility of the 6111–T4 alloy, obtained after rolling may be compared to that of tailor-welded blanks, obtained after laser welding. Venkat et al.

(1997) used a 3 kW laser beam to weld two pieces of the alloy of the same thickness. The ductility of the weld was measured in traditional tension tests. The longitudinal elongation varied, depending on the velocity of the laser beam, from a low of 8.6% to as high as 18.7%, much higher than in the present study, emphasizing again the importance of annealing after flexible rolling. The results of tension tests on the welds of tailor-welded aluminum samples of the 5000 designation were given by Davies et al. (2001). The strain hardening exponents of the alloys, taken at the onset of instability, varied from a low of 0.22 to a high of 0.257. Some, but not major, losses of ductility were reported, with the fracture strains of the welds approaching that of the parent metal.

Min and Kang (2000) tested the effect of flash welding on an S35C steel of 2 mm in thickness. The S35C steel is similar to the AISI 1030 in both chemical composition and tensile strength; it is somewhat more ductile in the as-received state, demonstrating an elongation of 53%. After welding, the elongation reduced to 15%.

The comparison of the formability of the tailor-welded blanks and that of the tailor-rolled blanks is not easy. The welded blanks were not subjected to cold working, so the loss of formability is the direct function of the welding process. The rolled samples lose ductility through the rolling passes, which can, however, be restored, at least to some extent, by well-designed annealing.

Two further aspects became clear and must form the topic of the next study of tailor-rolled blanks. These concern the stress concentration at the thickness change and its effect on the fatigue strength of the rolled blanks, especially on samples that will eventually be used to produce load-carrying components. These results should then be compared to that of tailor-welded blanks. The results will allow a clear choice concerning the superiority or otherwise of tailor-rolled blanks over tailor-welded blanks.

10.6 CONCLUSIONS

Flexible rolling, a recently developed technology to create sheet metal with varying thickness, was discussed. Details of a set of experiments, examining the effect of the process on the rolling variables, were presented.

The effect of the reduction on the roll separating forces, resistance to deformation and subsequent strain to fracture during the production of tailor-rolled blanks was considered. An aluminum alloy, used in the automotive industry, and two low carbon steels were subjected to single- and two-stage, continuous rolling passes. The reduction in the first stage was progressively increased, while the total reduction was kept to 55%. The metals' strength increased as a result of cold working and their ductility decreased drastically. Some of the ductility was regained after annealing. It became apparent that in order for the tailor-rolled blanks to be competitive with tailor-welded blanks, post-rolling annealing is unavoidable.

CHAPTER 11

Problems and Solutions

Abstract

In the first part of Chapter 11 assignments are presented, listed according to the Chapters in the book. The problems in the first section, General Background, are designed to recall the basic information needed in the understanding of the ideas presented in the discussion of the flat rolling process, concerning mechanics, strains, incompressibility, theory of plasticity, plastic deformation and flow, heat transfer and boundary conditions. This is followed by a set of problems for each of the nine remaining chapters. Some of these problems require extensive calculations, using the derivations and the expressions given in the text. While some of them are designed to illustrate the material by simply applying the formulas as given, some will require in-depth thinking, as well. Several of them request that the equations, stated in the text be derived from first principles. Many problems will be tractable only after a visit to a library and/or the Internet.

In the second section, the solutions to most, but not all, of the problems are listed. These are given in various ways. For some, detailed step-by-step solutions are shown. For some others, hints and advice that indicate a possible, but not necessarily unique, approach towards the solutions are offered. Numerical answers to some of the questions are also provided. It is possible that in some instances the answers the readers will obtain may differ from those given here as the set of assumptions may well be different from those made by the author.

PART 1: PROBLEMS

Chapter 1: Introduction

Problem 1. The figure below shows the undeformed and the deformed shapes of a plate. The strain distribution in the deformed plate is given by

$$\varepsilon_x = 0.005xy + 0.005y^2 \quad \text{and} \quad \varepsilon_y = 0.006yx^3 + 0.007\,(y + x)$$

where x and y are in meters.

- Determine the displacement components u and v of point D;
- Determine the shear strain at point C;
- Assuming a state of plane stress determines the stress component at point C.

Let $E = 207 \times 10^3$ MPa and $\nu = 0.3$.

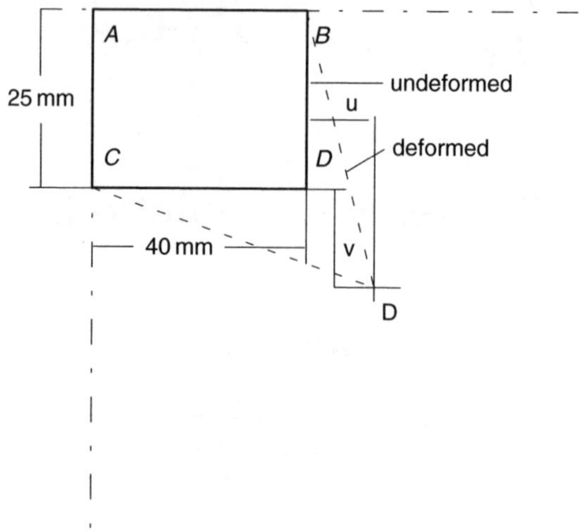

Figure 11.1 The undeformed and the deformed shapes of a plate.

Problem 2. Considering logarithmic strains, show that

$$\varepsilon_{\mathrm{I}} + \varepsilon_{\mathrm{II}} + \varepsilon_{\mathrm{III}} = 0$$

indicates incompressibility. Note that the Roman numeral subscripts indicate the principal strains.

Problem 3. The radial and circumferential stresses, respectively, in a rotating, hollow disk are given by the expressions

$$\sigma_r = \left(\frac{3+\nu}{8}\right)\rho\omega^2\left[a^2 + b^2 - (ab/r)^2 - r^2\right]$$

$$\sigma_\vartheta = \left(\frac{3+\nu}{8}\right)\rho\omega^2\left[a^2 + b^2 + (ab/r)^2 - \frac{1+3\nu}{3+\nu}r^2\right]$$

where the internal radius is $a = 25\,\mathrm{mm}$, the external radius is $b = 125\,\mathrm{mm}$, $\nu = 0.3$, the yield strength is $\sigma_y = 300\,\mathrm{MPa}$ and the density is $7800\,\mathrm{kg/m^3}$. Determine the rotational speed and the location at which yielding will begin, according to the maximum shear stress criterion.

Problem 4. Consider the problem of tube eversion, – in which a thin-walled tube is compressed longitudinally by two dies and in the process the ends of the tube are everted – shown schematically below. Show that if the material behaves according to the Tresca criterion of plastic flow, the wall thickness is not expected to change.

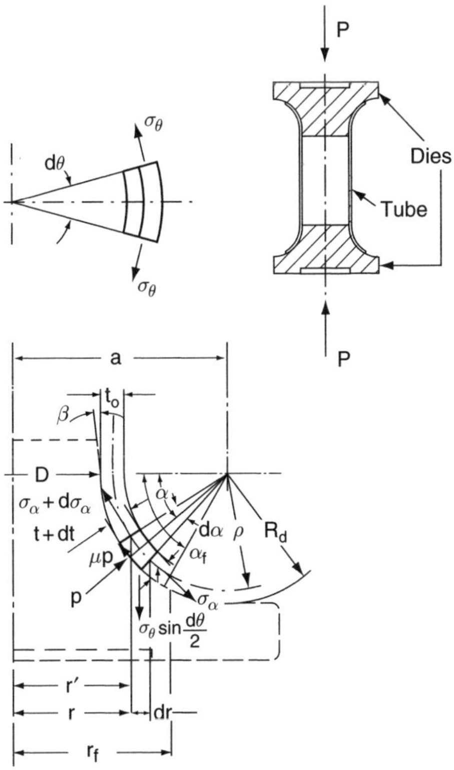

Figure 11.2 A schematic diagram of the tube-eversion process.

Problem 5. In open die forging of a right, circular cylinder, the work piece is compressed by two, rigid, rough platens. Define mathematically the boundary conditions for the stresses and velocities.

Problem 6. Write down the strain tensor for a state of plane stress.

Problem 7. Give the definitions for the three modes of heat transfer: conduction, convection and radiation.

Problem 8. Does the volume of a piece of metal, deforming elastically, remain constant? If not, how does it change?

Problem 9. Estimate the dimensional error of the rolled strip's thickness across the 2000 mm width when the roll axes make 0.5° with one-another while the rolls don't flatten or bend.

Problem 10. Check the World Bank's Yearbook for the amount of steel produced in the last few years.

Problem 11. Look for statistics that indicate the number of people employed in steel production over the years. Has the number increased or decreased during the last few decades?

Problem 12. Contact used equipment sales companies to find out the cost of various used rolling mills.

Problem 13. The work rolls are connected to the driving motor through flexible joints. It is essential that the angular velocity of the rolls be constant within each turn. What type of flexible coupling, with its flexible joints, will allow this to happen?

Chapter 2: Flat Rolling

Problem 1. Sketch the free-body diagram[1] of one of the driven back-up rolls of a four-high rolling mill.

Problem 2. Show that the minimum coefficient of friction is given by $\mu_{min} = \tan \phi_1$ where ϕ_1 is the bite angle.

Problem 3. Consider a heavy reduction of a fairly thick plate, such that the planes-remain-planes assumption is not valid any longer. How would an originally straight cross-sectional plane change by the time it reaches the exit?

Problem 4. The wide strip to be rolled is initially of uniform thickness across its width. It is rolled using a non-cambered, small diameter work roll, in a two-high mill. How would the strip's velocity at the exit vary across the width?

Problem 5. Estimate the radial strain experienced by the steel work roll during temper rolling. Assume that the original roll diameter is 400 mm. The entry thickness of the strip is 1 mm, and it is reduced by 2%, to a final thickness of 0.98 mm. Also assume that the roll becomes completely flat in the contact zone. Is this strain large enough to cause plastic deformation of the high carbon steel roll?

Problem 6. Consider a two-high mill. The work roll diameter is 200 mm and 1500 mm long. Estimate the magnitude of the force in the direction of rolling that would cause the work roll to deflect by 2 mm in that direction. Let $E = 200\,000$ MPa.

Problem 7. The entry speed of the transfer bar into the first stand of the finishing mill is 1000 mm/s. The entry thickness is 20 mm while the exit thickness from the last stand is 2 mm. Estimate the exit velocity of the strip.

---

[1] The free-body diagram shows *all* forces acting on a body.

Problem 8. Estimate the time of contact during the pass in the last stand of the finishing mill. Take the entry thickness as 2.7 mm and the exit thickness as 2 mm. The work roll diameter is 800 mm and its rotational speed is 200 rpm.

Problem 9. Show that in flat rolling, assuming a state of plane-strain plastic flow of a Huber-Mises material, the stress in the direction of vanishing strain is given by

$$\sigma_3 = \frac{\sigma_1 + \sigma_2}{2}.$$

Problem 10. Sketch how the actual entry of a strip into the roll gap may look like, including the thickening of the leading edge of the strip and the slight compression of the roll.

Chapter 3: Mathematical Modelling

Problem 1. Show that the projected contact length is given by $L = \sqrt{R'\Delta h}$, where R' is the radius of the flattened roll and $\Delta h = h_{\text{entry}} - h_{\text{exit}}$.

Problem 2. Compare the predicted power requirements of a rolling mill, using equations 3.8, 3.9 and 3.62. Under what conditions do they compare well? Comment critically on the predictions.

Problem 3. Estimate the magnitude of the mill stretch in a small laboratory rolling mill. Compare it to the stretch of a full size industrial mill. Take realistic assumption for the dimensions of both mills. Consider rolling a thin, low carbon steel strip.

Problem 4. The entry speed of the transfer bar into the first stand of the finishing mill is 1000 mm/s. There are seven mill stands and the reduction at each is 25%. The entry thickness is 20 mm. Estimate the exit velocity of the strip.

Problem 5. Derivations

Show that the following equations can be derived from first principles:

- Derive equation 3.18, using a balance of forces on a slab of the deforming material in the roll gap;
- Derive equation 3.19, using the Huber-Mises yield criterion and the condition of plane strain plastic flow;
- Derive equation 3.20, using geometry of the deforming strip;
- Derive equation 3.32, considering the independent variable to be the distance in the direction of rolling;
- Derive equation 3.48 for the stress due to the force of inertia.

Problem 6. A strip of 6061-T6 aluminum is to be produced. The dimensions of the rolled strip are to be: 0.25 mm thickness, 25 mm width and 1000 mm length. The starting material is available in strips of 10 mm thickness, 25 mm width and 2000 mm length.

The plates are to be hot rolled, at 500°C, to a thickness of 2 mm. The thickness of 0.25 mm is to be obtained subsequently by cold rolling, in which a light mineral oil base is used as the lubricant, with oleic acid as the boundary additive.

Design the flat rolling process by prescribing the reductions to be done in both the hot and the cold rolling processes.

The details of the two rolling mills are:

Hot mill:

Roll radius: 150 mm
Roll speed: 50 rpm
Sticking friction
Motor power: 45 kW

Cold mill:

Roll radius: 125 mm
Roll speed: 100 rpm
Coefficient of friction: 0.1
Motor power: 100 kW

The true stress–true strain relations, obtained in uniaxial tests, are:

For hot deformation: $\sigma = 37\dot{\varepsilon}^{0.17}$ MPa
For cold deformation: $\sigma = 450\varepsilon^{0.03}$MPa.

- What are the reductions to be in each of the hot and the cold rolling processes?
- How would the introduction of a more viscous oil in the roll gap affect the roll force and the roll torque?
- How would increasing the speed of rolling affect the roll force and the roll torque, when rolling with a lubricant? Why?
- Should a lubricant be applied in the hot rolling process? What would you expect it to achieve?

Problem 7. Derive the equation of motion for the metal in the roll gap. Under what conditions is the inertia term significant? (Give your answer in numerical terms of the roll speed and the reduction; assume a set of parameters for your calculations)

Problem 8. The roll pressure distribution, obtained experimentally during rolling of a 1100 H14 aluminum strip, is shown below. Determine the changes of the coefficient of friction along the roll gap. Let $\sigma_{fm} = 50$ MPa; $R = 125$ mm.

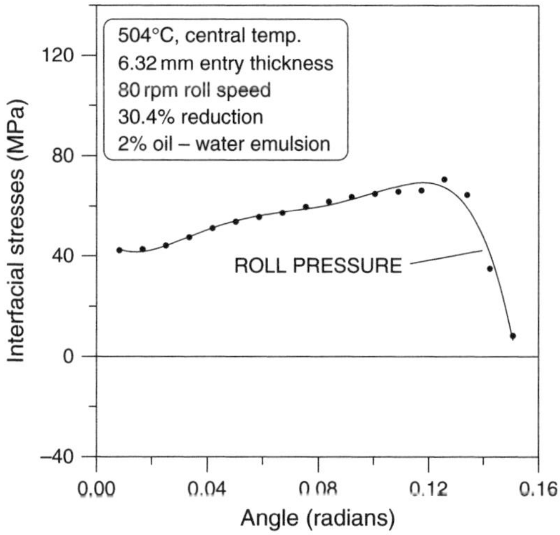

Figure 11.3 The roll pressure distribution.

Problem 9. A two-high rolling mill, for both hot and cold rolling of steel strips, is to be designed. The diameter of the work roll is 250 mm. Specifically, the dimensions for three components are needed:

- diameter of *one* of the drive spindles (recall that the shear stress in a spindle, subjected to a torque of T is given by: $\tau = Tc/J$; $J = \pi d^4/32$; c is the radius and J is the polar moment of inertia of the cross-section). The maximum allowable shear stress in the spindle is 100 MPa.
- load carrying area (A) of the frame such that the mill stretch is limited to 1 mm (recall that the elongation of a member is given by $\Delta = PL/AE$; the effective height, L, of the frame is 1 m; its modulus of elasticity is 200 000 MPa and P is the total load carried)
- power of the driving motor.

The requirements are, as follows.

- Hot rolling (sticking friction):
  Low carbon steel strips (5 mm thick, 50 mm wide, 500 mm long) are to be rolled at 1000°C, at 0.5 m/s entry velocity, to a reduction of 60%; the true stress–true strain rate curve of the metal, obtained in uniaxial tension, is given by $\sigma = 100\dot{\varepsilon}^{0.1}$ MPa. The rolls are turning at 45 rpm.
- Cold rolling ($\mu = 0.05$):
  Low carbon steel strips (5 mm thick, 50 mm wide, 500 mm long) are to be rolled, at 22°C, at 0.5 m/s entry velocity, to a reduction of 60%; the true stress–true strain curve of the metal, obtained in uniaxial tension, is given by $\sigma = 600\varepsilon^{0.25}$ MPa. The rolls are turning at 45 rpm.

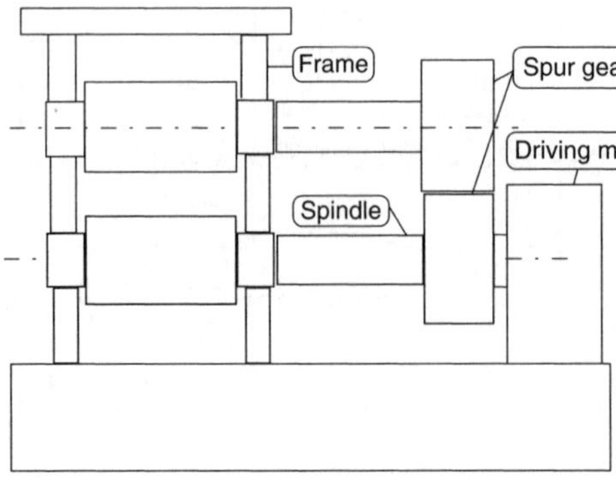

Figure 11.4 The schematic diagram of a rolling mill.

Problem 10. Show that, as long as the roll remains rigid, the strain rate in a flat rolling pass is given by

$$\dot{\varepsilon} = \frac{2R\sin\phi\dot{\phi}}{h_2 + 2R\,(1 - \cos\phi)}$$

Problem 11. The roll separating forces and the roll torques, measured during lubricated cold rolling of a low carbon steel strip, are given in the figures below.

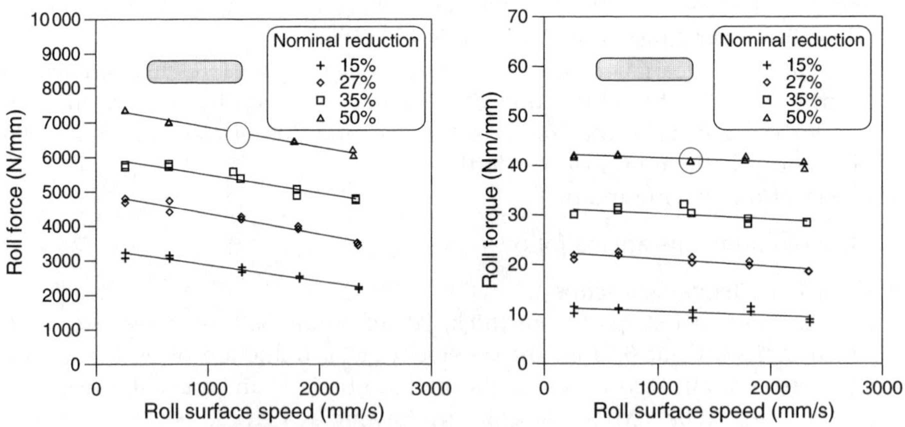

Figure 11.5 The roll force and the torque.

The details of the rolling mill: The roll diameter is 250 mm and its length is 150 mm. The centre-to-centre distance of the bearings on either side of the rolls is 250 mm. The rolls are made of D2 tool steel, hardened to $Rc = 63$. The surface roughness of the roll is $R_a = 0.2\,\mu m$. The mill is driven by a 42-kW, constant torque, DC motor.

The rolled metal: Cold rolled strips containing 0.05% C steel are used, having a true stress–true strain curve $\sigma = 150(1 + 234\varepsilon)^{0.251}$ MPa, obtained in uniaxial tension. The metal's strength is essentially independent of the rate of strain. The initial surface roughness of the samples is approximately $R_a = 0.8\,\mu m$, in both the rolling and transverse directions. Note that $\int (a + bx)^n\,dx = \frac{(a+bx)^{n+1}}{(n+1)b}$.

The sample dimensions: The to-be-rolled samples are 1 mm thick, 25 mm wide and 300 mm long.

Respond to the following:

- A pair of force and torque data are indicated on the figures, by small circles. Using an assumed value of the coefficient friction of 0.1, calculate the forces and the torques. Are the measurements and the calculation different? Why? (Account for at least three phenomena)
- Using the measured force, infer what the coefficient of friction should be.
- The roll separating forces decrease as the rolling speed increases. Why?
- The strip, rolled to a reduction of 50% in the first pass, is rolled once more. In the second pass it is reduced by another 30%. Calculate the roll separating force in the second pass. Let the coefficient of friction be 0.1.

Problem 12. Steel strips of 0.5 mm thickness are to be produced by cold rolling in three passes. The strip is supplied in 5 mm thickness, 50 mm width and 400 mm length. The true stress–true strain curve of the steel is given by $\sigma = 950\varepsilon^{0.12}$ MPa, obtained in uniaxial tension. The steel is reduced to 3 mm in the first pass at a roll speed of 50 rpm; to 1 mm in the second pass and to 0.5 mm in the third pass, both at 100 rpm. A light lubricant is used in all passes and the coefficient of friction is to be taken as 0.1 in all three passes. The rolls are 250 mm in diameter. The mill power is 100 kW.

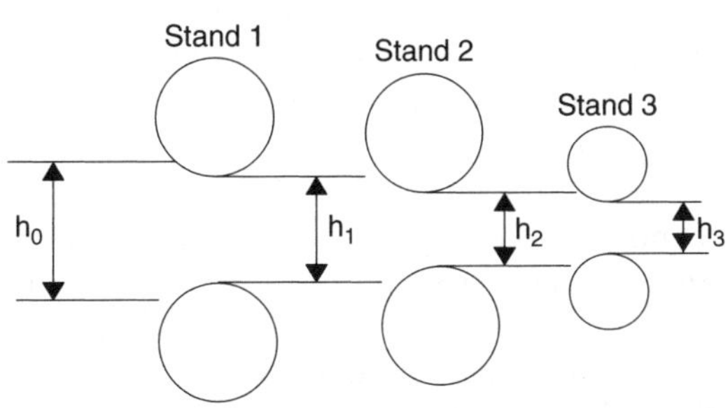

Figure 11.6 Three-pass rolling schedule.

- Is the motor power sufficient for all three passes?
- If not, what do you suggest to remedy the problem? (list at least two possibilities; *no calculations are needed*)
- How long will the strip be after the third pass?

Problem 13. 6061 – T6 aluminum strips are to be rolled in two passes to produce the 0.2 mm final thickness. The available strips are of 2 mm thickness, 25 mm width and 300 mm length. In the first pass, the strips are to be rolled to 1 mm, using roll speed of 100 rpm and an entry velocity of 1250 mm/s. In the second pass, the final thickness of 0.2 mm is produced. The roll speed is 200 rpm and the entry velocity of the strip is 2500 mm/s. The rolling mill is of a two-high design, having work rolls of 125 mm radius. The surface roughness of the work roll is 0.2 μm and the surface hardness is $Rc = 64$. During the two passes a light mineral oil with oleic acid as the boundary additive is used as the lubricant.

The true stress–true strain curve of the aluminum, *obtained in a uniaxial tension test*, is given by $\sigma = 450\,\varepsilon^{0.03}$ MPa.

- Determine the required power in the two passes. Take the coefficient of friction in the first pass to be 0.1 and in the second, 0.05.
- Ignoring the heat loss in between the two passes, what will be the temperature of the strip after the second pass? Let the specific heat of the aluminum be 900 J/kg K and its density be 2800 kg/m^3. The initial temperature of the strip is 22°C.
- Do you expect the width of the rolled strip to be different from the width at entry to the first pass? If there is any change, would the final width be larger or smaller than the initial width?
- In another pass, the light mineral oil is exchanged for one with a higher viscosity. Do you expect the roll separating force and the roll torque to change? How?

Problem 14. 1 mm thick and 25 mm wide AISI 1008 low carbon steel strips are to be produced by cold rolling. The metal is available in 8 mm thick, 25 mm wide bars so several passes will be necessary to reduce it to the final thickness. Its true stress–true strain curve, obtained in *uniaxial tension* is represented by $\sigma = 600\varepsilon^{0.25}$ MPa. The same two-high rolling mill, having 250 mm diameter rolls, will be used in all passes.

When rolling the strips, the time between passes is to be as short as possible, so there would be no heat losses. A low viscosity lubricant is used so the coefficient of friction in each pass is 0.1. The lubricant breaks down at a temperature of 200°C. The power of the driving motor is 150 kW. Assume that in all three passes the thickness of the strip at the neutral point equals the average of the entry and exit thickness of that pass. Let the specific heat of the steel be 500 J/kg K and its density be 7830 kg/m^3.

- Three consecutive passes are planned to produce the 1 mm thick strips. In the first, the strips are reduced to 4 mm at a roll speed of 100 rpm; in the second to 2 mm, at 180 rpm, and in the third, to the final thickness of 1 mm at 240 rpm. Is the equipment capable of carrying out the task?
- Will the lubricant provide the needed lubricating action in all of the passes?

Problem 16. The experimentally determined roll torques per unit width (for both rolls), during two-pass cold rolling of 3 mm thick carbon steel strips, are given below:

Pass #	rpm	Torque (Nm/mm)	h_{in} (mm)	h_{out} (mm)
1	100	60	3	1.8
2	100	100	1.8	0.6

The roll diameter is 200 mm. The strips are 50 mm wide. The true stress– true strain curve of the metal, obtained in uniaxial tension is $\sigma = 600\varepsilon^{0.25}$ MPa. The torque is measured by torque transducers, located in the drive spindles, as shown in the schematic diagram of the rolling mill. The schematic diagram of the two-high rolling mill is given below:

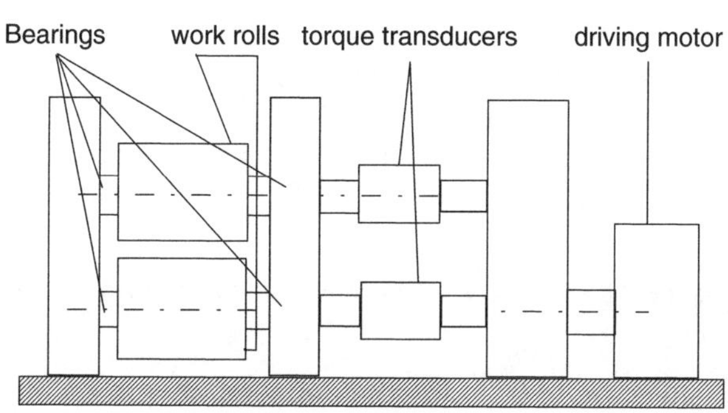

Figure 11.7 The rolling mill.

Determine:

- The coefficient of friction in each of the two passes;
- Can the second pass be performed without front pull? Why or why not?
- The temperature rise of the sample in the second pass. The density is 7800 kg/m^3 and the specific heat is 500 J/kg K.
- The necessary motor power.

Chapter 4: Materials

Problem 1. True stress–true strain curves, obtained during constant strain rate hot compression of a carbon steel show the characteristic behaviour of steels during hot deformation, indicating that the peak stresses are reached at progressively higher strains as the strain rates are increased. Why? What metallurgical phenomena are active in the steady-state region?

Problem 2. True stress–true strain curves, obtained in compression at 900°C for a Nb-V, are shown below. Determine the strain rate-hardening coefficient at a true strain of 0.4. At high temperatures the true stress–strain curve is given by an equation $\sigma = C\dot{\varepsilon}^m$.

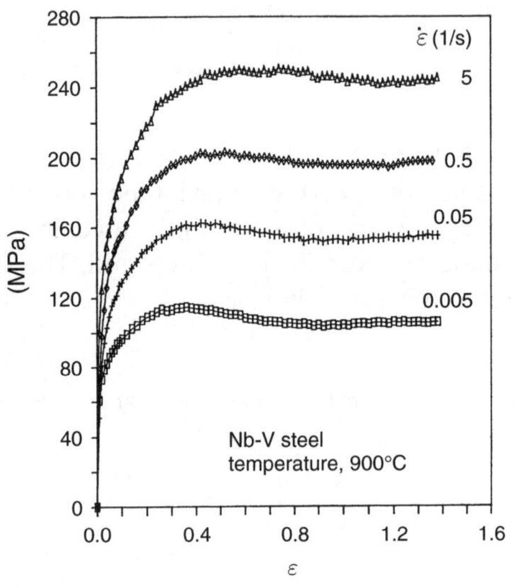

Figure 11.8 True stress–true strain curves of an Nb-V steel.

Problem 3. Washers are to be made of commercially pure aluminum by open die forging. Cylinders of 10 mm diameter and 15 mm length are available. Two possible approaches are possible: fast forging and slow forging. In both cases, the force of compression is the same. Which of the two processes will result in larger reduction?

The true stress-strain-strain rate curve is given by $\sigma = 90\varepsilon^{0.2}\dot{\varepsilon}^{0.1}$ MPa; it may be assumed that during the deformation, the strain rate remains constant. In the first forging process, the average rate of strain is $1\,\mathrm{s}^{-1}$ and in the next it is $100\,\mathrm{s}^{-1}$. The force reaches a maximum of 20 000 N during the deformation process and it increases linearly, from zero to its maximum value. The density of the aluminum is $2500\,\mathrm{kg/m^3}$ and its specific heat is 900 J/kg K.

Problem 4. Which of the two empirical relations represents the stress–strain curve of a steel better:

$$\sigma = K\varepsilon^n \quad \text{or} \quad \sigma = Y(1+B\varepsilon)^n$$

Why? Would the same argument hold for soft aluminum?

Problem 5. A cylindrical sample is subjected to torsion at high temperatures. Would dynamic recrystallization occur at the same time across the cross-section? Why or why not? Would static recrystallization occur at the same time across the cross-section as the cylinder is being twisted?

Problem 6. The true stress-true strain curves of a Nb-V steel, obtained in uniaxial, isothermal compression, are shown below. Determine the activation energy for dynamic recrystallization.

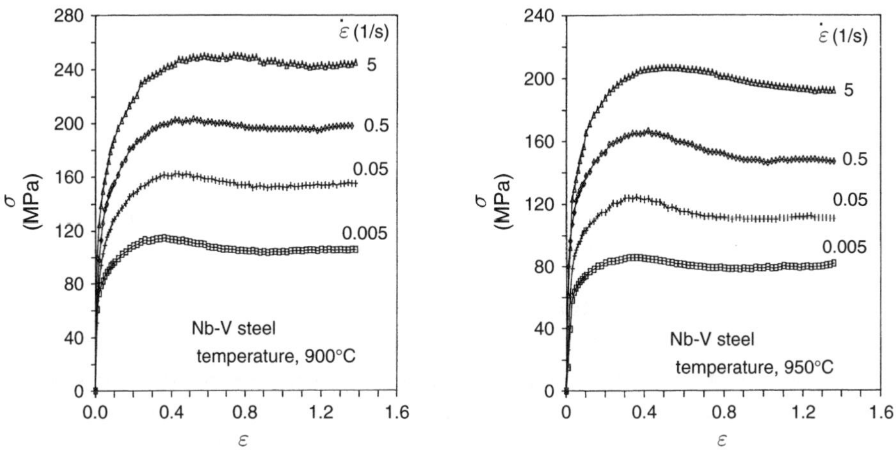

Figure 11.9 The true stress–true strain curves.

Problem 7. A drop hammer for open-die forging of brass cylinders of originally 25 mm height and 28 mm diameter, is to be designed. Specifically, the height from which the 1000 N hammer is to be dropped, is to be determined. The true stress–true strain equation of the brass is given by $\sigma = 800\varepsilon^{0.33}$ MPa. Its ductility, expressed in terms of total elongation to fracture $[e_f = (l_f - l_0)/l_0]$ is 0.5.

- Determine the height from which the hammer is to be dropped, such that the maximum permissible reduction of the brass cylinder is achieved;
- What will be the temperature rise of the cylinder after the drop? Let the density of the brass be 8530 kg/m^3 and its specific heat be 385 J/kg K;
- Will the distribution of the temperature be uniform? If not, how will it vary?

Problem 8. A cylindrical part (10 mm high, 158 mm diameter) is to be produced by open die forging using the drop tower, which has a weight of 20 000 N to be dropped on the sample from a height of 10 m. The stock from which it is to be made is 100 mm high and 50 mm in diameter.

The metal is a low carbon steel and its true stress–true strain curve in cold forming is given by $\sigma = 600\varepsilon^{0.25}$ MPa.

- Can the part be made in one drop?
- If a second drop is necessary, anneal the part first. How high should the weight be dropped from to produce the 10 mm final height?
- What should be the ductility of the metal to avoid tensile cracking?
- How is the velocity to change during the compression process for the strain rate to remain constant? (*do not calculate*).

Problem 9. An applicant is being interviewed for the position of master forger in a company and is given a test. The assignment is to hot forge a rail car wheel, the schematic diagram of which is shown below. The block from which the wheel is to be made is also shown in the figure.

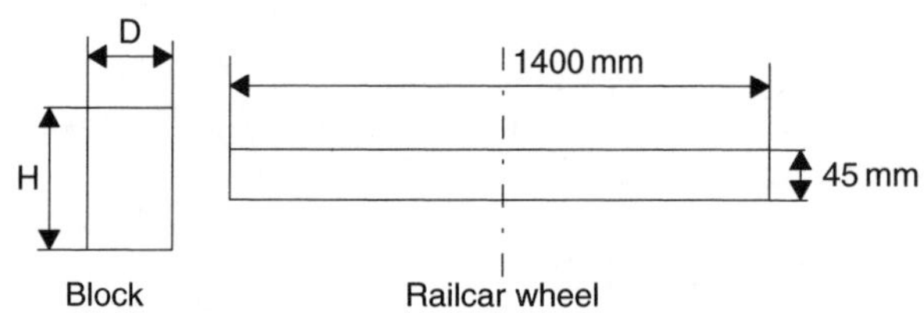

Figure 11.10 The rail car wheel.

The wheel is made of a medium carbon steel whose true stress–true strain curve is given by $\sigma = 120\varepsilon^{0.10}$ MPa, while formed at 1000°C. (Note that since the true stress–true strain curve is given, there is no need to consider the effects of the rate of strain.) The ductility of the steel, at 1000°C is 0.6, based on the tensile strain. The equipment available includes a high temperature furnace, and a drop forge which can drop a weight of 10 000 N from a height, adjustable from a minimum of 0.5 m to a maximum of 10 m. Several more high-powered forges are also available. The density of the steel is 7800 kg/m^3 and the specific heat is 500 J/kg K.

The applicant decides to use the drop forge and to forge the wheel, following the steps, given below:

Pre-heat the furnace to 1000°C and place the block in the furnace when the 1000°C is reached. The wheel is then held there for 4 hours, in air, during which it reaches a uniform temperature of 1000°C. The wheel is then taken to the drop forge; the transportation takes 2 minutes and the rate of cooling during the move is 1°C/s. The weight is raised to a height of 10 m and is dropped on the wheel.

- What should the height and the diameter of the block be to produce the rail car wheel?
- Is the 10 m height sufficient to forge the wheel in one drop?
- The first drop took 15 seconds. The cooling rate of the wheel, in contact with the cold dies is 2°C/s. If a second drop is necessary, should the wheel be placed back in the furnace for reheating?
- Would you recommend that the air in the furnace be replaced by argon or helium (both inert gases)? Why?
- Would you hire the applicant?

Problem 10. A drop hammer is available to produce, by cold forging, flat, low carbon steel washers of circular shape. The dimensions of the washers are to be: 2 mm thickness, 50 mm in diameter. The maximum height from where the 1000 N weight of the drop hammer may be dropped is 4 m.

The true stress–true strain rate curve of the steel, obtained in uniaxial tests, is $\sigma = 600\varepsilon^{0.25}$ MPa. The specific heat of the steel is 500 J/kg K and its density is 7800 kg/m^3.

- Determine the initial size of the sample;
- The initial temperature of the work piece is 22°C. What will be the temperature at the end of the compression process? Would it be possible to drop the weight from a height such that the sample would melt? ($T_{melt} \cong 1600$°C) What would that height be?
- The original sample was machined from a cold rolled sheet of steel to a right, circular cylinder. Would you expect that the flattened, deformed sample is circular? Why or why not? Sketch the true stress–true strain curve you would expect to see for the low carbon steel, obtained in uniaxial tension. How would the curve change if the steel is subjected to 40% cold working before the test?

Problem 11. An extensometer, attached to the sample, is used to monitor the extension in the tension test (see Figure 4.7). An LVDT (linear variable differential transformer), attached to the two compression platens, is used to monitor the deformation of a sample, subjected to compression (see Figure 4.8). How and why would the deformation of either sample be affected by the deformation of the tension tester or the compression tester?

Problem 13. Open die forging of a low carbon steel cylindrical sample (50 mm diameter, 75 mm high) is being performed on a hydraulic press, at 900°C. The compression platens are not heated. The metal's true stress–true strain curve, at 900°C, is given by $\sigma = 100\varepsilon^{0.1}$ MPa; its density is 7800 kg/m^3 and its specific heat is 500 J/kg K. Its ductility at 900°C, as measured by the fracture strain, is 45%. The procedure is as follows:

- Preheat the furnace to 1000°C;
- Hold the sample in the furnace, in air, for 1 hour, until its temperature distribution is uniform;

- Move the sample to the lower compression platen, position it in the centre and wait until it reaches the 900°C;
- Reduce its height by 75%, using a compression velocity of 1000 mm/s.

Determine

- The work done during the compression process
- The temperature rise during the compression process
- The strain rate at the end of the compression process
- Will tensile cracking limit the 75% compression process?
- The cylindrical sample is expected to deform as shown below, in (a). Instead, the sample deformed non-symmetrically, as shown in (b). What possible phenomena may have caused the distortion?

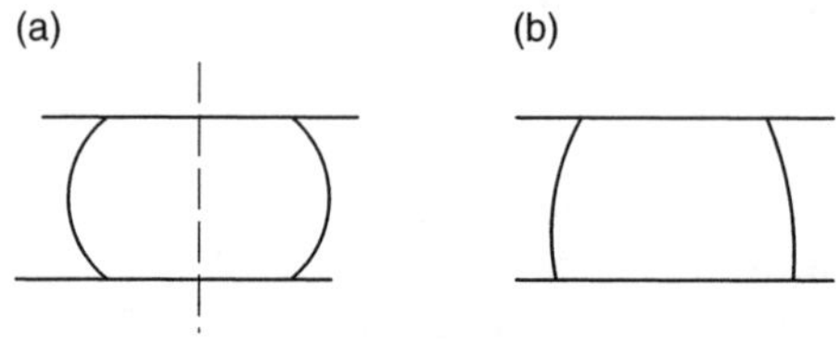

Figure 11.11 The distortion of an axially symmetrical, cylindrical sample, subjected to compression.

Problem 14. An open-die upsetting operation is shown in the figure below. The cylinder is made of AISI 1015 steel. The upsetting operation is performed at an initial temperature of 800°C and at a constant rate of strain of $10\,s^{-1}$. A glass-alcohol suspension is used as a lubricant in between the compression platens and the cylinder and it may be assumed that perfect lubrication exists.

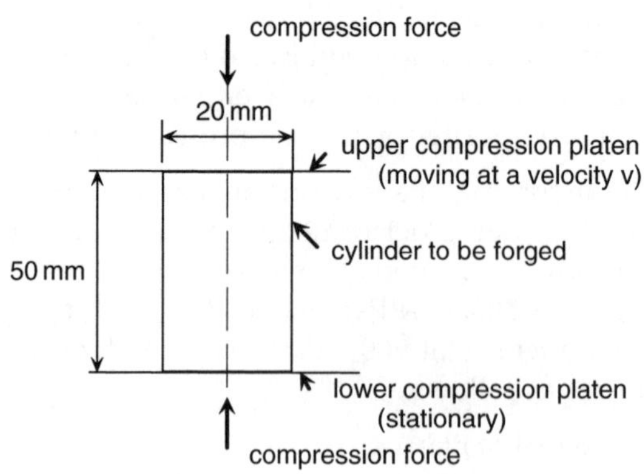

Figure 11.12 The open die forging process.

During the compression process the compression force and the change in height of the sample were monitored. The true stresses and strains were calculated and they are given in the table below.

- Determine the force of compression and the current height of the sample at a true strain of 0.25;
- Determine the temperature of the cylinder at a true strain of 0.6. Let the density be 7800 kg/m^3 and the specific heat to be 500 J/kg K;
- The ductility of the metal, at the temperature of 800°C is 0.6. Will the cylinder, compressed to a true strain of 0.8, experience tensile cracking on its outside circumference?
- Under what conditions will the cylinder remain cylindrical during the compression process?

True strain	True stress (MPa)
0	0
0.05	75
0.1	150
0.15	180
0.2	190
0.25	200
0.3	198
0.35	195
0.4	191
0.45	190
0.5	188.84
0.6	188.84
0.7	188.84
0.8	188.84

Problem 15. The force-deformation data, obtained during compression of a super purity aluminum cylinder are given below. The original dimensions are: 10 mm diameter and 15 mm high.

Determine:

- The engineering stress–engineering strain and the true stress–true strain curve of the metal; no more than two points in each are required;
- The constants in the relation $\sigma = K\varepsilon^n$ where σ is the true stress and ε is the true strain;
- The temperature rise of the sample at a true strain of 0.6; let the density be 2600 kg/m^3 and the specific heat to be 900 J/kg K;
- As the compression is proceeding, what phenomenon would limit the deformation process?

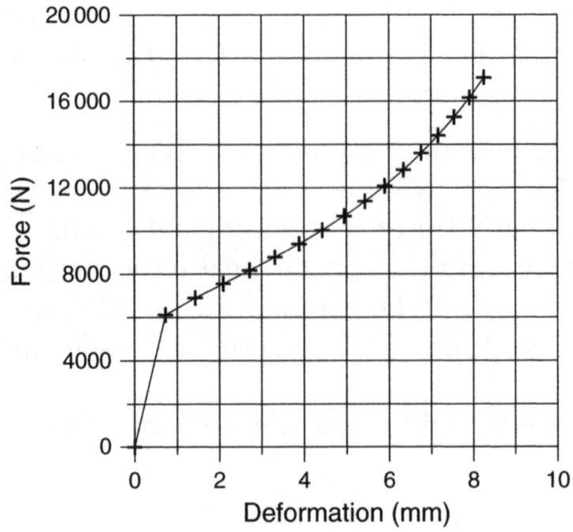

Figure 11.13 The force-deformation data.

Problem 16. The circumferential engineering strain (defined as the difference of the current and original circumference divided by the original circumference) during hot compression (at 1000°C) of a low carbon steel cylinder has been measured and is given in the figure below as a function of the force needed for the compression. The initial dimensions of the cylinder are also given in the figure. The compression velocity is constant, at 10 mm/s. The specific heat of the metal is 500 J/kg K and the density is 7830 kg/m^3.

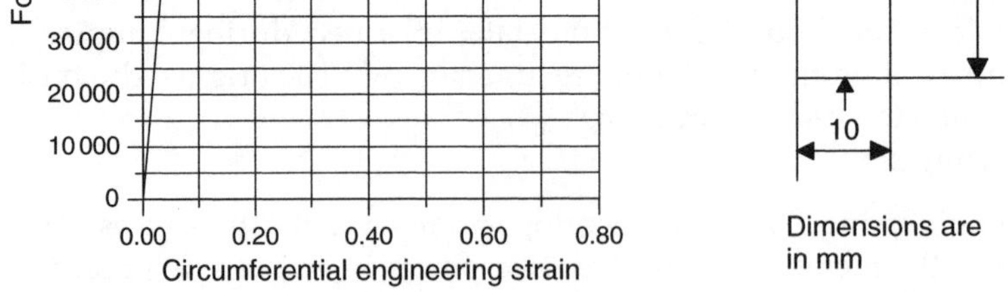

Figure 11.14 The force-deformation data and the sample dimensions.

- Determine the true stress–true strain curve, in the direction of the compression, of the alloy; no more than four pairs of stresses and strains need to be calculated – note that care is to be exercised in the choice of these points!

- What is the height of the sample at the end of the compression process?
- What is the rate of strain at a circumferential engineering strain of 0.4?
- A thermocouple, located at mid-height of the sample, indicated a tempera ture rise of 125 K. What is the temperature rise based on the conversion of the total work/volume?
- Are the measured and the predicted temperatures different? Why?

Problem 17. Let the stress–strain relation for a metal be $\sigma = Y(1+B\varepsilon)^n$; determine the average flow strength: $\sigma_{fm} = \frac{1}{\varepsilon_{max}} \int_0^{\varepsilon_{max}} \sigma(\varepsilon)\, d\varepsilon$

Chapter 5: Tribology

Problem 1. Two pieces of flat, strain-hardening metal are in non-lubricated contact under increasing normal pressure, high enough to cause plastic deformation. As well, there is relative, constant velocity motion in between the two. What are the mechanisms, parameters and variables that affect the coefficient of friction at the contact surface? How do they affect the coefficient?

Problem 2. Define the friction factor and the coefficient of friction. Under what conditions does the coefficient of friction loose meaning? Which one would you recommend for use in bulk metal forming? Why?

Problem 3. The figure below shows the true stress–true strain curves of a Nb-V steel, obtained at 950°C and at a constant true rate of strain of $0.05\,s^{-1}$. Curve #1

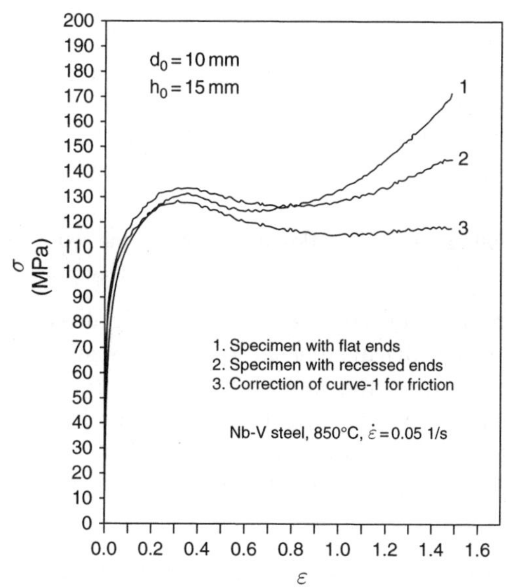

Figure 11.15 The stress–strain curves.

was obtained using a cylindrical sample with flat ends and with no lubrication. Curve #3 was obtained after corrections for friction were introduced. Determine the friction factor (m) necessary to get the corrected curve.

Problem 4. The figure below shows experimental results, obtained while cold rolling a low carbon steel, whose true stress–true strain curve is given by $\sigma = 150\,(1+234\varepsilon)^{0.251}$ MPa. The letters A–F refer to different lubricants used during the rolling passes. The roll diameter was 250 mm; the strips' dimensions were 0.96 mm thick, 25 mm wide, 300 mm long. The roll separating forces are given as a function of the surface speed of the roll, the lubricant and the reduction. Choose one of the lubricants and one of the reductions. Use Hill's relation to determine the coefficient of friction for each of the five speeds. Note that $\sigma_{fm} = \frac{1}{\varepsilon}\int_0^{\varepsilon} \sigma d\varepsilon$ and $R' = R = 125$ mm.

Plot the variation of the coefficient of friction as a function of the speed for the two reductions.

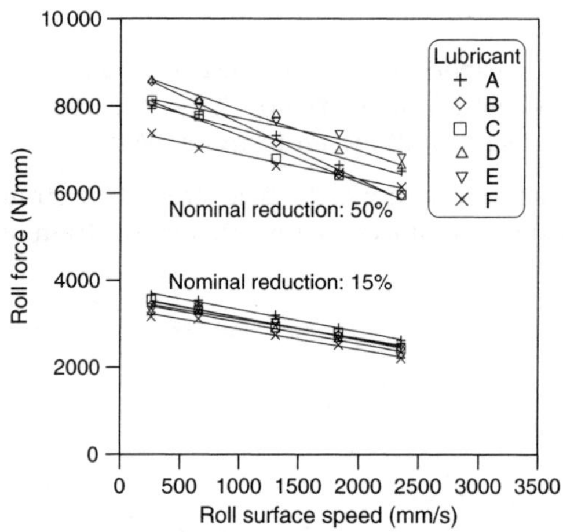

Figure 11.16 The roll force as a function of the speed, the reduction and the lubricant.

Problem 5. A drop hammer – in which a large weight is dropped on a sample – is used to compress cylindrical steel samples. Two tests are performed. In both, the weight is dropped from the same height, first on a completely dry sample and then on a sample, covered with highly viscous lubricant. Assume that in the second test friction was eliminated completely.

Which of the samples will be compressed more? Why?

Problem 6. The viscosity – pressure relationships indicate that as the pressure increases, the viscosity increases. In the boundary and in the mixed lubrication regimes, the Stribeck diagram indicates that the coefficient of friction decreases as the viscosity increases.

When rolling soft aluminum without lubrication the coefficient of friction was found to increase with reduction. The opposite was found during rolling of hard steel. How would you explain the apparent contradiction?

Problem 7. In a lubricated contact, the coefficient of friction is expected to decrease as the load is increased. What is the most important mechanism that contributes to this phenomenon?

Problem 8. Give the definition of the true area of contact.

Problem 9. How do the lubrication regimes affect the resulting surface roughness of the product?

Problem 10. Lubricating oils lower the coefficient of friction. Would water do the same?

Problem 11. Why does increasing viscosity lead to lower friction in the mixed lubrication regime but higher friction in the hydrodynamic regime?

Problem 12. A steel strip is rolled, successively in two passes. The reduction in both passes is 50%. All process parameters are kept to be the same. Would you expect the coefficient of friction to remain the same in the two passes? If not, how would it change and why?

Problem 13. A layer of scale is an insulator. Would that increase or decrease the coefficient of heat transfer?

Problem 14. The figure below shows the temperature distribution, obtained during hot rolling of a low carbon steel slab. The locations of the measuring thermocouples are also shown. Assuming that the roll temperature is 40°C, determine the coefficient of heat transfer at the roll-slab contact surface.

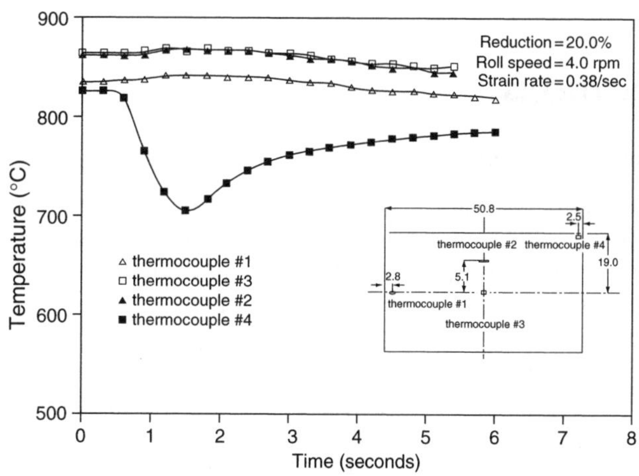

Figure 11.17 Temperature–time data.

Chapter 6: Applications and Sensitivity Studies

Reproduce the figures in the Chapter.

Chapter 7: Temper Rolling

Class discussion of the following topics

- What are the differences between temper rolling and conventional flat, cold rolling?
- Why is the coefficient of friction high in temper rolling?

Chapter 8: Severe Plastic Deformation – Accumulative Roll Bonding

Class discussion: What are the advantages and disadvantages of the accumulative roll-bonding process?

Chapter 9: Roll Bonding

Class discussion: Discuss the paper of Gilbreath (1967)

Chapter 10: Flexible Rolling

Class discussion: Would tailor-rolled blanks replace tailor-welded blanks in the automotive industry?

PART 2: SOLUTIONS

Chapter 1: Introduction

1. Integrate the expressions for the strains, realizing that they are the partial derivatives of the displacements so integration functions will result after the integration. Using the boundary conditions that $u = 0$ at $x = 0$ and $v = 0$ at all y, the integration functions are shown to vanish.

At point D therefore $u_D = 2.25 \times 10^{-7}$ m and $v_D = 9.187 \times 10^{-6}$ m.

The shear strain at point C is obtained by using its original definition in terms of the sum of partial derivatives of the displacements; the shear strain then is obtained as 1.75×10^{-4} radians.

The stress components at point C are obtained using the version of Hooke's law as $\sigma_x = 12.65$ MPa, $\sigma_y = 40$ MPa and $\tau_{xy} = 13.93$ MPa.

2. Consider a cube. For no volume change, the ratio of the original and the current volumes equals unity. The ratio can also be expressed as ratio of the product of the three sides of the undeformed cube to the product of the three sides of the deformed cube which will, of course, also equal unity. Take natural logarithm of both sides; on the left side one gets zero while on the right side the sum of the three logarithmic strains appears. Hence, the vanishing of the trace of the strain tensor (i.e. the sum of the three logarithmic strains) implies constancy of volume.

3. According to the maximum shear-stress criterion the largest principal stress will be in the circumferential direction and the smallest in the axial direction (perpendicular to the plane of the disk). So yielding will begin at $r = \mathrm{a}$ (the inside radius) at a rotational speed when the stress in the circumferential direction equals the yield strength, at a speed of 1720 r/s.

4. In the tube eversion process, the largest principal stress will be in the circumferential direction and the smallest in the longitudinal direction. The intermediate principal direction therefore is perpendicular to the wall of the tube. According to the flow rule, associated with the maximum shear-stress criterion of Tresca, the strain rate in the intermediate principal direction vanishes; the strain in that direction will therefore be constant and the wall thickness will remain unchanged. (Note: this answer may be arrived at by considering that while the tensile stress in the circumferential direction causes thinning of the wall, the compressive stress in the longitudinal direction causes thickening. The two processes therefore may cancel one-another).

5. At $r = R : \sigma_r = 0;\ \tau_{rz} = 0$ at all y; $\ 2\pi \int_0^R \sigma_z\, r dr = F$ at $y = \pm L/2$ and $\dot{w} = 0$ at $y = -L/2$ and $\dot{w} = \nu$ at $y = L/2$.

6. $\varepsilon_{xz} = \varepsilon_{yz} = 0$; all others components are non-zero.

7. Conduction is heat transfer by means of molecular agitation within a material without any motion of the material as a whole. Convection is the flow of heat through a bulk, macroscopic movement of matter from a hot region to a cool region. Radiation is heat transfer by the emission of electromagnetic waves which carry energy away from the emitting object.

8. No; the volume changes under elastic loading, controlled by the bulk modulus.

9. 8.73 mm

10. Check the web-site of the World Bank for the necessary information (worldbank.org)

11. Find the information from Statistics Canada or from the web-site of the American Iron and Steel Institute.

12. Use Google or any search engine to find addresses for used companies that sell used rolling mills.

13. A three-component flexible coupling with constant input speed will ensure constant output speed.

Chapter 2: Flat Rolling

1.

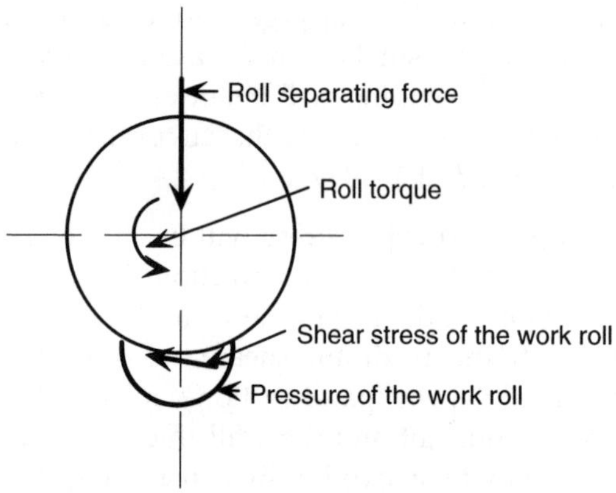

2. $\mu = \frac{F}{P}$; for entry: $F \cos \phi_1 = P \sin \phi_1$; so $\mu = \frac{\sin \phi_1}{\cos \phi_1} = \tan \phi_1$;

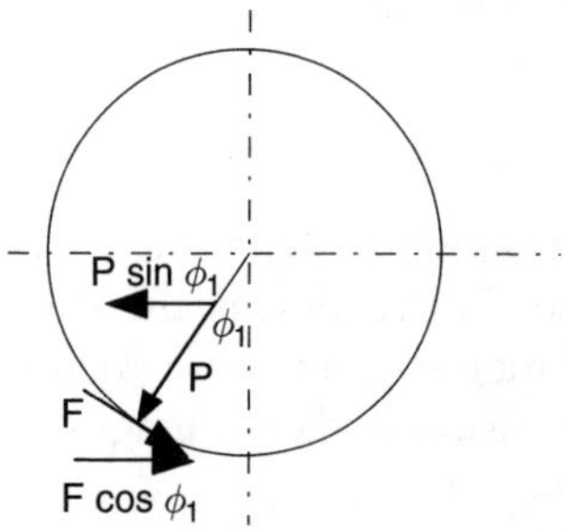

3. See Figure 2.4, Chapter 2.

4. The not-very-stiff work roll would bend under the forces it experiences during the pass, creating a strip which is thicker in the centre than at the sides; hence the edges would move faster than the centre.

5. For complete flattening, the radius would decrease by 0.01 mm. The radial strain is then estimated to be $0.01/200 = 0.00005$; much less than necessary to cause plastic flow.

6. Consider the work roll to be a simply supported beam, loaded by a uniformly distributed force along the 1500 mm length. The force is obtained as 714 887 N, distributed uniformly over the 1500 mm.

7. Using mass conservation, the exit velocity would be 10 000 mm/s.

8. The roll's surface speed is 8377.58 mm/s. The projected contact length is 16.73 mm so the time of contact is 0.001998 s. (The result is reached using the assumptions of rigid rolls and that the strip's speed equals the roll's surface speed.)

9. The plane-strain flow condition indicates that the perpendicular to the direction of rolling, ε_3, (the width direction) vanishes. According to the flow rule associated with the Huber-Mises flow criterion then

$$\varepsilon_3 = 0 = \sigma_3 - \frac{\sigma_1 + \sigma_2}{2}$$

10. While the work roll's radius of curvature would increase, an indentation is not likely to occur. The leading edge of the strip would thicken somewhat.

Chapter 3: Mathematical Modelling

1. The projected length of contact, from the geometry of the pass, is $L = \sqrt{R'^2 - \left[R' - \frac{1}{2}\Delta h\right]^2}$; as long as $\frac{\Delta h^2}{4}$ is much less than unity, true in strip rolling, the relationship is valid.

2. See Chapter 3, Section 3.10 for typical calculations.

3. Consider the frames of both the laboratory and the industrial mills to consist of four columns, subjected to a force along their lengths. For the laboratory mill take the load carrying area of one of the columns to be 100 × 100 mm, 800 mm long and the force to be 400 000 N. The stretch would then be 0.04 mm.

For the industrial mill, the load carrying area is to be 1 m^2; the length to be 4 m. Let the roll force be 60×10^6 N. The extension is then 0.3 mm.

4. Using mass conservation, the exit velocity from the seventh stand will be 7490.6 mm/s.

5. Derivations

Eq. 3.18: Refer to Figure 3.2. After some simplifications, the balance of the forces on the slab leads to

$$(d\sigma_x h) \mp 2\tau R' \cos\phi \, d\phi + 2pR' \sin\phi \, d\phi = 0;$$

Note that from geometry $\cos\phi = \dfrac{dx}{R'd\phi}$ and $\sin\phi = \dfrac{dh}{2R'd\phi}$;

Use $\tau = \mu p$ to obtain the required relation.

Eq. 3.19: The Huber-Mises flow criterion is written in terms of the principal stresses as $(\sigma_1 - \sigma_2)^2 + (\sigma_2 - \sigma_3{}^2) + (\sigma_1 - \sigma_3)^2 = 6k^2$; let $\sigma_1 = \sigma_x$; $\sigma_2 - p$; $\sigma_3 = \frac{\sigma_x - p}{2}$; substitute; simplify to get the required equation.

Eq. 3.20: Refer to Figure 3.3. The required equation is obtained immediately. The simplified version is valid as long as the angles are much smaller than unity.

Eq. 3.32: Take eq. 3.18 as the starting point. Examine the conditions of equilibrium in the direction perpendicular to the rolled metal, at the roll/strip interface to get $\sigma_y = p \mp \tau \tan\phi$; use the yield criterion to get $\sigma_y - \sigma_x = 2k$; hence $\sigma_x = p - 2k \mp \tau \tan\phi$; substitute this into eq. 3.18; recognize that $\tan\phi = \frac{dy}{dx}$ to get the required equation.

Eq. 3.48: See Section 3.5.2 for the detailed derivation.

6. *Hot rolling*: Use Schey's approach. Check if there is sufficient motor power to reduce the strip from 10 mm to 2 mm in one pass. The flow strength is obtained as 68.2 MPa, at a strain rate of $36.49\,s^{-1}$. The contact length is 34.64 mm; the pressure intensification factor is 2.4; so the roll force is 163 000 N and the required power is 29.6 kW; one hot rolling pass is sufficient, but just barely.

Cold rolling: Repeat the above steps; the flow strength is now 446.6 MPa; the contact length is 14.79 mm; the factor is 2 and the power needed is 58.8 kW so again, one pass is sufficient. The reductions are then: 80% in the hot rolling pass; 87.5% in the cold rolling pass. (Note: tensile cracking may be present at such high reductions; the ductility of each metal should be checked to see if the reductions need to be reduced.)

The lubricant would reduce the power requirements;

Increasing the speed would increase the strain rate and the metal's resistance to deformation; at the same time more oil would be drawn into the contact; the two contradictory events may cancel one-another;

Yes; lower power consumption would result.

7. Refer to Figure 3.2. Add the force of inertia, mass x acceleration, to the free body diagram of the slab. Balance of the forces in the direction of rolling will result in eq. 3.45. It is now necessary to express the mass and the acceleration in terms of the independent variables so the result can then be integrated. Mass conservation requires that $d(vh) = 0$ so the acceleration is then $a = dv/dt = -({}^{v}/_{h})\frac{dh}{dt}$. The time rate of change of the strip's thickness is obtained from the simplified version of eq. 3.20, as $\frac{dh}{dt} = \frac{2xv}{R'}$ where the velocity of the strip is $v = v_r \frac{h_{np}}{h}$. The mass of the slab, per unit width is $m = \rho dxh$. Use the Huber-Mises criterion to write: $\sigma_x = 2k - p$; substitute into eq. 3.45; the required equation 3.46 will result. See Section 3.5.2 for the numerical work.

8. Invert the equation of equilibrium so the coefficient of friction is on one side and all other terms are on the other side. All of these terms can be obtained numerically from the figure.

Angle	Pressure	h (mm)	dp/dh	k/h	μ
0.012567	45.65	4.979741	315.8733	5.020341	0.42614
0.016756	50.5	4.995095	−103.844	5.004909	−0.1804
0.020945	48.45	5.014837	−78.7463	4.985207	−0.18152
0.025134	46.55	5.038965	−61.3711	4.961337	−0.18047
0.029323	44.8	5.06748	−48.6293	4.933419	−0.17766
0.033512	43.2	5.100382	0	4.901594	−0.01939
0.037701	43.2	5.137671	−2.39947	4.866018	−0.03258
0.04189	43.1	5.179347	20.62403	4.826864	0.079522
0.046079	44.05	5.225409	58.47409	4.784314	0.293474
0.050268	47	5.275859	60.17875	4.738565	0.312832
0.054457	50.3	5.330696	57.40959	4.689819	0.304259
0.058646	53.7	5.389919	43.23186	4.638288	0.227175
0.062835	56.45	5.45353	16.91241	4.584187	0.074837
0.067024	57.6	5.521527	37.99164	4.527733	0.215003
0.071213	60.35	5.593911	13.67699	4.469145	0.060779
0.075402	61.4	5.670683	14.78593	4.40864	0.072266
0.079591	62.6	5.751841	18.11909	4.346435	0.10072
0.08378	64.15	5.837386	23.90693	4.282739	0.149608
0.087969	66.3	5.927318	25.97568	4.217759	0.171116
0.092158	68.75	6.021637	12.66388	4.151695	0.068709
0.096347	70	6.120343	−9.21499	4.084738	−0.11204
0.100536	69.05	6.223436	11.09516	4.017074	0.064136
					$\mu_{\text{average}} = 0.078932$

9. In hot rolling, for a reduction of 60% at a speed of 45 rpm, the flow strength is 139.5 MPa; the roll force is 373 576 N and the torque is 7247 Nm. The motor power to reduce the strip is 34.15 kW. In cold rolling, the required power is 54 kW; since only one mill is being designed, use the larger power.

The diameter of one of the spindles is then 68 mm; the load carrying area is 0.0031 m^2; and the motor power is 57 kW. Note that no allowance was made either for friction losses or for a generous factor of safety.

10. Start with eq. 3.20. The strain is $\varepsilon = \ln \frac{h_{\text{entry}}}{h}$; differentiate with respect to time to get required relation.

11. Read the force and the torque data off the figure; $P_r \approx 6600$ N/mm and $M \approx 41$ Nm/mm. Using the given data, the force is calculated to be 7080 N/mm and the torque is found as 56 Nm/mm. The differences are most likely due to a too high value of the coefficient of friction;

Using the measured data and inverse approach the coefficient of friction is estimated to be 0.08. Note that Schey's figure (see 3.1) allows only approximations, not exact values;

The roll force will drop with the speed as more oil is drawn into the contact zone;

In the second pass, the flow strength becomes 570 MPa and the roll force is 5110 N/mm.

12. *Pass #1*: The roll force is 853512 N; the power is 70.7 kW < 200 kW; OK.

Pass #2: The flow strength is now 952 MPa; the roll force is 1.38×10^6 N; the required power is 229 kW; the available motor is not powerful enough.

Remedy: only the installation of a more powerful motor will solve the problem; extra lubricant would help but it would not be sufficient.

13. *Pass #1*: The roll force is 208300 N so the power needed is 24.4 kW;

Pass #2: The power needed now is 27.4 kW

Temperature rise: Use equation 3.12 to determine the temperature rise. In the first pass, the rise is obtained as 154.9°C and in the second, 173.9°C. The strip temperature will be then 350.8°C.

The width will change some but not in any significant manner; the plane-strain flow assumption is quite close to reality.

Higher viscosity lubricant will likely lower the loads on the mill.

14. Design the schedule for three passes.

Pass #1: Reduce the strip from 8 mm to 4 mm. The flow strength is then 438 MPa; the roll force is 337 800 N and the power is 79 kW < 150 kW; OK.

Pass #2: Another 50% reduction; account for the effect of cumulative strain hardening so the flow strength is now 603.8 MPa; the force is 356 700 N and the power is 106.3 kW;

Pass #3: 50% reduction; the strength is 687.5 MPa; the power is 86.9 kW.

All three passes are possible.

In order to check the lubricant for possible breakdown, the temperature rise in the three passes needs to be calculated. As suggested, assume that there is no heat loss in between the passes. Use eq. 3.12 to get $\Delta T_{\text{pass1}} = 102.9$°C; $\Delta T_{\text{pass2}} = 153.6$°C so the lubricant will break down in the second pass.

15. Inverse calculations are needed.

Pass #1: The torque is given so the roll force can be obtained as 5477 N/mm (use eq. 3.6). The only unknown in eq. 3.1 is now the pressure intensification factor; it is obtained as 1.42; so the coefficient of friction is then 0.15.

Pass #2: Similarly, using the idea of cumulative strain hardening, the coefficient of friction is obtained as 0.04; note that this is below the minimum necessary coefficient to initiate bite, so in the second pass the strip needs some front pull to assist entry.

The rise of temperature in the second pass is calculated as 214 °C.

The motor power is 104 kW; twice what was used to calculate the temperature rise of the strip, needed to overcome friction losses in the bearings.

Chapter 4: Material Attributes

1. The metallurgical phenomena are time dependent. To have sufficient time available for the hardening and softening rates to reach equilibrium as the rate of strain is increased, a larger strain – hence, a little more time – is needed to get to the peak stress, corresponding to the peak strain.

2. Read the stress values at a strain of 0.4, at various strain rates. Prepare a ln-ln plot of the stresses at that strain versus the strain rates. The slope of the line is the strain-rate hardening exponent; obtained as 0.11.

3. The strain in slow forging is $\varepsilon_s = \ln \frac{15}{h_s}$ and in fast forging $\varepsilon_f = \ln \frac{15}{h_f}$. The stress in slow forging is $\sigma_s = \frac{F}{A_s} = 90\left(\ln \frac{15}{h_s}\right)^{0.2}$ where $A_s = \frac{\frac{\pi}{4}10^2 x15}{h_s}$.

The stress in fast forging is $\sigma_f = \frac{F}{A_f} = 90\left(\ln \frac{15}{h_f}\right)^{0.2} 100^{0.1}$ where $A_f = \frac{\frac{\pi}{4}10^2 x15}{h_f}$.

The force is the same in both cases, at 20 000 N. Hence, in slow forging $h_s = 5.334$ mm and in fast forging the final height it is 7.8 mm. The slow forging creates higher reduction.

4. The relation $\sigma = Y(1 + B\varepsilon)^n$ represents the behaviour of the metal better, since it indicates that at a strain of zero, the metal possesses some strength. The same argument holds for either steel or aluminum.

5. No. Dynamic recrystallization would occur sooner on the outer fibres since the stresses and strains there are the highest. No; static recrystallization would not occur since the sample is loaded dynamically (this is intended as a trick question.)

6. The procedure is outlined below:

- perform a number of stress–strain tests at several temperatures and rates of strain;
- obtain the peak stresses and prepare a log–log plot of the peak stresses versus the temperatures;
- at an arbitrary stress level, obtain from the plot, two temperatures and the corresponding rates of strain; and
- determine the activation energy from the slope $Q \approx \frac{\Delta(\ln \dot{\varepsilon})}{\Delta(-1/RT)}$.

Non-linear regression analysis is used to obtain the relationships of the strain rate in terms of the peak stress, using the lines from Figure 3.11.

For 950°C

$$\dot{\varepsilon} = \exp\left[(\ln \sigma_p - 5.244)/0.1178\right]$$

and at 900°C

$$\dot{\varepsilon} = \exp\left[(\ln \sigma_p - 5.4006)/0.10411\right]$$

The average activation energy for dynamic recrystallization is then obtained as 483 kJ/mole.

7. The limit of deformation is decided by the circumferential strain which cannot exceed 0.5.

The diameter after the drop is then obtained from $\varepsilon_\theta = \frac{\pi d_f - 28\pi}{28\pi} = 0.5$ so the maximum diameter will be 42 mm. The sample's height after the drop is obtained from mass conservation as 11.11 mm. The work done per volume is the integral of the stress–strain curve over the strain ($\varepsilon = \ln \frac{25}{11.11} = 0.811$) and is 455.26 MPa. The potential energy/volume at the start of the drop must match that magnitude; so the height is to be 7 m (excluding friction losses).

The rise of the temperature is 138°C.

Since the strain distribution will not be uniform within the sample, the temperature distribution will also not be uniform. The centre is expected to be warmer than the surfaces.

8. The available energy is 1018.6 MPa; the required work done is 1359.6 MPa so the part cannot be produced in one drop.

The height for the second drop: first determine the height from the first drop from $1018.6 \times 10^6 = 600 \times 10^6 \frac{\varepsilon^{1.25}}{1.25}$ so the height is 16.1 mm.

In the second drop, this is to be reduced to 10 mm. Since the sample was annealed, the original stress–strain curve may be used. Equate the potential energy to the required work/volume:

$$\frac{20\,000\,H}{0.000196} = \frac{600 \times 10^6}{1.25}\left(\ln \frac{16.1}{10}\right)^{1.25} \quad \text{so the height is } H = 1.86\,\text{m}.$$

9. The diameter and the height of the block:

There are two criteria: the volume remains constant; tensile cracking is to be avoided.

Constant volume indicates that $0.0627\ \text{m}^3 = \frac{\pi}{4}D^2H$; the diameter is obtained from the limiting circumferential strain: $0.6 = (1.4 - D)/D$; $D = 0.875$ m; the height is then 0.115 m.

Check if the 10 m drop height is sufficient:

The available potential energy/volume is $10 \times 10000/0.0627 = 1.4436 \times 10^6$ MPa; the required work/unit volume is the integral of the stress–strain curve over the strain of 0.938; it is obtained as 101.7×10^6 MPa; the wheel cannot be forged in one drop.

The temperature drop during transportation is 120°C; in the forge, it is 30°C. The wheel must be reheated for the second drop.

No need for protective atmosphere as the layer of scale will crack during the forging process;

Don't hire the person; a master forger should have known not to use the drop forge.

10. The available energy is 4000 Nm. The final volume of the sample is $3.927 \times 10^{-6}\,m^3$. Equate the available energy/volume to the required energy in terms of the strain; from the strain the original height is obtained as 12.41 mm.

The temperature rise will be 261°C; yes, in theory it is possible to drop the height so the sample will melt; in practice this may not be achievable, though. Taking the melting temperature to be 1600°C, dropping the weight from a height of 23.2 m may cause melting.

11. The extensometer in the tension test is attached directly to the sample so the deformation of the machine will have no effect on its readings. The LVDT in the compression test is attached to the compression platens so the deformation of the machine will have a significant effect on its readings.

12. F = 80760 N; h = 38.9 mm;

13. $\sigma = 100\varepsilon^{0.1}$ MPa; $\Delta T = 24.4$ C

Chapter 5: Tribology

1. The parameters that will affect the coefficient of friction are

- Pressure
- Surface roughness
- Surface hardness
- Relative velocity

The interactions of these parameters are shown in Figures 5.36 and 5.37.

2. The coefficient of friction is the ratio of the tangential and the normal forces. The friction factor is the ratio of the friction stress and the yield strength of the interface. The coefficient of friction loses meaning when the interfacial pressures are several times the strength of the material. The friction factor is recommended for bulk forming since its meaning does not depend on the pressure.

Read the data of the curves:

strain	p#1	p#3 = σ_f	h	d	m
0.2	128	125	12.28096	11.05406	0.138549
0.4	132	128	10.0548	12.21662	0.133645
0.6	128	120	8.232175	13.50145	0.211215
0.8	126	116	6.739934	14.92141	0.202334
1	131	115	5.518192	16.49071	0.241914
1.2	145	115	4.517913	18.22506	0.336027
					$m_{ave} = 0.210614$

3. Hill's equation is given in the text, see eq. 5.26. The results below are obtained for 50% reduction and lubricant F. The flow strength is 432 MPa. The calculations are given in the table:

h_{entry} (mm)	h_{exit} (mm)	speed (mm/s)	P_r (N/mm)	strain	σ_{fm}(MPa)	$R' =$ 125 mm	coefficient
1	0.5	260	7300	0.693147181	432.6655243	125	0.1563177
1	0.5	500	7000	0.693147181	432.6655243	125	0.1475527
1	0.5	1300	6700	0.693147181	432.6655243	125	0.1387877
1	0.5	1850	6500	0.693147181	432.6655243	125	0.1329444
1	0.5	2300	6050	0.693147181	432.6655243	125	0.1197969

4. The sample will be compressed more in the test with the lubricant as less of the available energy is spent on overcoming friction losses.

5. While rolling soft aluminum, the strip's asperities are flattened more easily than those of the steel, creating more sites for adhesive bonds to form, thus increasing the effort needed to cause relative motion.

6. Increasing loads cause the viscosity to increase. In the boundary and the mixed lubrication regimes this will lower the coefficient of friction.

7. The true area of contact is the sum of all areas in contact at the asperity tips.

8. In the boundary and the mixed regimes, the resulting surfaces will be smoother and will reflect the surface roughness of the rolls. In the hydrodynamic regime the strip will roughen as a result of the free deformation at the surface.

9. Yes; even though its viscosity is very low.

10. In the hydrodynamic regime, there is complete separation between the strip and the roll. Relative motion in this regime means that the oil film is being sheared. The force needed to shear a more viscous lubricant is higher than for one of lower viscosity.

In the second pass, the steel has hardened due to strain hardening. Its asperities will be somewhat harder to flatten so the creation of adhesive bonds will be harder; the coefficient of friction will be lower than in the first pass.

11. The coefficient of heat transfer would decrease.

REFERENCES

Ahmetoglu, M.A., Brouwers, D., Shulkin, L., Taupin, L., Kinzel, G.L. and Altan, T., (1995), "Deep Drawing of Round Cups From Tailor-Welded Blanks", J. Mat. Proc. Techn., 53, 684–694.

AISI website, "Tailor-Welded Blanks Expected to Soar Through 2000".

Alexander, J.M., (1972), "On the Theory of Rolling", R. Soc. London, 326, 535–563.

Alexandrov, I.V., Zhu, Y.T., Lowe, T.C., Islamgaliev, R.K. and Valiev, R.Z., (1998), Nanostructured Materials, 10, 49.

Al-Salehi, F.A.K., Firbank, T.C. and Lancaster, P.G., (1973), "An Experimental Determination of the Roll Pressure Distribution in Cold Rolling", Int. J. Mech. Sci., 15, 693–710.

Altan, T. and Boulger, F.W., (1973), "Flow Stress of Metals and its Application in Metal Forming Analysis", ASME, J. Eng. Ind., 95, 1009–1019.

Arnaud, J.C., (1995), "Trials of High Speed Steel Rolls in SOLLAC Hot Strip Mills", 37th Mechanical Working and Steel Processing Conference, Hamilton, 261–266.

ASM Handbook, (1985), American Society of Metals.

Auzas, J-D., Tice, D.K. and Caillaud, J-C., (1995), "New Studies and Developments in Eddy Current Testing of Rolling Mill Rolls", 37th Mechanical Working and Steel Processing Conference, Hamilton, 287–292.

Avitzur, B., (1968), "Metal Forming: Processes and Analysis", McGraw-Hill, New York.

Azarkhin, A. and Richmond, O., (1992), "Limits to Adhesive Friction", in Num. Meth. Ind. Form. Proc., eds, Chenot, J. L., Wood, R. D. and Zienkiewicz, O. C. 143–148.

Azushima, A., (1978), "Determination of Oil Film Thickness in Rolling from the Relationship between Surface Roughness of Strip and Roll", Bulletin of the JSME, 21, 1402–1407.

Azushima, A., (2000), "FEM Analysis of Hydrostatic Pressure Generated Within Lubricant Entrapped into Pocket on Workpiece Surface in Upsetting Process", ASME, J. Trib., 122, 822–827.

Backofen, W.A., (1972), "Deformation Processing", Addison-Wesley Publishing Company, Reading.

Banerji, A. and Rice, W.B., (1972), "Experimental Determination of Normal Pressure and Friction Stress in the Rolling Gap During Cold Rolling", Annals of the CIRP, 21, 53–54.

Baragar, D.L. and Crawley, A.F., (1984), "Frictional Effects on Flow Stress Determination at High Temperatures and Strain Rates", J. Mech. Working Techn., 9, 291–299.

Barber, J.R., (1991), "Is Modeling in Tribology a Useful Activity?", in Tribological Modeling for Mechanical Designers, ASTM STP 1105, eds, Ludema K.C. and Bayer, R.G., ASTM, PA, 165–172.

Barzan, D.R., (1995), "The Use and Performance of HSS Rolls at Inland's 80" Hot Strip Mill", 37th Mechanical Working and Steel Processing Conference, Hamilton, 271–274.

Batchelor, A.W. and Stachowiak, G.W., (1995), "Tribology in Materials Processing", J. Mat. Proc. Techn., 48, 503–515.

Bay, N., (1979), "Cold Pressure Welding – The Mechanisms Governing Bonding", ASME, J. Eng. Ind., 101, 121–127.

Bay, N., (1986), "Cold Welding Part 1: Characteristics, Bonding Mechanisms, Bond Strength", Metal Construction, 18, 369–372.

Benscoter, A.O. and Bramfitt, B.L., (2004), "Metallography and Microstructures of Low-Carbon and Coated Steels", Metallography and Microstructures, Vol 9, ASM Handbook, ASM International, 588–607.

Beynon, J.H. and Sellars, C.M., (1992), "Modeling Microstructure and Its Effects During Multipass Hot Rolling", ISIJ Int., 32, 339–367.

Beynon, J.H., Li. Y.H., Krzyzanowski, M. and Sellars, C.M., (2000), "Measuring, Modelling and Understanding Friction in the Hot Rolling of Steel", Metal Forming 2000, Kraków, Poland, eds, Pietrzyk, M., Kusiak, J., Majta, J., Hartley, P. and Pillinger, I., 3–10.

Bhattacharyya, D., (1981), "Tapering of Aluminium Rings During Ring-Compression Test", Proc. 9th NAMRC, Pennsylvania, 133–139.

Biglou, J., Lenard, J.G., Nelson, B.D. and Hall, D.R., (1995), "Recrystallization During Strip Rolling of Nb–Ti HSLA Steels", 37th Mechanical Working and Steel Processing Conference, October 1995, Hamilton, 661–668.

Bill, A.J. and Scriven, H., (1979), "Sendzimir Mills for Cold Rolling in BSC Stainless", in Flat Rolling: A Comparison of Rolling Mill Types, The Metals Society, 163–170.

Bland, D.R. and Ford, H., (1948), "The Calculation of Roll Force and Torque in Cold Strip Rolling with Tensions", Proc. I. Mech. E., 139, 144–133.

Blazevic, D.T., (1983a), "Rolled-in-Scale: the Continual Problem, Part I – Scale Formation and Rolling Characteristics", Hot Rolling Consultants, Homewood, IL.

Blazevic, D.T., (1983b), "Rolled-in-Scale: the Continual Problem, Part II – Rolled in Scale Descriptions, Causes, Cures", Hot Rolling Consultants, Homewood, IL.

Blazevic, D.T., (1985), "Rolled-in-Scale: the Continual Problem, Part IV – Red Oxide Scale", Hot Rolling Consultants, Homewood, IL.

Blazevic, D.T., (1996), "Tertiary Rolled in Scale – the Hot Strip Mill Problem of the 1990's", 37th MWSP Conf., ISS-AIME, XXXII, 33–38.

Bobig, P. and Stella, F., (2004), "Rolling of Ultra-Thin Gauges in the Finishing Line of a Minimill", Steel Grips, 5, 346–348.

Booser, E. R., (1984), Handbook of Lubrication, Volume II, CRC Press, Boca Raton, Florida.

Boratto, F., Barbosa, R., Yue, S. and Jonas, J.J., (1988), "Effect of Chemical Composition on the Critical Temperatures of Microalloyed Steels", Proc. THERMEC'88, ed., Tamura, I., Tokyo.

Bowden, F.P. and Tabor, D., (1950), "The Friction and Lubrication of Solids", Clarendon Press, Oxford.

Bowden, F.P. and Tabor, D., (1964), "The Friction and Lubrication of Solids, Part II", Clarendon Press, Oxford.

Bowden, F.P. and Tabor, D., (1973), "Friction – An Introduction to Tribology", Heinemann Educational Books, London.

Bower, A.F. and Johnson, K.L., (1989), "The Influence of Strain Hardening on Cumulative Plastic Deformation in Rolling and Sliding Contact", J. Mech. Phys. Solids, 37, 471–493.

Boyer, B. and Massoni, E., (2001), "Inverse Analysis for Identification of Parameters During Thermo-Mechanical Tests", Proceedings of NUMIFORM 2001, ed., Mori, K., A.A. Balkema, Rotterdam, 281–284.

Briant, J., Denis, J. and Parc, G., (1989), "Rheological Properties of Lubricants", Éditions Technip, Paris.

Bridgman, P.W., (1952), "Studies in Large Plastic Flow and Fracture", McGraw-Hill, New York.

Bryant, G.F. and Chiu, T.S.L., (1982), "Simplified Roll-Temperature Model: Convective Cooling", Metals Techn., 9, 478–484.

Caglayan, H. and Beynon, J.H., (1993), "SLIMMER: Sheffield Leicester Integrated Model for Microstructural Evolution in Rolling", 1st Int. Conf. on Modeling of Metal Rolling Processes, London, 274–282.

Callister, W.D., (2000), "Materials Science and Engineering", 5th edition, John Wiley and Sons, New York.

Cameron, A., (1966), "The Principles of Lubrication", John Wiley and Sons, New York.

Carter, W.T., (1994), "A Model for Friction in Metal Forming", ASME, J. Eng. Mat. Techn., 116, 8–13.

Chan, S.M., Chan, L.C. and Lee, T.C., (2003), "Tailor-Welded Blanks of Different Thickness Ratios Effects on Forming Limit Diagrams", J. Mat. Proc. Techn., 132, 95–101.

Chandra, S. and Dixit, U.S., (2004), "A Rigid-Plastic Finite Element Analysis of Temper Rolling Process", J. Mat. Proc. Techn., 152, 9–16.

Chen, W.C., Samarasekera, I.V. and Hawbolt, E.B., (1993), "Fundamental Phenomena Governing Heat Transfer During Rolling", Met. Trans. A, 24, 1307–1320.

Choquet, P., Fabregue, P., Giusti, J., Chamont, B., Pezant, J.N. and Blanchet, F., (1990), "Modeling of Forces, Structure and Final Properties During the Hot

Rolling Process on the Hot Strip Mill", Proc. Mathematical Modeling of Hot Rolling of Steel, ed., Yue, S., Hamilton, 34–43.

Chun, M.S., Biglou, J., Lenard, J.G. and Kim, G. (1999), "Using Neural Networks to Predict Parameters in Hot Working of Aluminium Alloys", J. Mat. Proc. Techn., 86, 245–251.

Clemenson, C., Juelstorp, O. and Bay, N., (1986), "Cold Welding Part 3: Influence of Surface Preparation on Bond Strength", Metal Construction, vol. 18, 625–629.

Coben, M. and Grossberg, S., (1983), "Absolute Stability of Global Pattern Formation and Parallel Memory Storage by Competitive Neural Networks", IEEE Trans. on Systems, Man and Cybernetics, SMC-13, 815–825.

Coldren, A.P., Biss, V. and Oakwood, T.G., (1981), "Microstructural Changes during Thermomechanical Processing of X80 Pipeline Steels", in Thermomechanical Processing of Microalloyed Austenite, eds, DeArdo, A.J., Ratz, G.A. and Wray, P.J., The Met. Soc. of AIME, Pittsburgh, 591–611.

Cole, I.M. and Sansome, D.H., (1968), "A Review of the Application of Pin Load-Cell Pressure Measurement Techniques to Metal Deformation Processes", Int. Machine Design Conf., 271–286.

Cook, P. and McCrum A., (1938), "The Calculation of Load and Torque in Hot Flat Rolling", The British Iron and Steel Research Association, Park Lane, London.

Cornette, G., Cugy, P., Hildebrand, A., Bouzekri, M. and Lovato, G., (2005), "Ultra High Strength FeMn TWIP Steel", La Revue de Métallurgy-CIT, 905–918.

Cuddy, L.J., (1981), "The Effect of Microalloy Concentration on the Recrystallization of Austenite During Hot Deformation", in Proc. Conf. Thermomechanical Processing of Microalloyed Austenite, Pittsburgh, eds, DeArdo, A.J., Ratz, G.A. and Wray. P.J., 129–140.

Czichos, H., (1993), "Wear Mechanisms in Tribological Systems Overview and Classification", in Contact Problems and Surface Interaction in Manufacturing and Tribological Systems, ASME, 239–241.

Dahi, E.D., (1987), "Accelerated Learning Using the Generalized Delta Rule", Proceedings of the IEEE Int. Conf. on Neural Networks, 2, 523–530.

Dake, L.S., Russel, J.A. and Debrodt, D.C., (1986), "A Review of DOE ECT Tribology Surveys", ASME, J. Trib., 497–501.

Davies, R.W., Grant, G.J., Eddie Oliver, H., Khaleel, M.A. and Smith, M.T., (2001), "Forming-Limit Diagrams of Aluminium Tailor-Welded Blank Weld Material", Met. Mat. Trans. A, 32A, 275–283.

Dayhoff, J., (1990), "Neural Network Architecture", Van Nostrand Reinhold, New York.

DePierre, V. and Gurney, F., (1974), "A Method for Determination of Constant and Varying Friction Factors During Ring Compression Tests", ASME, J. Lubr. Techn., 96, 482–488.

Devadas, C., Samarasekera, I.V. and Hawbolt, E.B., (1991), "The Thermal and Metallurgical State of Steel Strip During Hot Rolling: Part III. Microstructural Evolution", Metall. Trans., 22A, 335–349.

Dick, K. and Lenard, J.G., (2005), "The Effect of Roll Roughness and Lubricant Viscosity on the Loads on the Mill During Cold Rolling of Steel Strips", J. Mat. Proc. Techn., 168, 16–24.

Domanti, S.A., Edwards, W.J., Thomas, P.J. and Chefneux, L., (1994), "Application of Foil Rolling Models to Thin Steel Strip and Temper Rolling", Proc. 6th International Rolling Conference, Vol. 2, ed., VDEh, Düsseldorf, 422–429.

Dowson, D. and Higginson, G.R., (1977), "Elastohydrodynamic Lubrication", 2nd edition, Pergamon, Oxford.

Dubey, S.P., Sharma, G.K., Shishodia, K.S. and Sekhon, G.S., (2005), "A Study of Lubrication Mechanism of Oil-in-Water (O/W) Emulsions in Steel Cold Rolling", Ind. Lubr. Techn., 57, 208–212.

Duly, D., Lenard, J.G. and Schey, J.A., (1998), "Applicability of Indentation Tests Assess Ductility in Hot Rolling of Aluminium Alloys", J. Mat. Proc. Techn., 75, 143–151.

Dyment, J., (2004), "Effect of Bending Process on Hydroformability of Steel", thesis for the MASc degree , University of Waterloo, Canada.

Ehrhardt, B., Gerber, T., Hofmann, H. and Schauman, T.W., (2004), "Property Related Design of Advanced Cold Rolled Steels With Induced Plasticity", Steel Grips, 4, 247–255.

Ekelund, S., (1933), "Analysis of Factors Influencing Rolling and Power Consumption in the Hot Rolling of Steel", Steel, 93, 27–29.

El-Kalay, A.K.E.H.A. and Sparling, L.G.M., (1968)," Factors Affecting Friction and Their Effect upon Load, Torque, and Spread in Hot Flat Rolling", J. Iron Steel Inst., 43, 152–168.

Faires, V.M., (1955), "Design of Machine Elements", 3rd edition, Macmillan, New York.

Fang, X., Fan, Z., Ralph, B., Evans, P. and Underhill, R., (2002) "Effect of Temper Rolling on Tensile Properties of C-Mn Steels", Mat. Sci. Techno., 18, 285–288.

Fink, P. and Buch, E., (1979), "Platzer Planetary Mill for Hot High Reduction of Metal Strip", in Flat Rolling: A Comparison of Mill Types, The Metals Society, 73–81.

Fitzpatrick, J.J. and Lenard, J.G., (2001), "The Wear of Tool Steel Rolls During Hot Rolling of Low Carbon Steels", ASM, 45–49.

Flaxa, V. and Shaw, J., (2003), "Material Applications in ULSAB-AVC", Steel Grips, #4.

Fleck, N.A. and Johnson, K.L., (1987), "Towards a New Theory of Cold Rolling Thin Foil", Int. J. Mech. Sci., 29, 507–524.

Fleck, N.A., Johnson, K.L., Mear, M.E. and Zhang, L.C., (1992), "Cold Rolling of Foil", Proc. I. Mech. E., 206, 119–131.

Freeman, J.A. and Skapure, D.M., (1992), "Neural Networks, Algorithms, Application and Programming Techniques", Addison-Wesley Publishing Company, Reading, 93–102.

Fuchshumer, S. and Schlacher, K. (2000), "Mathematical Modelling of a Temper Rolling Mill", Metal Forming 2000, Kraków, A.A. Balkema, Rotterdam, 303–310.

Funke, P., Holland, J. and Kulbrok, R., (1978), "Einfluss unterschiedlicher Warmwalz–bedingungen beim Bandwalzen auf die Walzoberfläche und deren Auswirking auf das Walzgut", Stahl und Eisen, 98, 403–409.

Furukawa, M., Horita, Z., Nemoto, M. and Langdon, T.G., (1996), Scripta Materialia, 35, 2835.

Gaspard, C., Bataille, S., Batazzi, D. and Thonus, P., (2000), "Application of ESR Technology to the Manufacturing of Bimetallic HSS Rolls ", 42nd Mechanical Working and Steel Processing Conference, Hamilton, for "Cold and Hot Strip Mills", Toronto, 655–664.

Gavrus, A., Massoni, E. and Chenot, J.L., (1995), "An Inverse Analysis Using a Finite Element Model for Identification of Rheological Parameters", in Proc. Metal Forming'96, eds, Pietrzyk, M., Kusiak, J., Hartley, P. and Pillinger, I., J. Mat. Proc. Techn., 60, 447–454.

Gelin, J.C. and Ghouati, O., (1994), "The Inverse Method for Determining Viscoplastic Properties of Aluminium Alloys", in Proc. Metal Forming'94, Birmingham, eds, Hartley, P., Pillinger, I., Sturgess, C.E.N., Hall, R., Pietrzyk, M. and Kusiak, J., J. Mat. Proc. Techn., 34, 435–440.

Gelin, J.C. and Ghouati, O., (1999) "Recent Progresses in Optimal Design of Metal Forming Processes through Numerical Simulation", in Proceedings of the 6th ICTP, Advanced Technology of Plasticity, ed., Geiger, M. (Nurnberg), 1, 489–496.

Gigacher, G., Pierer, R. and Benhard, C., (2005), "High Temperature Properties and the Crack Susceptibility of New Steel Grades for Automotive Purposes", WIT Trans., 51, 163–168.

Gilbreath, P., (1967), "Definition and Evaluation of Parameters which Influence the Adhesion of Metals", ASTM STP 431, American Society of Testing of Materials, 128–148.

Ginzburg, V.B., (1989), "Steel Rolling Technology", Marcel Dekker, New York.

Ginzburg, V.B., (1993), "High-Quality Steel Rolling: Theory and Practice", M. Dekker, New York.

Gratacos, P. and Onno, F., (1994), "Elastoplastic Models for Cold Rolling, Application to Temper Rolling", Proc. 6th International Rolling Conference, Düsseldorf, 441–445.

Gratacos, P., Montmitonnet, P., Fromholz, C. and Chenot, J.L., (1992), "A Plane-Strain Model Elastoplastic Finite-Element Model for Cold Rolling of Thin Strip", Int. J. Mech. Sci., 34, 195–210.

Greenwood, J.A. and Williamson, J.B.P., (1966), "Contact of Nominally Flat Surfaces", Proc. Roy. Soc. London Series A, 295, 300–319.

Grimble, M.J., (1976), "A Roll Force Model for Tinplate Rolling", GECJ of Science and Techn., 43, 3–12.

Grimble, M.J., Fuller, M.A. and Bryant, G.F., (1978), "A Non-Circular Arc Roll Force Model for Cold Rolling", Int. J. Num. Meth. Eng., 12, 643–663.

Groover, M.P., (2002), "Fundamentals of Modern Manufacturing 2nd edition", John Wiley and Sons, New York.

Harding, R.A., (1976), "Temperature and Structural Changes in Hot Rolling", PhD Dissertation, University of Sheffield, Sheffield.

Hashimoto, M., Takigawa, H. and Kawakami, T., (1995), "Development and Application of High Speed Tool Steel (HSS) Rolls in Hot Strip Rolling", 37th Mechanical Working and Steel Processing Conference, Hamilton, 275–282.

Hatta, N., Kokado, J., Kikuchi, S. and Takuda, H., (1985), "Modelling on Flow Stress of Plain Carbon Steel at Elevated Temperature", Steel Research, 56, 575–582.

Heshmat, H., Godet M. and Berthier, Y., (1995), "On the Role and Mechanism of Dry Triboparticulate Lubrication", Lubr. Eng., 51, 557–564.

Hill, R. (1950), "The Mathematical Theory of Plasticity", Oxford University Press, Oxford.

Hill, W. and Kerr, E., (1995), "Tool Steel Work Roll Maintenance at Dofasco", 37th Mechanical Working and Steel Processing Conference, Hamilton, 283–286.

Hirano, F. and Ura A., (1970), "Effect of Difference in Hardness of Rubbing Surfaces on Abrasive Wear", Kyushu University, Japan, 164–216.

Hirt, G., Abratis, C., Ames, J., Jackel, F. and Meyer, A., (2005), "Manufacturing and Processing of Tailored Products", IX Conferencia Internacional de Forjamento (XXV SENAFOR) and VIII National Conference on Sheet Metal Forming, Porto Alegre, Brazil.

Hishikawa, S., Maeda, H., Nagakura, H., Hattori, S., Nakajima, M. and Katayama, Y., (1990), "New Control Techniques for Cold Rolling Mills – Applications to Aluminum Rolling", Hitachi Review, 39, 221–230.

Hitchcock, J.H., (1935), "Roll Neck Bearings", ASME Research Publication.

Hodgson, P.D. and Barnett, M.R., (2000), "The Thermomechanical Processing of Steels – Current Status and Future Trends", in Metal Forming 2000, Kraków, Poland, eds, Pietrzyk, M., Kusiak, J., Majta, J., Hartley, P. and Pillinger, I., 21–28.

Hodgson, P.D. and Gibbs, R.K., (1992), "A Mathematical Model to Predict the Mechanical Properties of Hot Rolled C-Mn and Microalloyed Steels", ISIJ Int. 32, 1329–1338.

Hodgson, P.D., McFarlane, D. and Gibbs, R.K., (1993), "The Mathematical Modeling of Hot Rolling: Accuracy vs. Utility", 1st Int. Conf. Modeling of Metal Rolling Processes, London, 2–15.

Hodgson, P.D., Hickson, M.R. and Gibbs, R.K., (1998), "The Production and Mechanical Properties of Ultrafine Ferrite", Mat. Sci. Forum, 35, 63.

Hoffman, O. and Sachs, G., (1953), "Introduction to the Theory of Plasticity for Engineers", McGraw-Hill, New York.

Horita, Z., Fujinami, T. and Langdon, T.G., (2001), "The Potential for Scaling ECAP: Effect of Sample Size on Grain Refinement and Mechanical Properties", Materials Science and Engineering, 318, 34–41.

Hosford, W.F. and Caddell, R.M., (1983), "Metal Forming – Mechanics and Metallurgy", Prentice Hall, Engelwood Cliffs, NJ.

Hum, B, Colquhoun, H.W. and Lenard, J.G., (1996), "Measurements of Friction During Hot Rolling of Aluminum Strips", J. Mat. Proc. Techn., 60, 331–338.

Hundy, B.B., (1955), "Inhomogeneous Deformation During the Temper Rolling of Annealed Mild Steel", J. Iron and Steel Inst., 181, 313–315.

Hush, D.R. and Salas, J.M., (1988), "Improving the Learning Rate of Back-Propagation with the Gradient Reuse Algorithm", Proceedings of the IEEE Int. Conf. on Neural Networks, 1, 441–448.

Hwang, Y.M., Hsu, H.H. and Hwang, Y.L. (2000), "Analytical and Experimental Study on Bonding Behaviour at the Roll Gap During Complex Rolling of Sandwich Sheets", Int. J. Mech. Sci., 42, 2417–2437.

Hwu, Y.J. and Lenard, J.G., (1998), "Phase Transformation Temperature of an Extra-Low Carbon Steel", ASME, J. Eng. Mat. Techn., 120, 19–25.

Imayev, R.M., Imayev, V.M. and Salishchev, G.A., (1992), J. Mat. Science, 27, 4465.

Inhaber, H., (1966), "The Coefficient of Friction in Cold Rolling", ASME, J. Eng. Ind., 88, 421–429.

Irvine, J. and Baker, T.N., (1984), "The Influence of Rolling Variables on the Strengthening Mechanisms Operating in Niobium Steels", Mat. Sci. Eng., 64, 123–134.

Irwing, W.R., (1993), "Continuous Casting of Steel", The Institute of Materials, London.

Januszkiewicz, K.R. and Sulek, H.H., (1988), "Effects of the Thiopenic Contamination of the Hot Rolling Lubricant in Breakdown Mill Rolling", Lubr. Eng., 44, 454–458.

Jeswiet, J., (1991), "A Friction Sensor for Sheet Metal Rolling", Annals of the CIRP, 40, 231–234.

Jin, W., Piereder, D. and Lenard, J.G., (2002), "A Study of the Coefficient of Friction During Hot Rolling of a Ferritic Stainless Steel", Trib. Trans., 13–21.

Johnson, K.L. and Bentall, R.H., (1969), "The Onset of Yield in the Cold Rolling of Thin Strip", J. Mech. Phys. Solids, 17, 253–264.

Johnson, W. and Mellor, P.B., (1962), "Plasticity for Mechanical Engineers", Van Nostrand, London.

Jonas, J.J. and Sakai, T., (1984), "A New Approach to Dynamic Recrystallization", in Deformation, Processing and Structure, ASM Materials Science Seminar, St. Louis, ASM, 185–243.

Jortner, D., Osterle, J.F. and Zorowski, C.F., (1960), "An Analysis of Cold Strip Rolling", Int. J. Mech. Sci., 2, 179–194.

Jost, H., (1966), "Lubrication (tribology) Education and Research", Department of Education and Science, HMSO, London.

Kampuš, Z. and Balič, J., (2003), "Deep Drawing of Tailored Blanks Without a Blankholder", J. Mat. Proc. Techn., 133, 128–133.

Karagiozis, A.N., (1986), "Experimental Determination of the Strip Temperature Distribution", MASC. Thesis, University of New Brunswick, Fredericton.

Karagiozis, A.N. and Lenard, J.G., (1985), "The Effect of Material Properties on the Coefficient of Friction in Cold Rolling", Proc. Eurotrib'95, Lyon, 1–7.

Karagiozis, A. N. and Lenard, J.G., (1987), "On Constitutive Relations for High Temperature Behaviour of Steels", Int. J. Mech. Eng. Ed., 15, 141–149.

Khoddam, S., Lam, Y.C. and Thomson, P.F., (1996), "Numerical Results Achieved with an Inverse Computational Method for Determining the Constitutive Parameters using the Hot Torsion Test Results", Steel Research, 67, 39–43.

Kimura, Y. and Okada, K., (1989), "Lubricating Properties of Oil-in-Water Emulsions", Trib. Trans., 32, 524–532.

Kobasa, D. and Schultz, R.A., (1968), "Experimental Determination of the Length of the Arc of Contact in Cold Rolling", Iron and Steel Eng., 1–6.

Kolmogorov, V.L. and Zalazinsky, A.G., (1998), "On Metal Joining and the Prediction of the Strength of Solid-Phase Joints", J. Mat. Proc. Techn., 75, 157–164.

Kopp, R., (1996), "Some Current Development Trends in Metal-Forming Technology", J. Mat. Proc. Techn., 60, 1–9.

Kopp, R., Böhlke, P., Hohmeier, P. and Wiedner, C., (2002), "Metal Forming of Lightweight Structures", Proc. 5th ESAFORM Conference, Kraków, Poland, 13–22.

Kopp, R., Wiedner, C. and Meyer, A., (2005), "Flexibly Rolled Sheet Metal and its Use in Sheet Metal Forming", Advanced Materials Research, 6–8, 81–92.

Korzekwa, D.A., Dawson, P.R. and Wilson, W.R.D., (1992), "Surface Asperity Deformation During Sheet Forming", Int. J. Mech. Sci., 34, 521–539.

Kosanov, J., Lenard, J.G., Uhrig, J. and Wallfarth, B., (2006), "The Effect of Lubricant Additives on the Coefficient of Friction in the Flat-Die Test", Mat. Sci. Eng., A, 427, 274–281.

Kosko, B., (1988), "Feedback Stability and Unsupervised Learning", Proc. IEEE Int. Conf. on Neural Networks, 1, 141–152.

Kragelsky, I.V., Dobychin, M.N. and Kombalov, V.S., (1982), "Friction and Wear Calculation Method", Pergamon Press, Oxford.

Krallics, G., Szeles, Z., Semenova, I.P., Dotsenko, T.V. and Alexandrov, I.V., (2002), "Experimental Investigations of the Al-Mg-Si Alloy Subjected to Equal Channel Angular Pressing", Proc. Conf. "Nanomaterials by Severe Plastic Deformation", NANOSPD2, Vienna, 183–189.

Krallics, G., Horváth, M. and Fodor, Á., (2004), "Influence of ECAP routes on Mechanical Properties of a Nanocrystalline Aluminium Alloy", Periodica Polytechnica Ser. Mech. Eng., 48, 145–150.

Krallics, G. and Lenard, J. G., (2004), "An Examination of the Accumulative Roll-Bonding Process", Journal of Materials Processing Technology, 152, 154–161.

Krzyzanowski, M. and Beynon, J.H., (2002), "Oxide Behaviour in Hot Rolling", in Metal Forming Science and Practice, ed., Lenard, J.G., Elsevier, Oxford.

Kudo, H., (1960), "Some Analytical and Experimental Studies of Axi-symmetric Cold Forging and Extrusion, Part I", Int. J. Mech. Sci., 2, 102–127.

Kumar, A., Schmid, S. and Wilson, W.R.D., (1997), "Particle Behavior in Two-Phased Lubrication", Wear, 206, 130–135.

Kunogi, M., (1954), "On Plastic Deformation of Hollow Cylinders under Axial Compressive Loading", J. Sci. Res. Inst., 2, 63 (in Japanese).

Kusiak, J., Pietrzyk, M. and Lenard, J.G., (1995), "Application of FE Simulation of the Compression Test to the Evaluation of Constitutive Equations for Steels at Elevated Temperatures", Proc. NUMIFORM'95, eds, Shen, S.F. and Dawson, P.R., Ithaca, 277–282.

Kuziak, R., Cheng, Y.-W., Glowacki, M. and Pietrzyk, M., (1997), "Modeling of the Microstructure and Mechanical Properties of Steels during Thermomechanical Processing", NIST Technical Note 1393, Boulder.

Laasraoui, A., and Jonas, J.J., (1991a), "Prediction of Temperature Distribution, Flow Stress and Microstructure During the Multipass Hot Rolling of Steel Plate and Strip", ISIJ Int., 31, 93–103.

Laasraoui, A., and Jonas, J.J., (1991b), "Recrystallization of Austenite after Deformation at High Temperature and Strain Rates - Analysis and Modeling", Metall. Trans. A, 22A, 131–160.

Larke, E.C., (1965), "The Rolling of Strip, Sheet and Plate, 2nd edition", Chapman and Hall, London.

Le Bon, A.B. and Saint Martin de, L.N., (1976), "Using Laboratory Simulation to Improve Rolling Schedules", Microalloying'75, Union Carbide Corp., New York, 90–99.

Lee, S.H., Saito, Y., Tsuji, N., Utsunomiya, H. and Sakai, T., (2002a), "Role of Shear Strain in Ultragrain Refinement by Accumulative Roll-Bonding (ARB) Process", Scripta Materialia 46, 281–285.

Lee, S.H., Saito, Y., Sakai, T. and Utsunomiya, H., (2002b), "Microstructures and Mechanical Properties of 6061 Aluminum Alloy Processed by Accumulative Roll-Bonding", Mat. Sci. Eng., A325, 228–235.

Lenard, J.G., (1990), "An Experimental Study of Boundary Conditions in Hot and Cold Flat Rolling", Annals of the CIRP, 39, 279–282.

Lenard, J.G., (1991), "Measurements of Friction in Cold Flat Rolling", J. Mat. Shaping Techn., 9, 171–180.

Lenard, J.G., (1992), "Friction and Forward Slip in Cold Strip Rolling", Trib. Trans., 35, 423–428.

Lenard, J.G., (2000), "Tribology of Metal Rolling", Keynote presentation, CIRP ANNALS 49/2, 567–590.

Lenard, J.G., (2001), "Friction During Flat Rolling of Metals", in Friction and Flow Stress in Forming and Cutting, eds, Boisse, P., Altan, T. and van Luttervelt, K., Int. J. Forming Processes, 4, 23–42.
Lenard, J.G., (2003), "Workability and Process Design in Rolling", Chapter 18 of Handbook of Workability and Process Design, eds, Dieter, G.E., Kuhn, H.A., Semiatin, S.L., ASM International, 258–277.
Lenard, J.G., (2004), "The Effect of Roll Roughness on the Rolling Parameters During Cold Rolling of an Aluminum Alloy", J. Mat. Proc. Techn., 132, 144–133.
Lenard, J.G. and Barbulovic-Nad, L., (2002), "The Coefficient of Friction During Hot Rolling of Low Carbon Steel Strips", ASME, J. Trib., 124, 840–845.
Lenard J.G. and Z. Malinowski, (1993), "Measurements of Friction During Warm Rolling of Aluminium", J. Mat. Proc. Techn., 39, 357–371.
Lenard, J.G. and Pietrzyk, M., (1993), "Tribology of Hot Forming: Its Effect on Grain Development", Proc. Int. Symposium "Advances in Hot Deformation Textures", eds, Jonas, J.J., Bieler, T.R. and Bowman, K.J., Pittsburgh, TMS, 107–120.
Lenard, J.G. and Zhang, S., (1997), "A Study of Friction During Lubricated Cold Rolling of an Aluminum Alloy", J. Mat. Proc. Techn., 72, 293–301.
Lenard, J.G., Wang, F. and Nadkarni, G., (1987), "The Role of Constitutive Formulation in the Analysis of Hot Rolling", ASME, J. Eng. Mat. Techn., 109, 343–349.
Lenard, J.G., Pietrzyk, M. and Cser, L., (1999), "Mathematical and Physical Simulation of the Properties of Hot Rolled Products", Elsevier, Amsterdam.
Li, E.B., Tieu, A.K. and Yuen, W.Y.D., (2003), "Forward Slip Measurements in Cold Rolling by Laser Doppler Velocimetry: Uncertainty Analysis and Accuracy Improvement", J. Mat. Proc. Techn., 133, 348–352.
Li, G. and Kobayashi, S., (1982), "Rigid-Plastic Finite-Element Analysis of Plane Strain Rolling", ASME, J. Eng. Ind., 104, 33–64.
Li, G-Y., (1999), "Analytical Equations for On-line Modelling of Friction in Cold Rolling", Proceedings of the Conference Modelling of Metal Rolling Processes 3, eds, Beynon, J.T., Clark, M.T., Ingham, P., Kern P. and Waterson, K., IOM Communications, London, 370–379.
Li, Y.H. and Sellars, C.M., (1996), "Evaluation of Interfacial Heat Transfer and Friction Conditions and Their Effects on Hot Forming Processes", Proc. 37th MWSP Conf., Hamilton, 385–394.
Li, Y.H. and Sellars, C.M., (1998), "Comparative Investigations of Interfacial Heat Transfer Behaviour During Hot Forging and Rolling of Steel with Oxide Scale Formation", Proc. 7th Int. Metal Forming 98, Birmingham, 282–286.
Li, Y.H. and Sellars, C.M., (1999), "Modelling Surface Temperatures During Hot Rolling of Steel", Proc. Modelling of Metal Rolling Processes, London, 178–186.
Lim, L.S. and Lenard, J.G., (1984), "Study of Friction in Cold Strip Rolling", ASME J. Eng. Mat. Techn., 106, 139–146.

Lin, J.F., Huang, T.K. and Hsu, C.T., (1991), "Evaluation of Lubricants in Cold Strip Rolling", Wear, 147, 79–91.

Liu, Y. and Lee, W-H., (2001), "Application of the Preliminary Displacement Principle to the Temper Rolling Model", KSME International Journal, 15, 225–231.

Lo, S-W and Yang, T-S., (2001), "Determining the Pressure Dependence of Lubricant Viscosity by Roughness Measurements of Rolled Strip", Trib. Trans., 44, 118–124.

Loung, L.H.S. and Heijkoop, T., (1981), "The Influence of Scale on Friction in Hot Metalworking", Wear, 71, 93–102.

Lu C., Kiet Tieu, A., Bo Ma and Jiang, Z., (2002), "Friction Measurement in Hot Rolling of Steel", Proceedings of the 44th MWSP Conference, Orlando, Florida, XL, 603–612.

Lubliner, J., (1990), "Plasticity Theory", Macmillan, New York.

Lundberg, S-E. and Gustaffson, T., (1993), "Roll Force, Torque, Lever Arm Coefficient and Strain Distribution in Edge Rolling", J. Mat. Eng. Performance, 26, 873–879.

Madaah-Hosseini, H.R. and Kokabi, A.H., (2002), "Cold Roll Bonding of 5754-Aluminum Strips", Mat. Sci. Eng., A335, 186–190.

Majta, J., Lenard, J.G. and Pietrzyk, M., (1996), "A Study of the Effect of Thermomechanical History on the Mechanical Properties of a High Niobium Steel", Mat. Sci. Eng., A208, 249–259.

Majumdar, A. and Bhushan, B., (1991), "Fractal Model of Elastic-Plastic Contact between Rough Surfaces", ASME, J. Trib., 113, 1–11.

Maki, T., Akasaka, K. and Tamura, I., (1981), "Dynamic Recrystallization Behavior of Austenite in Several Low and High Alloy Steels", Proc. Conf. "Thermomechanical Processing of Microalloyed Austentite", Pittsburgh, eds. DeArdo, A.J., Ratz, G.A. and Wray, P.J., 217–236.

Male, A.T. and Cockroft, M.G., (1964), "A Method for Determination of the Coefficient of Friction of Metals Under Conditions of Bulk Plastic Deformation", J. Inst. Met., 93, 38–46.

Male, A.T. and DePierre, V., (1970), "The Validity of Mathematical Solutions for Determining Friction from the Ring Compression Test", ASME, J. Lubr. Techn., 92, 389–397.

Malinowski, Z., Kusiak, J. and Pietrzyk, M., (1995), "Application of the Inverse Techniques to the Experimental Tests", Proc. Conf. Application of Mathematical Methods in Science and Technique, Kraków, 2, 43–52.

Malinowski, Z., Pietrzyk, M. and Lenard, J.G., (1993), "Analysis of the Flat Rolling Process: One-Dimensional and Finite Element Models", J Mat. Proc. Techn., 39, 373–387.

Manesh, D.H. and Taheri, A.K., (2005), "An Investigation of Deformation Behavior and Bonding Strength of Bimetal Strip During Rolling", Mechanics of Materials, 37, 531–542.

Martin, D.C., Mäntylä, P. and Hodgson, P.D., (1999), "An Analysis of Friction in Finish Rolling of Hot Steel Strip", Proc. Modelling of Metal Rolling Processes, IOM Communications, London, 150–159.

McConnell, C. and Lenard, J.G., (2000), "Friction in Cold Rolling of a Low Carbon Steel with Lubricants", J. Mat. Proc. Techn., 99, 86–93.

Medovar, B.I., Medovar, L.B., Chernets, A.V., Shabanov, V.B. and Sviridov, O.V., (2000), "Ukrainian ESS LM HSS Rolls for Hot Strip Mills", 42nd Mechanical Working and Steel Processing Conference, Toronto, 647–654.

Merklein, M. and Geiger, M., (2002), "New Materials and Production Technologies for Innovative Lightweight Construction", J. Mat. Proc. Techn., 125–126, 532–536.

Michell, J.H., (1900), "Some Elementary Distribution of Stress in Three Dimensions", Proc. London Math. Soc., 32, 23–35.

Mielnik, E.M., (1991), "Metalworking Science and Engineering", McGraw-Hill, New York.

Min, K.B. and Kang, S.S., (2000), "A Study on Resistance Welding in Steel Sheets for Tailor-Welded Blank; Evolution of Flash Weldability and Formability (2nd Report), J. Mat. Proc. Techn., 103, 218–224.

Montmitonnet, P., Delamare, F. and Rizouliers, B., (2000), "Transfer Layer and Friction in Cold Metal Strip Rolling Processes", Wear, 245, 125–135.

Morrison, W.B., Cochrane, R.C. and Mitchell, P.S., (1993), "The Influence of Precipitation Mode and Dislocation Substructure on the Properties of Vanadium-Treated Steels", ISIJ Int., 33, 1095–1107.

Mróz, Z. and Stupkiewicz, S., (1998), "Constitutive Model of Adhesive and Ploughing Friction in Metal-Forming Processes", Int. J. Mech. Sci., 40, 281–303.

Mubea.com.

Munther, P.A., (1997), "The Effect of Material and Process Parameters on the Frictional Conditions in Hot Flat Rolling of Steels", PhD Thesis, University of Waterloo, Waterloo, Canada.

Munther, P.A. and Lenard, J.G., (1999), "The Effect of Scaling on Interfacial Friction in Hot Rolling of Steels", J. Mat. Proc. Techn., 105–113.

Murthy, A. and Lenard, J.G., (1982), "Statistical Evaluation of Some Hot Rolling Theories", ASME, J. Eng. Mat. Techn., 104, 47–52.

Nadai, A., (1939), "The Forces Required for Rolling Steel Strip Under Tension", ASME, A54–A62.

Nakahara, T., Makino, T. and Kyogoku, K., (1988), "Observations of Liquid Droplet Behaviour and Oil Film Formation in O/W Type Emulsion Lubrication", ASME, J. Trib., 110, 348–353.

Nanba, S., Kitamura, M., Shimada, M., Katsumata, M., Inoue, T., Imamura, H., Maeanda, Y. and Hattori, S., (1992), "Prediction of Microstructure Distribution in the Through-Thickness Direction During and after Hot Rolling in Carbon Steels", ISIJ Int., 377–386.

Nishida, Y., Arima, H., Kim, J.C. and Ando, T., (2001), Scripta Materialia, 261.

Nordic Steel and Mining Review, (1998), "Measuring Oil Film Thickness", Nordic Steel and Mining Review, 106.

Onions, R. A., and Archard, J. F., (1973), "The Contact of Surfaces Having a Random Structure," J. Phys. D, 6, 289–304.

Orowan, E., (1943), "The Calculation of Roll Pressure in Hot and Cold Flat Rolling", Proc. I. Mech. E., 150, 140–167.

Park, K-T., Kwon, H-J., Kim, W-J. and Kim, Y-S., (2001), "Microstructural Characteristics and Thermal Stability of Ultrafine Grained 6061 Al Alloy Fabricated by Accumulative Roll Bonding Process", Mat. Sci. Eng., A316, 145–152.

Parker, D.B., (1982), "Learning Logic, Invention Report S81–64, File 1", Office of Technology Licensing, Stanford University.

Parker, D.B., (1987), "Optimal Algorithms for Adaptive Networks: Second Order Back Propagation, Second Order Direct Propagation, and Second Order Hebbin Learning", Proceedings of the IEEE Int. Conf. on Neural Networks, 2, 593–600.

Pashley, M.D., Pethica, J.B. and Tabor, D., (1984), "Adhesion and Micromechanical Properties of Metal Surfaces", Wear, 100, 7–31.

Pawelski, H., (2000), "Modelling of Temper Rolling Considering Surface Change of Strip", Metal Forming 2000, eds, Pietrzyk, M., Kusiak, J., Majta, J. Hartley, P. and Pillinger, I., A.A. Balkema, Rotterdam, 285–301.

Peng, X.K., Heness, G. and Yeung, W.Y., (1999), "Effect of Rolling Temperature on Interface and Bond Development of Roll Bonded Copper/Aluminium Laminates", J. Mat. Sci., 34, 277–281.

Pietrzyk, M., (1982), "Analiza Metodl Elementów Skolczonych Procesu Walcowania Blach i Tasm", Hutnik, 49, 213–217 (in Polish).

Pietrzyk, M., (1994), "Numerical Aspects of the Simulation of Hot Metal Forming Using Internal Variable Method", Metall. Foundry Eng., 20, 429–439.

Pietrzyk, M., (2001), "Identification of Parameters in the History Dependent Constitutive Model for Steels", Annals of the CIRP, 50, 161–164.

Pietrzyk, M. and Lenard, J.G., (1988), "Experimental Substantiation of Modelling Heat Transfer in Hot Flat Rolling", Proc. 25th Nat. Heat Transfer Conf., Houston, ed., Jacobs, H.R., HTD-96, 3, 47–53.

Pietrzyk, M. and Lenard, J.G., (1991), "Thermal-Mechanical Modelling of the Flat Rolling", Process, Springer-Verlag, Berlin.

Pietrzyk, M. and Tibbals J.E., (1995), "Application of the Finite Element Technique to the Interpretation of the Plane Strain Compression Test for Aluminum", Proc. COMPLAS 4, eds, Owen, D.R.J., Onate, E. and Hinton, E., Pineridge Press, Barcelona, 889–900.

Pietrzyk, M., Lenard, J.G. and Dalton, G.M., (1993), "A Study of the Plane Strain Compression Test", Annals of the CIRP, 42, 331–334.

Pietrzyk, M., Kusiak, H., Lenard, J.G. and Malinowski, Z., (1994), "Heat Exchange between the Workpiece and the Tool in Metal Forming Processes", Conf. FORMABILITY'94, ed. Bartecek, J., Ostrava, 329–338.

Pleschiutschnigg, F-P., Singh, B.K., Jamnis, V.V., Talwar, S.R., Misra, A.K., Atluri, R.P.V., Singh, R.B., Shankar, P., Verma, R.K., Goyal, R.K., Mishra, B., Deepu B., Meierling, P. and Pleschiutschnig, J., (2004), "Quality-conscious Coil Production of Value-Added Steel Grades at Ispat Industries Ltd. India", Steel Grips, 2, 256–260.

Puchi-Cabrera, E.S., (2001), "Modeling Cold-Rolling Processes of Commercial Twin Roll Cast Aluminum Alloys", ASME, J. Eng. Mat. Techn., 123, 149–154.

Rabinowicz, E., (1965), "Friction and Wear of Materials", John Wiley and Sons, New York.

Rabinowicz, E., (1982), ASME/ASLE Lub. Conference, Washington.

Rastegaev, M.V., (1940), "New Method of Homogeneous Compression of Specimens for the Determination of Flow Stress and the Coefficient of Internal Friction", Zavod Lab, 354 (in Russian).

Reich, R., Panseri, N. and Bohaychick, J. (2001), "The Effects of Lubricant Starvation in the Cold Rolling of Aluminum Metal Using an Oil-in-Water Emulsion", Lubr. Eng., 57, 15–18.

Reichert, M., Stuwe, H.P., Richert, J., Pippan, R. and Motz, Ch., (2001), Materials Science and Engineering, A301, 237.

Reid, J. V. and Schey, J.A., (1977), "Full Fluid Film Lubrication in Aluminum Strip Rolling", ASLE Trans., 21, 191–200.

Rich, E. and Knight, K., (1991), "Artificial Intelligence", McGraw-Hill, New York.

Ricotti, L.P., Ragazzini, S. and Martinelli, G., (1988), "Learning of Word Stress in a Sub-optimal Second order Back-Propagation Neural Network", Proceedings of the IEEE Int. Conf. on Neural Networks, 1, 355–361.

Roberts, W.L., (1967), "Computing the Coefficient of Friction in the Roll Bite from Mill Data", Blast Furnace and Steel Plant, 55, 499–508.

Roberts, C.D., (1997), "Mechanical Principles of Rolling", Iron and Steelmaker, 24, 113–114.

Roberts, W.L., (1977), "Tribological Considerations in the Hot-Rolling of Low-Carbon Steels", Lubr. Eng., 575–580.

Roberts, W.L., (1978), "Cold Rolling of Steel", Marcel Dekker Inc., New York.

Roberts W.L., (1983), "Hot Rolling of Steel", Marcel Dekker Inc., New York.

Roberts, W.L., (1988), "Flat Processing of Steel", Marcel Dekker Inc., New York.

Roberts, W.L., Sandberg, A., Siwecki, T. and Welefors, T., (1983), "Prediction of Microstructure Development During Recrystallization Hot Rolling of Ti-V Steels", Proc. Conf. HSLA Steels, Technology and Applications, Philadelphia, ASM, 67–84.

Rowe, G.W., (1977), "Principles of Industrial Metalworking Processes", Edward Arnold Publishers, London.

Roychoudhury, R. and Lenard, J.G., (1984), "A Mathematical Model for Cold Rolling – Experimental Substantiation", Proc. 1st Int. Conf. Techn. Plast., Tokyo, 1138–1143.

Rumelhart, D.E. and McClelland, J.L., (1986), "Parallel Distributed Processing", MIT Press, Cambridge, MA.

Sa, C-Y. and Wilson, W.R.D., (1994), "Full Film Lubrication of Strip Rolling", ASME, J. Trib., 116, 569–576.

Saeki, K. and Hashimoto, Y., (1967), Journal of the JSTP, 34, 8–72.

Saito, Y., Tsuji, N., Utsunomiya, H., Sakai, T. and Hong, R.G., (1998), "Ultra-fine Grained Bulk Alumium Produced by Accumulative Roll-Bonding Process", Scripta Materialia 39, 1221–1227.

Saito, Y., Utsunomiya, H., Tsuji, N. and Sakai, T., (1999), "Novel Ultra-High Straining Process for Bulk Materials – Development of the Accumulative Roll-Bonding (ARB) Process", Acta Materialia 47, 579–583.

Sakai, T., (1995), "Dynamic Recrystallization Microstructures under Hot Working Conditions", J. Mat. Proc. Techn., 33, 349–361.

Saltavets, W.I., Kirjukhin, N.M., Luke, W.I. and Liddle, W.T., (2001), "Increase in Hot Mill Roll Life", 43rd Mechanical Working and Steel Processing Conference, Charlotte, 637–648.

Sargent, L.B., (1983), "Pressure-Viscosity Coefficients of Liquid Lubricants", ASLE Transactions, 26, 1–10.

Schey, J.A., (1983), "Tribology in Metalworking", ASM, Metals Park, Ohio.

Schey, J.A. (1984), "Workability in Rolling", in Workability Testing Techniques, ed., Dieter, G.E., ASM, 269–314.

Schey, J.A., (2000), "Introduction to Manufacturing Processes, 3rd edition", McGraw-Hill, New York.

Schmid, S.R., (1997), "Hydrodynamic Segregation, Entrainment and Rejection of Oil in Emulsion Lubrication Problems", ASME, J. Trib., 119, 342–348.

Schmid, S.R. and Wilson, W.R.D., (1995), "Lubrication of Aluminum Rolling by Oil-in-Water Emulsion", Trib. Trans., 38, 452–458.

Schmid, S.R. and Wilson, W.R.D., (1996), "Lubrication Mechanisms for Oil-in-Water Emulsions," Lubr. Eng., 52, 168–175.

Schunke, J.N., Sudarshan, T.S. and Srivatsan, T.S., (1988), "Effects of Oxygen in Influencing Friction Characteristics of Metals", Wear, 128, 211–221.

Segal, V.M., Reznikov, V.I., Drobyshevkij, A.E. and Kopylov, V.I., (1981), Metally (in Russian), 1, 99.

Segal, V.M., Reznikov, A.E., Kopylov, V.I., Pavlik, D.A. and Malyshev, V.F., (1984), in Process of Plastic Transformation of Metals (in Russian), Navuka i Teknika, 295.

Sellars, C.M., (1979), "Physical Metallurgy of Hot Working", in Hot Working and Forming Processes, eds, Sellars, C.M. and Davies, G.J., The Metals Soc., London, 3–15.

Sellars, C.M., (1990), "Modelling Microstructural Development During Hot Rolling", Mat. Sci. Techn., 6, 1072–1081.

Sellars, C.M. and Beynon, J.H., (1984), in High Strength Low Alloy Steels, University of Wollongong, Wollongong, NSW, eds, Dunne, D.P. and Chandra, T., 142–150.

Seredynski, T., (1973), "Prediction of Plate Cooling During Rolling Mill Operation", J. Iron and Steel Inst., 211, 197–203.

Shaw, L., Miracle, D. and Abbaschian, R., (1995), "Microstructure and Mechanical Properties of Metal/Oxide and Metal/Silicide Interfaces", Acta Metall. Mater, 43, 4267–4279.

Sheppard, T. and Duan, X., (2002), "A New Spread Formula for Hot Flat Rolling of Aluminium Alloys", Mod. Sim. Mat. Sci. Eng, 10, (6), 597–610.

Shida, S., (1969), "Empirical Formula of Flow-Stress of Carbon Steels — Resistance to Deformation of Carbon Steels at Elevated Temperatures", J. Jpn. Soc. Techn. Plasticity, 10, 610–617.

Shida, S. and Awazuhara, H., (1973), "Rolling Load and Torque in Cold Rolling", Jpn Soc. Techn. Plasticity, 14, 267–278.

Shirizly, A. and Lenard, J.G., (2000), "Emulsions Versus Neat Oils in the Cold Rolling of Carbon Steel Strips", ASME J. Trib., 122, 550–556.

Shirizly, A., Lenard, J.G., Sauer, J. and Januszkiewicz, K., (2002), "Lubricant Capture During Hot Rolling of an Aluminum Alloy", Trib. Trans., 45, 205–210.

Siebel, E. and Lueg, W., (1933), "Investigation into the Distribution of Pressure at the Surface of the Material in Contact with the Rolls", Mitt. K.W. Inst. Eisenf. 13, 1–14.

Siegert, K., Wagner, S. and Dogan, N., (1999), "Development of a Portable Sensor for the Three-Dimensional Measurement of Sheet and Tool Surfaces", Proc. Advances in Sheet Metal Stamping (SP-1435), Detroit, 1–10.

Silk, N.J. and Li, Y.H., (1999), "Friction Measurements in a Tandem Finishing Mill by the Forward Slip Method", Proc. 3rd Int. Conf. on Modelling of Metal Rolling Processes, London, 400–409.

Silvonen, A., Malinen, A. and Korhonen, A.S., (1987), "A Finite Element Study of Plane Strain Hot Rolling", Scand. J. Metal., 16, 103–108.

Simpson, P. K., (1990), "Artificial Neural Systems-Foundations, Paradigms, Applications and Implementations", McGraw-Hill, New York, 7–22.

Sims, R.B., (1952), "The Forward Slip in Cold Strip Rolling", Sheet Metal Industries, 29, 869–877.

Sims, R.B., (1954), "The Calculation of Roll Force and Torque in Hot Rolling Mills", Proc. I. Mech. E., 168, 191–200.

Sims, R.B. and Arthur, D.F., (1952), "Speed Dependent Variables in Cold Rolling", JISI, B.I.S.R.A., 285–295.

So, B.Y.C. and Klaus, E.E., (1980), "Viscosity-Pressure Correlation of Liquids", ASLE Transactions, 23, 409–421.

Stachowiak, G.W. and Batchelor, A.W., (2005), "Engineering Tribology", Elsevier, Oxford.

Stancu-Niederkorn, S. Engel, U. and Geiger, M., (1993), "A New Method for Investigating the Relation Between Real Contact Area and Friction in Bulk Metal Forming Processes", Wear, 43, 381–386.

Starling, C. W., (1962), "The Theory and Practice of Flat Rolling", University Press, London.

Stevens, P.G., Ivens, K.P. and Harper, P.J., (1971), "Increasing Work Roll Life by Improved Roll Cooling Practice", J. Iron and Steel Inst., 209, 1–11.

Stone, M.D., (1953), "The Rolling of Thin Strip – Part I", Iron and Steel Engineer Year Book, 115–128.

Sutcliffe, M.P.F., (1990), "Experimental Measurements of Lubricant Film Thickness in Cold Strip Rolling", Proc. Inst. Mech. Eng, 204, 263–273.

Sutcliffe, M.P.F., (2000), "Flattening of Random Rough Surfaces in Metal-Forming Processes", ASME, J. Trib., 121, 433–440.

Sutcliffe, M.P.F., (2002), "Surface Finish and Friction in Cold Metal Rolling", in Metal Forming Science and Practice, ed., Lenard, J.G., Elsevier, Oxford, Chapter 4, 19–59.

Sutcliffe, M.F.P. and Johnson, K.L., (1990), "Lubrication in Cold Strip Rolling in the 'Mixed' Regime", Proc. Instn. Mech. Engrs., Part B, 204, 249–261.

Sutcliffe, M.P.F. and Rayner, P.J., (1998), "Experimental Measurements of Load and Strip Profile in Thin Strip Rolling", Int. J. Mech. Sci., 40, 887–899.

Suzuki, H., Hashizume, S., Yabuki, Y., Ichihara, Y., Nakajima, S. and Kenmochi, K., (1968), "Studies on the Flow Stress of Metals and Alloys", Report of the Institute of Industrial Science, The University of Tokyo (Serial No. 117), 18.

Svietlichnyy, D. and Pietrzyk, M., (1999), "On-line Model for Control of Hot Plate Rolling", Modeling of Metal Rolling Processes 3, London, 62–71.

Szeri, A.Z., (1998), Fluid Film Lubrication, Cambridge University Press, Cambridge.

Szyndler, D., Lenard, J.G., Fyke, S.H. and Pietrzyk, M., (2000), "Inverse Analysis Applied to the Evaluation of Friction and Rheological Parameters in Hot Forming of Steels", Metal Forming 2000, Kraków, 101–106.

Tabary, P.E., Sutcliffe, M.F.P., Porral, F. and Deneuville, P., (1994), "Measurements of Friction in Cold Metal Rolling", in Tribology in Manufacturing Processes, ASME.

Tamano, T. and Yanagimoto, S., (1978), "Theory of Rolling for the Range of Mixed Friction", Bulletin of the JSME, 21, 592–599.

Tamura I., Sekine H., Tanaka T., Ouchi C., (1988), "Thermomechanical Processing of High-Strength Low-Alloy Steels", Butterworth, London.

Tan, X., Martins, P.A.F., Bay, N. and Zhang. W., (1998), "Friction Studies at Different Normal Pressures with Alternative Ring-Compression Tests", 7th Int. Conf. on Metal Forming, Birmingham, 292–297.

Tarnovskii, I. Ya., Pozdeyev, A.A. and Lyashkov, V.B., (1965), "Deformation of Metals during Rolling", Pergamon, Oxford.

Thaller, G., Djumlija, G., Gruber, W., Champion, N. and Marples, A. (2005), "VAI Steckel Mill Technology: Economical and Flexible Production of Hot-Rolled Strip and Plates", Rev. Met. Paris, 3, 251–261.

Tiley, J.B. and Lenard, J.G., (2003), "A Study of the Temperature at the Interface of the Work Roll and the Thin Scale Layer on a Low Carbon Steel During Hot

Rolling", The ISS 45th Mechanical Working and Steel Processing Conference, Chicago, Illinois, 103–113.

Tirosh, J., Iddan, D. and Pawelski, O., (1985), "The Mechanics of High-Speed Rolling of Viscoplastic Materials", ASME, J. Applied Mechanics, 52, 309–318.

Tong, K. and Sachs, G., (1957), "Roll Separating Force and Minimum Thickness of Cold-Rolled Strips", J. Mech. Phys. Solids, 6, 35–46.

Touloukian, Y.S. and Buyco, E.H., (1970), "Thermomechanical Properties of Matter", vol. 4, Specific Heat – Metallic Elements and Alloys, IFI Plenum, New York – Washington.

Trijssenaar, M., (2002), "Emulsion Lubrication in Cold Rolling", Doctoral Dissertation, Technische Universiteit Delft, The Netherlands.

Tselikov, A., (1967), "Stress and Strain in Metal Rolling", MIR, Moscow.

Tsuji, N., Shiotsuki, K., Utsunomiya, H. and Saito, Y., (1999a), "Low Temperature Superplasticity of Ultra-Fine Grained 5083 Aluminum Alloy Produced by Accumulative Roll-Bonding", Materials Science Forum 304-306, 73–78.

Tsuji, N., Saito, Y., Utsunomiya, H. and Tanigawa, S., (1999b), "Ultra-fine Grained Bulk Steel Produced by Accumulative Roll-Bonding (ARB) Process", Scripta Materialia 40, 795–800.

Tsuji, N., Saito, Y., Lee, S-H. and Minamino, Y., (2002), "ARB (Accumulative Roll-Bonding) and Other New Techniques to Produce Bulk Ultrafine Grained Materials", Proc. 2nd Int. Conf. on Nanomaterials by Severe Plastic Deformation, Vienna.

Tzou, G.Y., Tieu, A.K., Huang, M.N., Lin, C.Y. and Wu, E.Y., (2002), "Analytical Approach to the Cold-and-Hot Bond Rolling of Sandwich Sheet with Outer Hard and Inner Soft Layers", J. Mat. Proc. Techn., 125–126, 664–669.

Underwood, L.R., (1950), "The Rolling of Metals", John Wiley and Sons, New York.

Valiahmetov, O.R., Galeyev, R.M. and Salishchev, G.A., (1990), Fizika Metallov i Metallovedenije (in Russian), 10, 204.

Valiev, R.Z., (1996), "Synthesis and Processing of Nanocrystalline Powder", The Minerals and Materials Society, 153.

Valiev, R.Z., Krasilnikov, N.A. and Tsenev, N.K., (1991), Materials Science and Engineering, A137, 35.

van Rooyen, G.T. and Backofen, W.A., (1960), "A Study of Interface Friction in Plastic Compression", Int. J. Mech. Sci., 1–27.

Venkat, S., Albright, C.E., Ramasamy, S. and Hurley, J.P., (1997), "CO_2 Laser beam Welding of Aluminium 5754-O and 6111-T4 Alloys", Welding Research Supplement, 76, 275–282.

Venter, R. and Abd-Rabbo, A., (1980), "Modelling of the Rolling Process – I Inhomogeneous Deformation Model", Int. J. Mech. Sci., 22, 83–92.

Wagoner R.H. and Chenot, J-L., (1996), "Fundamentals of Metal Forming", John Wiley and Sons, New York.

Wang, F., (1989), "Simulation of Hot Strip Rolling by Compression Testing", PhD Thesis, The University of New Brunswick, Fredericton.

Wang, F. and Lenard, J.G., (1991), "The Constitutive Behaviour of a Nb-V HSLA Steel in the Temperature Range of 900 to 975°C", Steel Research, 2, 66–71.

Wang, F. and Lenard, J.G., (1992), "An Experimental Study of Interfacial Friction – Hot Ring Compression", ASME, J. Eng. Mat. Techn., 114, 13–18.

Wanheim, T., (1973), "Friction at High Normal Pressures", Wear, 25, 225–244.

Wanheim, T. and Bay, N., (1978), "A Model for Friction in Metal Forming Processes", Annals of the CIRP, 27, 189–194.

Wanheim, T., Bay, N. and Petersen, A.S., (1974), "A Theoretically Determined Model for Friction in Metal Working Processes", Wear, 28, 251–258.

Wankhede, U. and Samarasekera, I.V., (1997), "Thermal Behavior of High Speed Steel Work Rolls in the Finishing Stands of a Hot Strip Mill", Iron and Steelmaker, 24, 55–60.

Werbos, P., (1974), "New Tools for Prediction and Analysis in the Behaviour Science", Ph D Thesis, Harvard University, Cambridge, MA.

Whetzel, J.C. and Rodman, S., (1959), Iron and Steel Engineering, 123.

White, H., (1989), "Neural-Networks", AI Expert, 48–52.

Whitehouse, D.J. and Archard, J.F., (1970), "The Properties of Random Surfaces of Significance in Their Contact", Proc. Roy. Soc. London, Series A, 316, 97–121.

Wiklund, O., (1996a) in Proceedings of 5th International Conference "Steel Strip '96", eds, Boruta, J., Kisa, A., Kure, T., Pasternak, B., Zamarsky, G. and Zela, P., Spolecnost OCELOVE PASY, Opava, Czech Republic, 136.

Wiklund, O., (1996b) in Proceedings of 2nd International Conference on "Modelling of Metal Rolling Processes", eds, Beynon J.H., Ingham P., Teichert H. and Waterson, K., The Institute of Materials, London, 400.

Wiklund, O., (1999), "Improved Modelling of Temper Rolling Using Finite Elements to Fit Fast 'Empirical' Models Combined with Artificial Neural Networks", in Proceedings of International Conference "Modelling of metal rolling processes 3", eds, Beynon J.H., Clark, M.T., Kern, P., Ingham P. and Waterson, K., The Institute of Materials, London, 72.

Wiklund, O. and Sandberg, F., (2002), "Modelling and Control of Temper Rolling and Skin Pass Rolling", Chapter 15, in Metal Forming Science and Practice, ed., Lenard, J.G., Elsevier, Oxford.

Wilson, W.R.D. and Chang, D-F., (1994), "Low Speed Mixed Lubrication of Bulk Metal Forming Processes", in Tribology in Manufacturing, eds, Dohda, K., Jahanmir, S. and Wilson, W.R.D., ASME, 159–168.

Wilson, W.R.D. and Walowit, J.A., (1972), "An Isothermal Hydrodynamic Lubrication Theory for Strip Rolling with Front and Back Tension", Proc. 1971 Tribology Convention, I. Mech. E., London, 164–172.

Wilson, W.R.D., Sakaguchi, Y. and Schmid, S.R., (1994), "A Mixed Flow Model for Lubrication with Emulsions", Trib. Trans., 37, 543–551.

Worswick, M.J., (2002), "Numerical Simulation of Sheet Metal Forming", in Metal Forming Science and Practice, ed., Lenard, J.G., Elsevier, Oxford.

Wu, C.S., Klaus E.E. and Duda, J.L, (1989), "Development of a Method for the Prediction of Pressure-Viscosity Coefficients of Lubricating Oils Based on free-Volume Theory", ASME, J. Trib., 111, 121–128.

Wu, H-Y., Lee, S. and Wang, J-Y., (1998), "Solid-state Bonding of Iron-Based Alloys, Steel, C Brass and Aluminium alloys", J. Mat. Proc. Techn., 75, 173–179.

Wusatowski, Z., (1969), "Fundamentals of Rolling", Pergamon, Oxford.

Xing, Z.P., Kang, S.B. and Kim, H.W., (2002), "Structure and Properties of AA3003 Alloy Produced by Accumulative Roll Bonding Process", J. Mat. Sci., 37, 717–722.

Yada, H, (1987), "Prediction of Microstructural Changes and Mechanical Properties in Hot Strip Rolling", Proc. Symp. Accelerated Cooling of Rolled Steel, eds, Ruddle, G.E. and Crawley, A.F., Pergamon Press, Winnipeg, 103–119.

Yan, S., and Kuroda, S., (1997a), "Lubrication with Emulsion: First Report, the Extended Reynolds Equation," Wear, 206, 230–237.

Yan, S., and Kuroda, S., (1997b), "Lubrication with Emulsion II. The Viscosity Coefficients of Emulsions," Wear, 206, 238–243.

Yan, H. and Lenard, J.G., (2004), "A Study of Warm and Cold Roll-Bonding of an Aluminium Alloy", Mat. Sci. Eng., A 385, 419–428.

Yoneyama, T. and Hatamura, Y., (1987), "The Development of a Die Sensor", JSME International, 30, 670–677.

Yuen, W.Y.D., (2003), "On-line and Off-line Models for the Rolling Process", Scan. J. Metall., 32, 86–93.

Zhang, S. and Lenard, J.G., (1994), "Use of Lubricants During Cold Rolling of Aluminium", Synthetic Lubrication, 11, 99–109.

Zhang, W. and Bay, N., (1997), "Cold Welding – Theoretical Modeling of the Weld Formation", Welding Research Supplement, 76, 417–420.

Zhang, W., Peeters, M.J.P.A. and Bay, N., (1995), "Numerical Modelling of Cold Rolling of Metal Plate Adopting a General Friction Model", in Simulation of Materials Processing, eds, Dawson, P. and Shen, H., A.A. Balkema, Rotterdam.

Zhang, Y., (2005), "How Does Dry Contact Occur in Elastohydrodynamic Lubrication?", Ind. Lubr. Trib., 57, 176–180.

Zhu, D., Biresaw, G., Clark, S.J. and Kasun, T.J., (1994), "Elastohydrodynamic Lubrication with O/W Emulsions", ASME, J. Trib., 116, 310–320.

AUTHOR INDEX

SUBJECT INDEX